GROUNDWATER CHEMICALS FIELD GUIDE

JOHN H. MONTGOMERY

LEWIS PUBLISHERS

Introduction

The compounds profiled include all the organic Priority Pollutants promulgated by the U.S. Environmental Protection Agency (U.S. EPA) under the Clean Water Act of 1977 [1]. Many of these organic Priority Pollutants were included among the Target Compounds promulgated by the U.S. EPA under the Comprehensive Environmental Response, Compensation and Liability Act (CERCLA) in 1980 and the Superfund Amendments and Reauthorization Act (SARA) of 1986. Most of the organic compounds most commonly found in the work environment according to the National Institute of Occupational Safety and Health are also included. In addition, aliphatic and aromatic hydrocarbons common in fuel mixtures are included.

This pocket guide was designed to provide a quick and easy access to information on a compound's physical properties, health and exposure data. Information for this pocket guide was obtained from Groundwater Chemicals Desk Reference [2], Groundwater Chemicals Desk Reference - Volume 2 [3] and more recent journal articles. Chemical structures and manufacturers are not included in this book, however, symptoms of chemical exposure have been added. The reader is referred to these books for identifying competing values and detailed literature citations.

The compound headings are those commonly used by the U.S. EPA and are arranged alphabetically. Positional and/or structural prefixes set in italic type are not an integral part of the chemical name and are disregarded in alphabetizing. These include *asym*-, *sym*-, *n*-, *sec*-, *cis*-, *trans*-, *α*-, *β*-, *γ*-, *o*-, *m*-, *p*-, *N*-, etc.

Synonyms: These are listed alphabetically following the convention used for the compound headings. Compounds in boldface type are the Chemical Abstracts Service (CAS) names listed in the eighth Collective Index. If no synonym appears in boldface type, then the compound heading is the CAS assigned name. Synonyms include chemical names, common or generic names, trade names, registered trademarks, government codes and acronyms. All synonyms found in the literature are listed except foreign, slang and obsolete names.

Although synonyms were retrieved from several references, most of them were retrieved from the Registry of Toxic Effects of Chemical Substances [4].

Chemical Designations

Chemical Abtracts Service (CAS) Registry Number: This is a unique identifier assigned by the American Chemical Society to chemicals recorded in the CAS Registry System. This number is used to access various chemical databases such as the Hazardous Substances Data Bank (HSDB), CAS Online, Chemical Substances Information Network and many others. This entry is also useful to conclusively identify a substance regardless of the assigned name.

Department of Transportation (DOT) Designation: This is a four-digit number assigned by the U.S. Department of Transportation (DOT) for hazardous materials and is identical to the United Nations identification number (which is preceded by the letters UN). This number is required on shipping papers, on placards or orange panels on tanks, and on a label or package containing the material. These numbers are widely used for personnel responding to emergency situations, e.g., overturned tractor trailers, in which the identification of the transported material is quickly and easily determined. Additional information may be obtained through the U.S. Department of Transportation, Research and Special Programs Administration, Materials Transportation Bureau, Washington, D.C. 20590.

Molecular Formula (mf): This is arranged by carbon, hydrogen and remaining elements in alphabetical order in accordance with the Hill system [5]. Molecular formulas are useful in identifying isomers (i.e., compounds with identical molecular formulas) and are required if one wishes to calculate the formula weight of a substance.

Formula Weight (fw): This is calculated to the nearest hundredth using the empirical formula and the 1981 Table of Standard Atomic Weights as reported in Weast [6]. Formula weights are required for many calculations, such as converting weight/volume units, e.g., mg/L or g/L, to molar units (mol/L); with density for calculating molar volumes; and for estimating Henry's law constants.

Registry of Toxic Effects of Chemical Substances (RTECS) Number: Many compounds are assigned a unique accession number consisting of two letters followed by seven numerals. This number is needed to quickly and easily locate additional toxicity and health-based data which are cross-referenced in the Registry of Toxic Effects of Chemical Substances (RTECS) [4]. Contact the National Institute for Occupational Safety and Health (NIOSH), U.S. Department of Health and Human Services, 4676 Columbia Pkwy., Cincinnati, Ohio 45226 for additional information.

Physical and Chemical Properties

Appearance and Odor: The appearance, including the physical state (solid, liquid, or gas) of a chemical at room temperature (20-25 °C) is provided. If the compound can be detected by the olfactory sense, the odor is noted. Unless noted otherwise, the information provided in this category is for the pure substance and was obtained from many sources [7-12].

Boiling Point (bp): This is defined as the temperature at which the vapor pressure of a liquid equals the atmospheric pressure. Unless otherwise noted, all boiling points are reported at one atmosphere pressure (760 mmHg). Although not used in environmental assessments, boiling points for aromatic compounds have been found to be linearly correlated with aqueous solubility [13]. Boiling points are also useful in assessing entry of toxic substances into the body. Body contact with high-boiling liquids is the most common means of entry into the body whereas the inhalation route is the most common for low-boiling liquids [14].

Dissociation Constant (K_a): In an aqueous solution, an acid (HA) will dissociate into the carboxylate anion (A^-) and hydrogen ion (H^+) and may be represented by the general equation:

$$HA_{(aq)} \rightleftharpoons H^+ + A^- \quad (1)$$

At equilibrium, the ratio of the products (ions) to the reactant (non-ionized electrolyte) is related by the equation:

$$K_a = \frac{[H^+]\,[A^-]}{[HA]} \quad (2)$$

where K_a is the dissociation constant. This expression shows that K_a increases if there is increased ionization and vice versa. A strong acid (weak base) such as hydrochloric acid ionizes readily and has a large K_a, whereas a weak acid (or stronger base) such as benzoic acid ionizes to a lesser extent and has a lower K_a. The dissociation constants for weak acids are sometimes expressed as K_b, the dissociation constant for the base, and both are related to the dissociation constant for water by the expression:

$$K_w = K_a + K_b \quad (3)$$

where

K_w = dissociation constant for water (10^{-14} at 25 °C)
K_a = acid dissociation constant
K_b = base dissociation constant

The dissociation constant is usually expressed as $pK_a = -\log_{10}K_a$. The above equation becomes:

$$pK_w = pK_a + pK_b \quad (4)$$

When the pH of the solution and the pK_a are equal, 50% of the acid will have dissociated into ions. The percent dissociation of an acid or base can be calculated if the pH of the solution and the pK_a of the compound are known [15]:

For organic acids: $$\alpha_a = \frac{100}{1 + 10^{(pH-pKa)}} \quad (5)$$

For organic bases: $$\alpha_b = \frac{100}{1 + 10^{(pKw-pKb-pH)}} \quad (6)$$

where

α_a = percent of the organic acid that is non-dissociated
α_b = percent of the organic base that is non-dissociated
pK_a = log dissociation constant for acid
pK_w = log dissociation constant for water (14.00 at 25 °C)
pK_b = log dissociation constant for base ($pK_b = pK_w - pK_a$).
pH = hydrogen ion activity (concentration) of the solution

Since ions tend to remain in solution, the degree of dissociation will affect processes such as volatilization, photolysis, adsorption and bioconcentration [16].

Henry's Law Constant (K_H): Sometimes referred to as the air-water partition coefficient, the Henry's law constant is defined as the ratio of the partial pressure of a compound in air to the concentration of the compound in water at a given temperature under equilibrium conditions. If the vapor pressure and solubility of a compound are known, this parameter can be calculated at 1 atm (760 mmHg) as follows:

$$K_H = \frac{PFW}{760S} \quad (7)$$

where

K_H = Henry's law constant ($atm \cdot m^3/mol$)
P = pressure (mm)
S = solubility (mg/L)
FW = formula weight (g/mol)

Henry's law constant can also be expressed in dimensionless form and may be calculated using one of the following equations:

$$K_H' = \frac{K_H}{RK} \quad \text{or} \quad K_H' = \frac{S_a}{S} \quad (8)$$

where

K_H = Henry's law constant ($atm \cdot m^3/mol$)
K_H' = Henry's law constant (dimensionless)
R = ideal gas constant (8.20575×10^{-5} $atm \cdot m^3/mol \cdot K$)

K = temperature of water (K)
S_a = solute concentration in air (mol/L)
S = aqueous solute concentration (mol/L)

It should be noted that estimating Henry's law constant assumes that the gas obeys the ideal gas law and the aqueous solution behaves as an ideally dilute solution. The solubility and vapor pressure data inputted into the equations are valid only for the pure compound and must be in the same standard state at the same temperature.

The major drawback in estimating Henry's law constant is that both the solubility and the vapor pressure of the compound are needed in equation 7. If one or both these parameters are unknown, an empirical equation based on quantitative structure-activity relationships (QSAR) may be used to estimate Henry's law constants [17]. In this QSAR model, only the structure of the compound is needed. From this, connectivity indexes (based on molecular topology), polarizability (based on atomic contributions) and the propensity of the compound to form hydrogen bonds can easily be determined. These parameters, when regressed against known Henry's law constants for 180 organic compounds, yielded an empirical equation that explained more than 98% of the variance in the data set having an average standard error of only 0.262 logarithm units. Henry's law constant may also be estimated using the bond or group contribution method developed by Hine and Mookerjee [18]. The constants for the bond and group contributions were determined using experimentally determined Henry's law constants for 292 compounds. The authors found that those estimated values significantly deviating from observed values (particularly for compounds containing halogen, nitrogen, oxygen, and sulfur substituents) could be explained by "distant polar interactions", i.e., interactions between polar bonds or structural groups.

Henry's law constants provided an indication of the relative volatility of a substance. According to Lyman and others [19], if $K_H < 10^{-7}$ atm·m^3/mol, the substance has a low volatility. If K_H is greater than 10^{-7} but less than 10^{-5} atm·m^3/mol, the substance will volatilize slowly. Volatilization becomes an important transfer mechanism in the range $10^{-5} < H < 10^{-3}$ atm·m^3/mol. Values of K_H exceeding 10^{-3} atm·m^3/mol indicate volatilization will proceed rapidly.

Ionization Potential (IP): The ionization potential of a compound is defined as the energy required to remove a given electron from the molecule's atomic orbit (outermost shell) and is expressed in electron volts (eV). One electron volt is equivalent to 23,053 cal/mol.

Knowing the ionization potential of a contaminant is required in determining the appropriate photoionization lamp for detecting that contaminant or family of contaminants. Photoionization instruments are equipped with a radiation source (ultraviolet lamp), pump, ionization chamber, an amplifier and a recorder (either digital or meter). Generally, compounds with ionization potentials smaller than the radiation source (UV lamp rating) being used will readily ionize and will be detected by the instrument. Conversely, compounds with ionization potentials higher than the lamp rating will not ionize and will not be detected by the instrument.

Log K_{oc}: The soil/sediment partition or sorption coefficient is defined as the ratio of adsorbed chemical per unit weight of organic carbon to the aqueous solute concentration. This value provides an indication of the tendency of a chemical to partition between particles containing organic carbon and water. Compounds that bind strongly to organic carbon have characteristically low solubilities, whereas compounds with low tendencies to adsorb onto organic particles have high solubilities.

Non-ionizable chemicals that sorb onto organic materials in an aquifer (i.e., organic carbon), are retarded in their movement in groundwater. The sorbing solute travels at linear velocity that is lower than the groundwater flow velocity by a factor of R_d, the retardation factor. If the K_{oc} of a compound is known, the retardation factor may be calculated using the following equation from Freeze and Cherry [20] for unconsolidated sediments:

$$R_d = \frac{V_w}{V_c} = 1 + \frac{BK_d}{n_e} \qquad (9)$$

where

R_d = retardation factor (unitless)
V_w = average linear velocity of groundwater (e.g., ft/day)
V_c = average linear velocity of contaminant (e.g., ft/day)
B = average soil bulk density (g/cm^3)
n_e = effective porosity (unitless)
K_d = distribution (sorption) coefficient (cm^3/g)

By definition, K_d is defined as the ratio of the concentration of the solute on the solid to the concentration of the solute in solution. This can be represented by the Freundlich equation:

$$K_d = \frac{VM_S}{MM_L} = \frac{C_S}{C_L{}^n} \qquad (10)$$

where

V = volume of the solution
M_S = mass of the sorbed solute
M = mass of the porous medium
M_L = mass of the solute in solution
C_S = concentration of the sorbed solute
C_L = concentration of the solute in the solution
n = a constant

Values of n are normally between 0.7 and 1.1 although values of 1.6 have been reported [19]. If n is unknown, it is assumed to be unity and a plot of C_S versus C_L will be linear. The distribution coefficient is related to K_{oc} by the equation:

$$K_{oc} = \frac{K_d}{f_{oc}} \tag{11}$$

where f_{oc} is the fraction of naturally occurring organic carbon in soil.

Sometimes K_d is expressed on an organic-matter basis and is defined as:

$$K_{om} = \frac{K_d}{f_{om}} \tag{12}$$

where f_{om} is the fraction of naturally occurring organic matter in soil. The relationship between K_{oc} and K_{om} is defined as:

$$K_{om} = 0.58K_{oc} \tag{13}$$

where the constant 0.58 is assumed to represent the fraction of carbon present in the soil or sediment organic matter [21].

For fractured rock aquifers in which the porosity of the solid mass between fractures is insignificant, Freeze and Cherry [20] report the retardation equation as:

$$R_d = \frac{V_w}{V_c} = 1 + \frac{2K_A}{b} \tag{14}$$

where

K_A = distribution coefficient (cm)
b = aperture of fracture (cm)

To calculate the retardation factors for ionizable compounds such as acids and bases, the fraction of unionized acid (α_a) or base (α_b) needs to be determined (see **Dissociation Constant**). According to Guswa and others [15], if it is assumed only the unionized portion of the acid is adsorbed onto the soil, the retardation factor for the acid becomes:

$$R_a = 1 + \frac{\alpha_a BK_d}{n_e} \tag{15}$$

However, for a base they assume that the ionized portion is exchanged with a monovalent ion and the unionized portion of the base is adsorbed hydrophobically. Therefore, the retardation factor for the base is:

$$R_b = 1 + \frac{\alpha_b B K_d}{n_e} + \frac{C_{ec} B}{100 \Sigma z^+ n_e}(1-\alpha_b) \quad (16)$$

where

C_{ec} = cation exchange capacity of the soil (cm^3/g)
Σz^+= sum of all positively charged particles in the soil (milliequivalents/cm^3)

Guswa and others [15] report that the term Σz^+ is $\approx$ 0.001 for most agricultural soils.

Correlations between K_{oc} and bioconcentration factors in fish and beef have shown a log-log linear relationship [22] as well as solubility of organic compounds in water [23,24]. Moreover, the log K_{oc} has been shown to be related to molecular connectivity indices [25-27] and high performance liquid chromatography (HPLC) capacity factors [28-31].

In instances where experimentally determined K_{oc} values are not available, they can be estimated using recommended regression equations as cited in Lyman and others [19]. All the K_{oc} estimations are based on regression equations in which the solubility or the K_{ow} of the substance is known.

Log K_{ow}: The K_{ow} of a substance is the *n*-octanol/water partition coefficient and is defined as the ratio of the solute concentration in the water-saturated *n*-octanol phase to the solute concentration in the *n*-octanol-saturated water phase. Values of K_{ow} are therefore unitless.

The partition coefficient has been recognized as a key parameter in predicting the environmental fate of organic compounds. The log K_{ow} has been shown to be linearly correlated with bioconcentration factors (BCF) in aquatic organisms [32,33], in fish [22,32-36], soil/sediment partition coefficients (K_{oc}) [37,38], solubility of organic compounds in water [32,39-46], molecular surface area [45-48], molar refraction [49], reversed-phase liquid chromatography (RPLC) retention factors [50], high performance liquid chromatography (HPLC) capacity factors [51-55], HPLC retention times [56-58], distribution coefficients [59], solvatochromic parameters [60], biological responses [61], molecular descriptors and physicochemical properties [62] and substituent constants which are based on empirically derived atomic or group constants and structural factors [63,64].

For ionizable compounds such as acids, amines and phenols, K_{ow} values are a function of pH. Unfortunately, many investigators have neglected to report the pH of the solution at which the K_{ow} was determined. If a K_{ow} value is used for an ionizable compound for which the pH is known, both values should be noted.

Melting Point (mp): The melting point of a substance is defined as the temperature at which a solid substance undergoes a phase change to a liquid. The reverse process, the temperature at which a liquid freezes to a solid, is called the freezing point. For a given substance, the melting point is identical to the freezing point.

Unless noted otherwise, all melting points are reported at the standard pressure of 1 atmosphere. Although the melting point of a substance is not directly used in predicting its behavior in the environment, it is useful in determining the phase in which the substance would be found under typical conditions.

Solubility in Organics (S_0): The presence of small quantities of solvents can enhance a compound's solubility in water [65]. Consequently, its fate and transport in soils, sediments and groundwater will be changed due to the presence of these cosolvents. For example, soils contaminated with compounds having low water solubilities tend to remain bound to the soil by adsorbing onto organic carbon and/or by interfacial tension with water. A solvent introduced to an unsaturated soil environment (e.g., a surface spill, leaking aboveground tank, etc.) may come in contact with existing soil contaminants. As the solvent interacts with the existing contamination, it may mobilize it, thereby facilitating its migration. Consequently, the organic solvent can facilitate the leaching of contaminants from the soil to the water table. Therefore, the presence of cosolutes must be considered when predicting the fate and transport of contaminants in the unsaturated zone and the water table.

Solubility in Water (S_w): The water solubility of a compound is defined as the saturated concentration of the compound in water at a given temperature and pressure. This parameter is perhaps the most important factor in estimating a chemical's fate and transport in the aquatic environment. Compounds with high water solubilities tend to desorb from soils and sediments (i.e., they have low K_{oc} values), are less likely to volatilize from water, and are susceptible to biodegradation. Conversely, compounds with low solubilities tend to adsorb onto soils and sediments (have high K_{oc}), volatilize more readily from water, and bioconcentrate in aquatic organisms. The more soluble compounds commonly enter the water table more readily than their less soluble counterparts.

The water solubility of a compound varies with temperature, pH (particularly, ionizable compounds such as acids and bases), and other dissolved constituents, e.g., inorganic salts (electrolytes) and organic chemicals including naturally occurring organic carbon, such as humic and fulvic acids. At a given temperature, the variability/discrepancy of water-solubility measurements documented by investigators may be attributed to one or more of the following: (1) purity of the compound, (2) analytical method employed, (3) particle size (for solid solubility determinations only), (4) adsorption onto the container and/or suspended solids, (5) time allowed for equilibrium conditions to be reached, (6) losses due to volatilization, and (7) chemical transformations (e.g., hydrolysis).

The water solubility of chemical substances has been related to bioconcentration factors (BCF), soil/sediment partition coefficients (K_{oc}), [23,24], *n*-octanol/water partition coefficients (K_{ow}) [40-43,45,46], molecular descriptors and physicochemical properties [62], soil organic matter K_{om} [66], total molecular surface area [67-69], the compound's molecular structure [70,71], boiling points [13], and for homologous series of hydrocarbons or classes of organic compounds - carbon number [72-75] and molar volumes [76]. With the exception of the molecular structure-solubility relationship, regression equations generated from the other relationships have demonstrated a log-log linear relationship for these properties. The reported regression equations are useful in estimating the solubility of a compound in water if experimental values are not available. In addition, the solubility of a compound may be estimated if an experimentally determined Henry's law constant is available [77].

Unless otherwise noted, all reported solubilities were determined using distilled water. For some compounds, solubilities were determined using well water, surface water, natural seawater or artificial seawater.

Specific Density (*d*): The specific density, also known as relative density, is defined as:

$$\text{Specific density} = \frac{d}{d_w} \qquad (17)$$

where

d = density of a substance (g/mL or g/cm^3)
d_w = density of distilled water (g/mL or g/cm^3)

Values of specific density are unitless and are reported in the form:

d at T_s/T_w

where

d = specific density of the substance
T_s = temperature of substance at the time of measurement (°C)
T_w = water temperature (°C).

For example, the value 1.1750 at 20/4 °C indicates a specific density of 1.1750 for the substance at 20 °C with respect to water at 4 °C. At 4 °C, the density of water is exactly 1.0000 g/mL (g/cm^3). Therefore, the specific density of a substance is equivalent to the density of the substance relative to the density of water at 4 °C.

The density of a hydrophobic substance enables it to sink or float in water. Density values are especially important for liquids migrating through the unsaturated zone and encountering the water table as "free product." Generally, liquids that are less dense than water "float" on the water table. Conversely, organic liquids that are more dense than water commonly "sink" through the water table, e.g., dense non-aqueous phase liquids (DNAPLs) such as chloroform, dichloroethane, and tetrachloroethylene.

Hydrophilic substances, on the other hand, behave differently. Acetone, which is less dense than water, does not float on water because it is freely miscible with it in all proportions. Therefore, the solubility of a substance must be considered in assessing its behavior in the subsurface.

Transformation Products: Chemicals released in the environment are susceptible to several degradation pathways. These include chemical (i.e., hydrolysis, oxidation, dehydrochlorination, reduction, isomerization, and conjugation), photolysis or photooxidation and biodegradation. Compounds transformed by one or more of these processes may result in the formation of more toxic or less toxic substances. In addition, the transformed product(s) will behave differently than the parent compound due to changes in their physicochemical properties. Many researchers focus their attention on transformation rates rather than the transformation products. Consequently, only limited data exist on the resultant end products. Where available, compounds that are transformed into identified products are listed.

In addition to chemical transformations occurring under normal environmental conditions, abiotic degradation products are also included. Types of abiotic transformation processes or treatment technologies fall into two categories -- physical and chemical. Types of physical processes used in removing or eliminating hazardous wastes include sedimentation, centrifugation, flocculation, oil/water separation, dissolved air flotation, heavy media separation, evaporation, air stripping, steam stripping, distillation, soil flushing, chelation, liquid-liquid extraction, supercritical extraction, filtration, carbon adsorption, reverse osmosis, ion exchange and electrodialysis. This information can be useful in evaluating abiotic degradation as a possible remedial measure. Chemical processes include neutralization, precipitation, hydrolysis (acid or base catalyzed), photolysis, oxidation-reduction, oxidation by hydrogen peroxide, alkaline chlorination, electrolytic oxidation, catalytic dehydrochlorination, and alkali metal dechlorination.

Most of the abiotic chemical transformation products reported in this book are limited to only three processes: hydrolysis, photooxidation and chemical oxidation-reduction. These processes are the most widely studied and reported in the literature. Detailed information describing the above technologies, their availability/limitation and company sources is available [78].

Vapor Density (vap d): The vapor density of a substance is defined as the ratio of the mass of vapor per unit volume. An equation for estimating vapor density is readily derived from a varied form of the ideal gas law:

$$PV = \frac{\mathcal{M}RK}{FW} \tag{18}$$

where

P = pressure (atm)
V = volume (L)

M = mass (g)
R = ideal gas constant (8.20575×10^{-2} atm·L/mol·K)
K = temperature (K)
FW = formula weight

Recognizing that the density of a substance is defined as:

$$d = \frac{M}{V} \tag{19}$$

Substituting this equation into the equation 18, rearranging, and simplifying results in an expression to determine the vapor density (g/L):

$$p = \frac{PFW}{RK} \tag{20}$$

At standard temperature (293.15K) and pressure (1 atm), equation 20 simplifies to:

$$p = \frac{FW}{24.47} \tag{21}$$

The specific vapor density of a substance relative to air is determined using:

$$p_v = \frac{FW}{24.47 p_{air}} \tag{22}$$

where

p_v = specific vapor density of a substance
p_{air} = vapor density of air (g/L)

The specific vapor density, p_v, is simply the ratio of the vapor density of the substance to that of air under the same pressure and temperature. According to Weast [6], the vapor density of dry air at 20 °C and 760 mmHg is 1.204 g/L. At 25 °C, the vapor density of air decreases slightly to 1.184 g/L. Calculated specific vapor densities are reported relative to air (set equal to 1) only for compounds which are liquids at room temperature (i.e., 25 °C). These are reported in addition to the calculated vapor densities.

Vapor Pressure (vp): The vapor pressure of a substance is defined as the pressure exerted by the vapor (gas) of a substance when it is under equilibrium conditions.

It provides a semi-quantitative rate at which it will volatilize from soil and/or water. The vapor pressure of a substance is a required input parameter for calculating the air-water partition coefficient (see **Henry's Law Constant**), which in turn is used in volatilization rate constant calculations.

Fire Hazards

Flash Point (fl p): The flash point is defined as the minimum temperature at which a substance releases ignitable flammable vapors in the presence of an ignition source. e.g., spark or flame. Flash points may be determined by two methods --- Tag closed cup (ASTM method D56) or Cleveland open cup (ASTM method D93). Unless otherwise noted, all flash point values represent closed cup method determinations. Flash point values determined by the open cup method are slightly higher than those determined by the closed cup method; however, the open cup method is more representative of actual conditions.

According to Sax [7], a material with a flash point of 100 °F or less is considered dangerous, whereas a material having a flash point greater than 200 °F is considered to have a low flammability. Substances with flash points within this temperature range are considered to have moderate flammabilities.

Lower Explosive Limit (LEL): The minimum concentration (vol % in air) of a flammable gas or vapor required for ignition or explosion to occur in the presence of an ignition source (see also **Flash Point**).

Upper Explosive Limit (UEL): The maximum concentration (vol % in air) of a flammable gas or vapor required for ignition or explosion to occur in the presence of an ignition source (see also **Flash Point**).

Health Hazard Data

Immediately Dangerous to Life or Health (IDLH): According to the National Institute of Occupational Safety and Health [79], the IDLH level ". . . for the purpose of respirator selection represents a maximum concentration from which, in the event of respirator failure, one could escape within 30 minutes without experiencing any escape-impairing or irreversible health effects." Concentrations are reported in parts per million (ppm) or milligrams per cubic meter (mg/m^3).

Exposure Limits: The permissible exposure limits (PEL) in air, set by the Occupational Health and Safety Administration (OSHA), can be found in the Code of Federal Regulations [80]. Unless noted otherwise, the PEL are 8-hour time-weighted average (TWA) concentrations. If NIOSH [79] and/or the American Conference of Governmental Industrial Hygienists (ACGIH) has published recommended exposure limits, these are also included. The ACGIH's recommended exposure limits, commonly known as threshold limit values (TLV),

Society of Agronomy, 1965), pp 1367-1378.
22. Kenaga, E.E. "Correlation of Bioconcentration Factors of Chemicals in Aquatic and Terrestrial Organisms with Their Physical and Chemical Properties," *Environ. Sci. Technol.*, 14(5):553-556 (1980).
23. Means, J.C., Wood, S.G., Hassett, J.J., and W.L. Banwart. "Sorption of Polynuclear Aromatic Hydrocarbons by Sediments and Soils," *Environ. Sci. Technol.*, 14(2):1524-1528 (1980).
24. Abdul, S.A., Gibson, T.L., and D.N. Rai. "Statistical Correlations for Predicting the Partition Coefficient for Nonpolar Organic Contaminants between Aquifer Organic Carbon and Water," *Haz. Waste Haz. Mater.*, 4(3):211-222 (1987).
25. Sabljić, A., and M. Protić. "Relationship between Molecular Connectivity Indices and Soil Sorption Coefficients of Polycyclic Aromatic Hydrocarbons," *Bull. Environ. Contam. Toxicol.*, 28(2):162-165 (1982).
26. Sabljić, A. "Predictions of the Nature and Strength of Soil Sorption of Organic Pollutants by Molecular Topology," *J. Agric. Food Chem.*, 32(2):243-246 (1984).
27. Sabljić, A. "On the Prediction of Soil Sorption Coefficients of Organic Pollutants from Molecular Structure: Application of Molecular Topology Model," *Environ. Sci. Technol.*, 21(4):358-366 (1987).
28. Hodson, J., and N.A. Williams. "The Estimation of the Adsorption Coefficient (K_{oc}) for Soils by High Performance Liquid Chromatography," *Chemosphere*, 19(1):67-77 (1988).
29. Szabo, G., Prosser, S.L., and R.A. Bulma. "Adsorption Coefficient (K_{oc}) and HPLC Retention Factors of Aromatic Hydrocarbons," *Chemosphere*, 21(4/5):495-505 (1990).
30. Szabo, G., Prosser, S.L., and R.A. Bulman. "Prediction of the Adsorption Coefficient (K_{oc}) for Soil by a Chemically Immobilized Humic Acid Column using RP-HPLC," *Chemosphere*, 21(6):729-739 (1990).
31. Szabo, G., Prosser, S.L., and R.A. Bulman. "Determination of the Adsorption Coefficient (K_{oc}) of Some Aromatics for Soil by RP-HPLC on Two Immobilized Humic Acid Phases," *Chemosphere*, 21(6):777-788 (1990).
32. Isnard, S., and S. Lambert. "Estimating Bioconcentration Factors from Octanol-Water Partition Coefficient and Aqueous Solubility," *Chemosphere*, 17(1):21-34 (1988).
33. Davies, R.P., and A.J. Dobbs. "The Prediction of Bioconcentration in Fish," *Water Res.*, 18(10):1253-1262 (1984).
34. Neely, W.B., Branson, D.R., and G.E. Blau. "Partition Coefficient to Measure Bioconcentration Potential of Organic Chemicals in Fish," *Environ. Sci. Technol.*, 8(13):1113-1115 (1974).
35. Oliver, B.G., and A.J. Niimi. "Bioconcentration Factors of Some Halogenated Organics for Rainbow Trout: Limitations in Their Use for Prediction of Environmental Residues," *Environ. Sci. Technol.*, 19(9):842-849 (1985).
36. Ogata, M., Fujisawa, K., Ogino, Y., and E. Mano. "Partition Coefficients as a Measure of Bioconcentration Potential of Crude Oil in Fish and Sunfish," *Bull. Environ. Contam. Toxicol.*, 33(5):561-567 (1984).
37. Kenaga, E.E., and C.A.I. Goring. "Relationship between Water Solubility, Soil Sorption, Octanol-Water Partitioning and Concentration of Chemicals in Biota,"

in *Aquatic Toxicology, ASTM STP 707*, Eaton, J.G., Parrish, P.R., and A.C. Hendricks, Eds. (Philadelphia: American Society for Testing and Materials, 1980), pp 78-115.

38. Chiou, C.T., Peters, L.J., and V.H. Freed. "A Physical Concept of Soil-Water Equilibria for Nonionic Organic Compounds," *Science*, 206:831-832 (1979).
39. Banerjee, S., Yalkowsky, S.H., and S.C. Valvani. "Water Solubility and Octanol/Water Partition Coefficients of Organics. Limitations of the Solubility-Partition Coefficient Correlation," *Environ. Sci. Technol.*, 14(10):1227-1229 (1980).
40. Chiou, C.T., Freed, V.H., Schmedding, D.W., and R.L. Kohnert. "Partition Coefficients and Bioaccumulation of Selected Organic Chemicals," *Environ. Sci. Technol.*, 11(5):475-478 (1977).
41. Chiou, C.T., Schmedding, D.W., and M. Manes. "Partitioning of Organic Compounds in Octanol-Water Systems," *Environ. Sci. Technol.*, 16(1):4-10 (1982).
42. Hansch, C., Quinlan, J.E., and G.L. Lawrence. "The Linear Free-Energy Relationship between Partition Coefficients and Aqueous Solubility of Organic Liquids," *J. Org. Chem.*, 33(1):347-350 (1968).
43. Miller, M.M., Wasik, S.P., Huang, G.-L., Shiu, W.-Y., and D. Mackay. "Relationships between Octanol-Water Partition Coefficient and Aqueous Solubility," *Environ. Sci. Technol.*, 19(6):522-529 (1985).
44. Tewari, Y.B., Miller, M.M., Wasik, S.P., and D.E. Martire. "Aqueous Solubility of Octanol/Water Partition Coefficient of Organic Compounds at 25.0 °C," *J. Chem. Eng. Data*, 27(4):451-454 (1982).
45. Yalkowsky, S.H., and S.C. Valvani. "Solubilities and Partitioning 2. Relationships between Aqueous Solubilities, Partition Coefficients, and Molecular Surface Areas of Rigid Aromatic Hydrocarbons," *J. Chem. Eng. Data*, 24(2):127-129 (1979).
46. Miller, M.M., Ghodbane, S., Wasik, S.P., Tewari, Y.B., and D.E. Martire. "Aqueous Solubilities, Octanol/Water Partition Coefficients, and Entropies of Melting of Chlorinated Benzenes and Biphenyls," *J. Chem. Eng. Data*, 29(2):184-190 (1984).
47. Camilleri, P., S.A. Watts, and J.A. Boraston. "A Surface Area Approach to Determination of Partition Coefficients," *J. Chem. Soc. Perkin Trans. II*, (September 1988), pp 1699-1707.
48. Funasaki, N., Hada, S., and S. Neya. "Partition Coefficients of Aliphatic Ethers - Molecular Surface Area Approach," *J. Phys. Chem.*, 89(14):3046-3049 (1985).
49. Yoshida, K., Tadayoshi, S., and F. Yamauchi. "Relationship between Molar Refraction and *n*-Octanol/Water Partition Coefficient," *Ecotoxicol. Environ. Safety*, 7(6):558-565 (1983).
50. Khaledi, M.G., and E.D. Breyer. "Quantitation of Hydrophobicity with Micellar Liquid Chromatography," *Anal. Chem.*, 61(9):1040-1047 (1989).
51. Eadsforth, C.V. "Application of Reverse-Phase H.P.L.C. for the Determination of Partition Coefficients," *Pestic. Sci.*, 17(3):311-325 (1986).
52. Brooke, D.N., A.J. Dobbs, and N. Williams. "Octanol:Water Partition Coefficients (P): Measurement, Estimation, and Interpretation, Particularly for

Chemicals with $P > 10^5$," *Ecotoxicol. Environ. Safety*, 11(3):251-260 (1986).

53. DeKock A.C., and D.A. Lord. "A Simple Procedure for Determining Octanol-Water Partition Coefficients using Reverse Phase High Performance Liquid Chromatography (RPHPLC)," *Chemosphere*, 16(1):133-142 (1987).
54. Harnisch, M., Mockel, H.J., and G. Schulze. "Relationship between Log P_{ow} Shake-Flask Values and Capacity Factors Derived from Reversed-Phase High Performance Liquid Chromatography for *n*-Alkylbenzene and some OECD Reference Substances," *J. Chromatogr.*, 282:315 (1983).
55. Carlson, R.M., Carlson, R.E., and H.L. Kopperman. "Determination of Partition Coefficients by Liquid Chromatography," *J. Chromatogr.*, 107:219-223 (1975).
56. Sarna, L.P., Hodge, P.E., and G.R.B. Webster. "Octanol-Water Partition Coefficients of Chlorinated Dioxins and Dibenzofurans by Reversed-Phase HPLC Using Several C_{18} Columns," *Chemosphere*, 13(9):975-983 (1984).
57. Burkhard, L.P., and D.W. Kuehl. "*n*-Octanol/Water Partition Coefficients by Reverse Phase Liquid Chromatography/Mass Spectrometry for Eight Tetrachlorinated Planar Molecules," *Chemosphere*, 15(2):163-167 (1986).
58. Webster, G.R.B., Friesen, K.J., Sarna, L.P., and D.C.G. Muir. "Environmental Fate Modeling of Chlorodioxins: Determination of Physical Constants," *Chemosphere*, 14(6/7):609-622 (1985).
59. Campbell, J.R., Luthy, R.G., and M.J.T. Carrondo. "Measurement and Prediction of Distribution Coefficients for Wastewater Aromatic Solutes," *Environ. Sci. Technol.*, 17(10):582-590 (1983).
60. Sadek, P.C., Carr, P.W., Doherty, R.M., Kamlet, M.J., Taft, R.W., and M.H. Abraham. "Study of Retention Processes in Reversed-Phase High-Performance Liquid Chromatography by the Use of the Solvatochromic Comparison Method," *Anal. Chem.*, 57(14):2971-2978 (1985).
61. Schultz, T.W., Wesley, S.K., and L.L. Baker. "Structure-Activity Relationships for Di and Tri Alkyl and/or Halogen Substituted Phenols," *Bull. Environ. Contam. Toxicol.*, 43(2):192-198 (1989).
62. Warne, M.St.J., Connell, D.W., Hawker, D.W., and G. Schüürmann. "Prediction of Aqueous Solubility and the Octanol-Water Partition Coefficient for Lipophilic Organic Compounds Using Molecular Descriptors and Physicochemical Properties," *Chemosphere*, 21(7):877-888 (1990).
63. Hansch, C., and S.M. Anderson. "The Effect of Intramolecular Hydrophobic Bonding on Partition Coefficients," *J. Org. Chem.*, 32:2853-2586 (1967).
64. Hansch, C., Leo, A., and Nikaitani. "On the Additive-Constitutive Character of Partition Coefficients," *J. Org. Chem.*, 37(20):3090-3092 (1972).
65. Nyssen, G.A., Miller, E.T., Glass, T.F., Quinn II, C.R., Underwood, J., and D.J. Wilson. "Solubilities of Hydrophobic Compounds in Aqueous-Organic Solvent Mixtures," *Environ. Monit. Assess.*, 9(1):1-11 (1987).
66. Chiou, C.T., Porter, P.E., and D.W. Schmedding. "Partition Equilibria of Nonionic Organic Compounds between Organic Matter and Water," *Environ. Sci. Technol.*, 17(4):227-231 (1983).
67. Amidon, G.L., S.H. Yalkowsky, S.T. Anik, and S.C. Valvani. "Solubility of Nonelectrolytes in Polar Solvents. V. Estimation of the Solubility of Aliphatic

Monofunctional Compounds in Water Using a Molecular Surface Area Approach," *J. Phys. Chem.*, 79(21):2239-2246 (1975).
68. Hermann, R.B. "Theory of Hydrophobic Bonding. II. The Correlation of Hydrocarbon Solubility in Water with Solvent Cavity Surface Area," *J. Phys. Chem.*, 76(19):2754-2759 (1972).
69. Lande, S.S., Hagen, D.F., and A.E. Seaver. "Computation of Total Molecular Surface Area from Gas Phase Ion Mobility Data and its Correlation with Aqueous Solubilities of Hydrocarbons," *Environ. Toxicol. Chem.*, 4(3):325-334 (1985).
70. Nirmalakhandan, N.N., and R.E. Speece. "Prediction of Aqueous Solubility of Organic Compounds Based on Molecular Structure," *Environ. Sci. Technol.*, 22(3):328-338 (1988).
71. Nirmalakhandan, N.N., and R.E. Speece. "Prediction of Aqueous Solubility of Organic Compounds Based on Molecular Structure. 2. Application to PNAs, PCBs, PCDDs, etc.," *Environ. Sci. Technol.*, 23(6):708-713 (1989).
72. Krzyzanowska, T., and J. Szeliga. "A Method for Determining the Solubility of Individual Hydrocarbons," *Nafta*, 28:414-417 (1978).
73. Bell, G.H. "Solubilities of Normal Aliphatic Acids, Alcohols and Alkanes in Water," *Chem. Phys. Lipids*, 10:1-10 (1973).
74. Robb, I.D. "Determination of the Aqueous Solubility of Fatty Acids and Alcohols," *Aust. J. Chem.*, 18:2281-2285 (1966).
75. Mitra, A., Saksena, R.K., and C.R. Mitra. "A Prediction Plot for Unknown Water Solubilities of Some Hydrocarbons and Their Mixtures," *Chem. Petro-Chem. J.*, 8:16-17 (1977).
76. Lande, S.S., and S. Banerjee. "Predicting Aqueous Solubility of Organic Nonelectrolytes from Molar Volumes," *Chemosphere*, 10:751:759 (1981).
77. Kamlet, M.J., Doherty, R.M., Abraham, M.H., Carr, P.W., Doherty, R.F., and R.W. Taft. "Linear Solvation Energy Relationships. 41. Important Differences between Aqueous Solubility Relationships for Aliphatic and Aromatic Solutes," *J. Phys. Chem.*, 91(7):1996-2004 (1987).
78. "A Compendium of Technologies Used in the Treatment of Hazardous Wastes," Office of Research and Development, U.S. EPA Report-625/8-87-014, 49 p.
79. "NIOSH Pocket Guide to Chemical Hazards," U.S. Department of Health and Human Services, U.S. Government Printing Office (1987), 241 p.
80. "General Industry Standards for Toxic and Hazardous Substances," U.S. Code of Federal Regulations, 29 CFR 1910.1000, Subpart Z (January 1977).
81. *Threshold Limit Values and Biological Exposure Indices for 1987-1988* (Cincinnati, OH: American Conference of Governmental Industrial Hygienists, 1987), 114 p.
82. Sittig, M. *Handbook of Toxic and Hazardous Chemicals and Carcinogens* (Park Ridge, NJ: Noyes Publications, 1985), 950 p.
83. Verschueren, K. *Handbook of Environmental Data on Organic Chemicals* (New York: Van Nostrand Reinhold Co., 1983), 1310 p.

Abbreviations and Symbols

Å	angstrom
α	alpha
@	at
α_a	percent of acid that is non-dissociated
α_b	percent of base that is non-dissociated
$\approx$	approximately equal to
abdom	abdominal
ACGIH	American Conference of Governmental Industrial Hygienists
album	albuminuria
anem	anemia
anes	anesthesia
anor	anorexia
appre	apprehension
aq	aqueous
arrhy	arrhythmias
asphy	asphyxia
asth	asthma
ASTM	American Society for Testing and Materials
asym	asymmetric
atax	ataxia
atm	atmosphere
avg	average
b	aperture of fracture
B	average soil bulk density (g/cm^3)
β	beta
BCF	bioconcentration factor
blur	blurred
bp	boiling point
BP	blood pressure
breath	breathing
bron	bronchitis
bronspas	broncospasm
C	ceiling
°	degrees Centigrade (Celsius)
cal	calorie
calcd	calculated
card	cardiac
CAS	Chemical Abstracts Service
C_{ec}	cation exchange capacity
CERCLA	Comprehensive Environmental Response, Compensation and Liability Act
chol	cholinesterase
CHRIS	Chemical Hazard Response Information System
cirr	cirrhosis

cm	centimeter
C_L	concentration of solution in solution
CNS	central nervous system
CO	carbon monoxide
CO_2	carbon dioxide
CoA	coenzyme A
coll	collapse
concn	concentration
conf	confusion
conj	conjunctivitis
convuls	convulsions
corn	cornea
cryst	crystalline, crystal(s)
C_S	concentration of sorbed solute
CVS	cardiovascular system
cyan	cyanosis
d	specific density
defat	defatting
dent	dental
depres	depressant/depression
derm	dermatitis
diarr	diarrhea
dil	dilated
dist	disturbance
dizz	dizziness
DOT	Department of Transportation (U.S.)
drow	drowsiness
d_s	density of a substance
d_w	density of water
dys	dysuria
dysp	dyspnea
δ	delta
enl	enlargement
epis	epistaxis
equi	equilibrium
eryt	erythema
est	estimated
euph	euphoria
eV	electron volts
extrem	extremities
° F	degrees Fahrenheit
fasc	fasiculation
fl p	flash point
f_{oc}	fraction of organic carbon
frost	frostbite
ftg	fatigue

fvr	fever
FW	formula weight
γ	gamma
g	gram
GC/MS	gas chromatography/mass spectrometry
GI	gastrointestinal
gidd	giddiness
>	greater than
≥	greater than or equal to
h	hour
H_2O_2	hydrogen peroxide
halu	hallucinations
head	headache
hema	hematuria
hemat	hematoma
hemog	hemoglobinuria
hemorr	hemorrhage
HPLC	high performance liquid chromatography
HSDB	Hazardous Substances Data Bank
ict	icterus
IDLH	immediately dangerous to life or health
inco	incoordination
inflamm	inflammation
inj	injury
IP	ionization potential
insom	insomnia
irrit	irritation
irrity	irritability
jaun	jaundice
K	kelvin (°C + 273.15)
K_a	acid dissociation constant
K_A	distribution coefficient (cm)
K_b	base dissociation constant
K_d	distribution coefficient (cm^3/g)
kg	kilogram
K_H	Henry's law constant ($atm \cdot m^3/mol \cdot K$)
K_H'	Henry's law constant (dimensionless)
kid	kidney
K_{oc}	soil/sediment partition coefficient (organic carbon basis)
K_{om}	soil/sediment partition coefficient (organic matter basis)
K_{ow}	*n*-octanol/water partition coefficient
kPa	kilopascal
K_w	dissociation constant for water (10^{-14} at 25 °C)
<	less than
≤	less than or equal to
L	liter

lac	lacrimation
lar	laryngeal
lass	lassitude
lel	lower explosive limit
leucyt	leukocytosis
li-head	lightheadedness
liq	liquid
liv	liver
low-wgt	weight loss
m	meter
m	meta (as in *m*-dichlorobenzene)
M	molarity (moles/liter)
M	mass
mal	malaise
mg	milligram
min	minute
misc	miscible
mL	milliliter
M_L	mass of sorbed solute
mm	millimeters of mercury
mmol	millimole
mo	month
mol	mole
monocy	monocytosis
mp	melting point
M_S	mass of solute in solution
muc memb	mucous membrane
musc	muscle
n, *N*	normal (as in *n*-propyl, *N*-nitroso)
narc	narcotic
narco	narcosis
nas	nose/nasal
nau	nausea
n_e	effective porosity
nec	necrosis
neph	nephritis
ner	nervousness
neur	neurologic
ng	nanogram
NIOSH	National Institute for Occupational Safety and Health
nm	nanometer
NO	nitric oxide
NO_x	nitrogen oxides
NO_2	nitrogen dioxide
N_2O_5	nitrogen pentoxide
numb	numbness

σ	sigma
o	ortho (as in *o*-dichlorobenzene)
O_3	ozone
OC	open cup
opac	opacity
OSHA	Occupational Safety and Health Administration
p	density
p	para (as in *p*-dichlorobenzene)
P	pressure
Pa	pascal
p_{air}	vapor density of air
pal	pallor
para	paralysis
pares	paresthesia
paresis	incomplete loss of muscular power; weakness of a limb
parox	paroxysm
PEL	permissible exposure limit
peri neur	peripheral neuropathy
pH	log hydrogen ion activity (concentration)
phar	pharyngeal
photo	photophobia
pK_a	log dissociation constant of an acid
pK_b	log dissociation constant of a base
pK_w	log dissociation constant of water
pleur	pleurisy
pneu	pneumonia
polyneur	polyneuropathy
ppb	parts per billion (μg/L)
ppm	parts per million (mg/L)
p_v	specific vapor density
pulm	pulmonary
pup	pupil
ρ	rho
QSAR	quantitative structure-activity relationships
R	ideal gas constant (8.20575×10^{-2} atm·L/mol·K)
R_a	retardation factor for an acid
R_b	retardation factor for a base
RBC	red blood cell
RCRA	Resource Conservation and Recovery Act
R_d	retardation factor
resp	respiratory
resp ar	respiratory arrest
retster	retrosternal
rhin	rhinorrhea
RT	room temperature
RTECS	Registry of Toxic Effects of Chemical Substances

S	solubility
S_a	solute concentration in air (mol/L)
salv	salivation
SARA	Superfund Amendments and Reauthorization Act
scotoma	an area of absent or depressed vision in the visual field
sec	secondary (as in *sec*-butyl)
sens	sensitization
sleep	sleepiness
sneez	sneezing
S_o	solubility in organics
S_w	solubility in water
sol	soluble
soln	solution
som	somnolence
sp.	species
spas	spasm
STEL	short-term exposure limit
subl	sublimes
sweat	sweating
swell	swelling
sym	symmetric
SYM	symptoms
t	tertiary (as in *t*-butyl; but *tert*-butyl)
T	temperature
$t_{1/2}$	half-life
tacar	tachycardia
tech	technical
temp	temperature
tend	tenderness
TLV	threshold limit value
TLV-C	threshold limit value ceiling
TR PROD	transformation products
T_s	temperature of a substance
T_w	temperature of water
TWA	time-weighted average
μ	micro (10^{-6})
μg	microgram
uel	upper explosive limit
unsym	unsymmetric
U.S. EPA	U.S. Environmental Protection Agency
UV	ultraviolet
V, vol	volume
vap d	vapor density
V_c	average linear velocity of contaminant (e.g., ft/day)
verti	vertigo
vesic	vesiculation

vis dist	visual disturbance
vomit	vomiting
vp	vapor pressure
V_w	average linear velocity of groundwater (e.g., ft/day)
W	watt
λ	wavelength
weak	weakness
wheez	wheezing
wk	week
wt	weight
yr	year
z^+	positively charged species (milliequivalents/cm^3)

Contents

GROUNDWATER CHEMICALS FIELD GUIDE

ACENAPHTHENE

SYNONYMS: 1,2-Dihydroacenaphthylene; Ethylene naphthalene; 1,8-Ethylene naphthalene; 1,8-Hydroacenaphthylene; Periethylene naphthalene.

CHEMICAL DESIGNATIONS: CAS: 83-32-9; DOT: none assigned; mf: $C_{12}H_{10}$; fw: 154.21; RTECS: AB 1255500.

PROPERTIES: White cryst solid. Mp: 89.9°, bp: 279°, *d*: 1.0242 @ 90/4°, pK_a: > 15, K_H: 1.5 x 10^{-4} atm·m^3/mol @ 20°, log K_{oc}: 1.25, log K_{ow}: 3.92, 4.33, S_o (g/L): methanol (17.86), ethanol (32.26), propanol (40), chloroform (400), benzene or toluene (200), glacial acetic acid (32), S_w: 3.47-4.16 mg/L @ 25°, vp: 1.55 x 10^{-3} mm @ 25°.

TRANSFORMATION PRODUCTS: Ozonation in water @ 60° produced 7-formyl-1-indanone, 1-indanone, 7-hydroxy-1-indanone, 1-indanone-7-carboxylic acid, indane-1,7-dicarboxylic acid and indane-1-formyl-7-carboxylic acid. Wet oxidation of acenaphthene @ 320° yielded formic and acetic acids.

USES: Manufacture of dye intermediates, pharmaceuticals, insecticides, fungicides and plastics; chemical research.

ACENAPHTHYLENE

SYNONYM: Cyclopenta[*de*]naphthalene.

CHEMICAL DESIGNATIONS: CAS: 208-96-8; DOT: none assigned; mf: $C_{12}H_8$; fw: 152.20; RTECS: AB 1254000.

PROPERTIES: Solid. Mp: 92-93°, bp: 280°, *d*: 0.8988 @ 16/2°, pK_a: > 15, K_H: 1.14 x 10^{-4} atm·m^3/mol @ 25°, IP: 8.29 eV, 8.73 eV, log K_{oc}: 3.68 (calcd), log K_{ow}: 4.07, S_o: sol in ethanol, ether and benzene, S_w: 3.93 mg/L @ 25°, vp: 2.90 x 10^{-2} mm @ 20°.

TRANSFORMATION PRODUCTS: Based on data for structurally similar compounds, acenaphthylene may undergo photolysis to yield quinones. In a toluene soln, irradiation of acenaphthylene at various temperatures and concentrations all resulted in the formation of dimers. Ozonation in water @ 60° produced 1,8-naphthalene dialdehyde, 1,8-naphthalene anhydride, 1,2-epoxyacenaphthylene, 1-naphthoic acid and 1,8-naphthaldehydic acid.

EXPOSURE LIMITS: No individual standards have been set, however, as a constituent in coal tar pitch volatiles, the following exposure limits have been established: NIOSH REL: 10-h TWA 0.1 mg/m^3 (cyclohexane-extractable fraction); OSHA PEL: TWA 0.2 mg/m^3 (benzene-soluble fraction); ACGIH TLV: TWA 0.2 mg/m^3 (benzene solubles).

USES: Research chemical. Derived from industrial and experimental coal gasification operations where maximum concn detected in gas, liq and coal tar streams were 28, 4.1 and 18 mg/m^3, respectively.

ACETALDEHYDE

SYNONYMS: Acetic aldehyde; Aldehyde; Ethanal; Ethyl aldehyde; NCI-C56326;

RCRA waste number U001; UN 1089.

CHEMICAL DESIGNATIONS: CAS: 75-07-0; DOT: 1089; mf: C_2H_4O; fw: 44.05; RTECS: AB 1925000.

PROPERTIES: Colorless fuming liq or gas with a penetrating, pungent, fruity odor. Mp: -121°, bp: 20.8°, *d*: 0.7834 @ 18/4°, fl p: -37.8°, lel: 4%, uel: 60%, pK_a: 14.15 @ 0°, K_H: 6.58 x 10^{-5} atm·m^3/mol @ 25°, IP: 10.21 eV, log K_{ow}: 0.43 (calcd), S_o: misc with acetone, alcohol, benzene, ether, gasoline, solvent naphtha, toluene, turpentine and xylene, S_w: misc, vap d: 1.80 g/L @ 25°, 1.52 (air = 1), vp: 750 mm @ 20°.

TRANSFORMATION PRODUCTS: Oxidation in air yields acetic acid. Photooxidation of acetaldehyde in NO_x-free air (λ = 2900-3500 Å) yielded H_2O_2, alkylhydroperoxides, CO and lower molecular weight aldehydes. In the presence of NO_x, photooxidation products included O_3, H_2O_2 and peroxyacyl nitrates. Anticipated products from the reaction of acetaldehyde with O_3 or hydroxyl radicals in the atmosphere are formaldehyde and CO_2. Reacts with NO_2 forming peroxyacyl nitrates, formaldehyde and methyl nitrate. Irradiation in the presence of chlorine yielded peroxyacetic acid, CO and CO_2. Synthetic air containing gaseous nitrous acid and exposed to artificial sunlight (λ = 300-450 nm) photooxidized acetaldehyde into formic acid, methyl nitrate and peroxyacetal nitrate. In the presence of sulfuric acid, polymerizes explosively to trimeric paraldehyde.

EXPOSURE LIMITS: NIOSH REL: IDLH 10,000 ppm; OSHA PEL: TWA 200 ppm; ACGIH TLV: TWA 100 ppm, STEL 150 ppm.

SYMPTOMS OF EXPOSURE: Eye, nose, throat irrit; conj; cough; CNS depres; eye, skin burns; derm; delayed pulm edem.

USES: Manufacture of acetic acid, acetic anhydride, aldol, aniline dyes, *n*-butyl alcohol, 1,3-butylene glycol, chloral, 2-ethylhexanol, paraldehyde, pentaerythritol, peracetic acid, pyridines, trimethylolpropane, flavors, perfumes, plastics, synthetic rubbers, disinfectants, drugs, explosives, antioxidants, yeast; silvering mirrors; hardening gelatin fibers.

ACETIC ACID

SYNONYMS: Acetic acid (aq soln); Ethanoic acid; Ethylic acid; Glacial acetic acid; Methanecarboxylic acid; Pyroligneous acid; UN 2789; UN 2790; Vinegar acid.

CHEMICAL DESIGNATIONS: CAS: 64-19-7; DOT: 2789, 2790 (10-80% soln); mf: $C_2H_4O_2$; fw: 60.05; RTECS: AF 1225000.

PROPERTIES: Colorless liq with a strong vinegar-like odor. Mp: 16.6°, bp: 117.9°, *d*: 1.0492 @ 20/4°, fl p: 40°, lel: 4.0%, uel: 16%, pK_a: 4.74, K_H: 1.23 x 10^{-3} atm·m^3/mol @ 25°, IP: 10.37 eV, 10.69 eV, log K_{ow}: -0.53 to -0.17, S_o: sol in acetone and benzene; misc with alcohol, carbon tetrachloride and glycerol, S_w: misc, vap d: 2.45 g/L @ 25°, 2.07 (air = 1), vp: 11.4 mm @ 20°.

TRANSFORMATION PRODUCTS: Near Wilmington, NC, organic wastes containing acetic acid (representing 52.6% of total dissolved organic carbon) were injected into an aquifer containing saline water to a depth of about 1,000 feet. The generation of gaseous components (hydrogen, nitrogen, hydrogen sulfide, CO_2 and methane) suggests acetic acid and possibly other waste constituents, were anaerobically degraded by microorganisms. Ozonolysis of acetic acid in distilled water @ 25° yielded glyoxylic acid which was oxidized readily to oxalic acid. Oxalic acid was further oxidized to CO_2. Ozonolysis accompanied by UV irradiation enhanced the removal of acetic acid.

EXPOSURE LIMITS: NIOSH REL: IDLH 1,000 ppm; OSHA PEL: TWA 10 ppm; ACGIH TLV: TWA 10 ppm, STEL 15 ppm.

SYMPTOMS OF EXPOSURE: Conj lac; irrit nose, throat; phar edema, chronic bron; burns eyes, skin; skin sens; dent erosion; black skin, hyperkeratosis.

USES: Preparation of acetates, acetic anhydride, acetone, acetyl compounds, cellulose acetates, chloroacetic acid, ethyl alcohol, ketene, methyl ethyl ketone, vinyl acetate, plastics and rubbers in tanning; laundry sour; acidulant and preservative in foods; printing calico and dyeing silk; solvent for gums, resins, volatile oils and other substances; manufacture of nylon and fiber, vitamins, antibiotics and hormones; production of insecticides, dyes, photographic chemicals, stain removers; latex coagulant; textile printing; food preservative.

ACETIC ANHYDRIDE

SYNONYMS: **Acetic acid anhydride**; Acetic oxide; Acetyl anhydride; Acetyl ether; Acetyl oxide; Ethanoic anhydrate; Ethanoic anhydride; UN 1715.

CHEMICAL DESIGNATIONS: CAS: 108-24-7; DOT: 1715; mf: $C_4H_6O_3$; fw: 102.09; RTECS: AK 1925000.

PROPERTIES: Colorless liq with a strong acetic acid odor. Mp: -73.1°, bp: 139.55°, *d*: 1.0820 @ 20/4°, fl p: 48.9°, lel: 2.9%, uel: 10.3%, K_H: 3.92 x 10^{-6} atm·m^3/mol @ 20° (calcd), S_o: sol in benzene and chloroform; misc with acetic acid, alcohol and ether, S_w: 12 wt % @ 20°, vap d: 4.17 g/L @ 25°, 3.52 (air = 1), vp: 3.5 mm @ 20°.

TRANSFORMATION PRODUCTS: Slowly reacts with moisture forming acetic acid.

EXPOSURE LIMITS: NIOSH REL: IDLH 1,000 ppm; OSHA PEL: C 5 ppm; ACGIH TLV: C 5 ppm.

SYMPTOMS OF EXPOSURE: Conj lac, corn edema; opac, photophobia, nas phar irrit, cough, dysp bron; skin burns, vesic, derm.

USES: Preparation of acetyl compounds and cellulose acetates; detection of rosin; dehydrating and acetylating agent in the production of pharmaceuticals, dyes, explosives, perfumes and pesticides; organic synthesis.

ACETONE

SYNONYMS: Chevron acetone; Dimethylformaldehyde; Dimethylketal; Dimethyl ketone; DMK; Ketone propane; β-Ketopropane; Methyl ketone; Propanone; **2-Propanone**; Pyroacetic acid; Pyroacetic ether; RCRA waste number U002; UN

ACGIH TLV: TWA 0.1 ppm, STEL 0.3 ppm.

SYMPTOMS OF EXPOSURE: Irrit eyes, skin, muc memb; abnormal pulm func; delayed pulm edema, chronic resp disease.

USES: Intermediate in the manufacture of many chemicals (glycerine, 1,3,6-hexanediol, β-chloropropionaldehyde, 1,2,3,6-tetrahydrobenzaldehyde, β-picoline, nicotinic acid), pharmaceuticals, polyurethane, polyester resins, liquid fuel, slimicide; herbicide; antimicrobial agent; control of aquatic weeds; warning agent in gases.

ACRYLAMIDE

SYNONYMS: Acrylamide monomer; Acylic amide; Ethylenecarboxamide; Propenamide; **2-Propenamide**; RCRA waste number U007; UN 2074.

CHEMICAL DESIGNATIONS: CAS: 79-06-1; DOT: 2074; mf: C_3H_5NO; fw: 71.08; RTECS: AS 3325000.

PROPERTIES: Colorless solid or flake-like cryst. Mp: 84-85°, bp: 125° @ 25 mm, *d*: 1.122 @ 30/4°, K_H: 3.03 x 10^{-9} atm·m^3/mol @ 20° (calcd), S_o: @ 30°: acetone (631 g/L), benzene (3.46 g/L), chloroform (26.6 g/L), ethanol (862 g/L), ethyl acetate (126 g/L), heptane (68 mg/L), methanol (1,550 g/L), S_w: 2,155 g/L @ 30°, vp: 7 x 10^{-3} mm @ 20°.

TRANSFORMATION PRODUCTS: Readily polymerizes at the melting point or under UV light. Acrylic acid and ammonium ions (NH_4^+) were reported as hydrolysis products. Under aerobic conditions, NH_4^+ is oxidized to nitrite ions and nitrate ions. The NH_4^+ produced in soil may volatilize as ammonia or accumulate as nitrite ions in sandy or calcareous soils.

EXPOSURE LIMITS: NIOSH REL: 10-h TWA 0.3 mg/m^3; OSHA PEL: TWA 0.3 mg/m^3; ACGIH TLV: TWA 0.03 mg/m^3.

SYMPTOMS OF EXPOSURE: Atax; numb limbs, pares; musc weak; absent deep tendon reflex; hand sweat; ftg; lethargy; irrit eyes, skin.

USES: Synthesis of dyes; flocculants; polymers or copolymers as plastics, adhesives, soil conditioning agents; sewage and waste treatment; ore processing; permanent press fabrics.

ACRYLONITRILE

SYNONYMS: Acritet; Acrylon; Acrylonitrile monomer; An; Carbacryl; Cyanoethylene; ENT 54; Fumigrain; Miller's fumigrain; Nitrile; Propenenitrile; **2-Propenenitrile**; RCRA waste number U009; TL 314; UN 1093; VCN; Ventox; Vinyl cyanide.

CHEMICAL DESIGNATIONS: CAS: 107-13-1; DOT: 1093; mf: C_3H_3N; fw: 53.06; RTECS: AT 5250000.

PROPERTIES: Clear, colorless, watery liq with a sweet irritating odo resembling peach pits. Mp: -83°, bp: 77.5-79°, *d*: 0.8060 @ 20/4°, fl p: -1°, 3.05%, uel: 17.0 ± 0.5%, K_H: 1.10 x 10^{-4} atm·m^3/mol @ 25°, IP: 10.91 eV, log

-1.13 (calcd), log K_{ow}: -0.92 to 1.20, S_o: sol in ethanol, ether, acetone, benzene, carbon tetrachloride, toluene; misc in alcohol and chloroform, S_w: 80 g/L @ 25°, vap d: 2.17 g/L @ 25°, 1.83 (air = 1), vp: 83 mm @ 20°.

TRANSFORMATION PRODUCTS: In an aq soln @ 50°, UV light converted acrylonitrile to CO_2. After 24 h, the concn of acrylonitrile was reduced 24.2%. Ozonolysis of acrylonitrile in the liquid phase yielded formaldehyde and the tentatively identified compounds glyoxal, an epoxide of acrylonitrile and acetamide. In the gas phase, cyanoethylene oxide was reported as an ozonolysis product. Anticipated products from the reaction of acrylonitrile with O_3 or hydroxyl radicals in air included formaldehyde, formic acid, HC(O)CN and cyanide ions. Degradation by the microorganism *Nocardia rhodochrous* yielded ammonium ion and propionic acid, the latter being oxidized to CO_2 and water. Wet oxidation of acrylonitrile @ 320° yielded formic and acetic acids.

EXPOSURE LIMITS: NIOSH REL: TWA 1 ppm, 15-min C 10 ppm; OSHA PEL: TWA 2 ppm, 15-min C 10 ppm; ACGIH TLV: TWA 2 ppm.

SYMPTOMS OF EXPOSURE: Asphy; irrit eyes; head; sneez; nau, vomit; weak, li-head; skin vesic, scaling derm.

USES: Copolymerized with methyl acrylate, methyl methacrylate, vinyl acetate, vinyl chloride, or 1,1-dichloroethylene to produce acrylic, modacrylic and high-strength fibers; ABS (acrylonitrile-butadiene-styrene) and acrylonitrile-styrene copolymers; nitrile rubber; cyanoethylation of cotton; synthetic soil block (acrylonitrile polymerized in wood pulp); manufacture of adhesives; organic synthesis; grain fumigant; pesticide; monomer for a semi-conductive polymer that can be used similar to inorganic oxide catalysts in dehydrogenation of *t*-butyl alcohol to isobutylene and water; pharmaceuticals; antioxidants; dyes and surfactants.

ALDRIN

SYNONYMS: Aldrec; Aldrex; Aldrex 30; Aldrite; Aldrosol; Altox; Compound 118; Drinox; ENT 15,949; Hexachlorohexahydro-*endo,exo*-dimethanonaphthalene; **1,2,3,4,10,10-Hexachloro-1,4,4a,5,8,8a-hexahydro-1,4:5,8-dimethanonaphthalene**; 1,2,3,4,10,10-Hexachloro-1,4,4a,5,8,8a-hexahydro-1,4-*endo,exo*-5,8-dimethanonaphthalene; 1,2,3,4,10,10-Hexachloro-1,4,4a,5,8,8a-hexahydro-*exo*-1,4-*endo*-5,8-dimethanonaphthalene; 1,4,4a,5,8,8a-Hexahydro-1,4-*endo,exo*-5,8-dimethanonaphthalene; HHDN; NA 2,761; NA 2,762; NCI-C00044; Octalene; RCRA waste number P004; Seedrin; Seedrin liquid.

CHEMICAL DESIGNATIONS: CAS: 309-00-2; DOT: 2761; mf: $C_{12}H_8Cl_6$; fw: 364.92; RTECS: IO 2100000.

PROPERTIES: White, odorless cryst when pure; tech grades are tan to dark brown with a mild chemical odor. Mp: 104°, bp: 145° @ 2 mm, *d*: 1.70 @ 20/4°, K_H: 4.96 x 10^{-4} atm·m^3/mol @ 25°, log K_{oc}: 2.61, 4.69, log K_{ow}: 5.17-7.4, S_o: sol in most organic solvents, S_w: 17-180 µg/L @ 25°, vp: 2.31 x 10^{-5} mm @ 20°.

TRANSFORMATION PRODUCTS: Dieldrin is the major metabolite from the microbial degradation of aldrin by oxidation or epoxidation. Dieldrin may further degrade to photodieldrin. Under oceanic conditions, aldrin may undergo dihydroxylation at the chlorine free double bond to produce aldrin diol. Photolysis of 0.33 ppb aldrin in San Francisco Bay water by sunlight produced photodieldrin ($t_{1/2}$ = 1.1 day). Aldrin on silica gel plates in the presence of photosensitizers and exposed to sunlight produced photoaldrin. Photoaldrin also formed when a benzene soln containing aldrin and benzophenone as a sensitizer was exposed to UV light (λ = 268-356 nm). Photoaldrin and photodieldrin formed when aldrin was codeposited on bean leaves and exposed to sunlight. Photodegradation of aldrin by sunlight yielded the following products after 1 mo: dieldrin, photodieldrin, photoaldrin and a polymeric substance. Photolysis of solid aldrin using a high pressure mercury lamp with a pyrex filter (λ > 300 nm) yielded a polymeric substance with small amounts of photoaldrin, dieldrin, hydrochloric acid and CO_2. Sunlight and UV light can convert aldrin to photoaldrin. Photooxidation of aldrin in water is accelerated when H_2O_2 is present. Products identified include dieldrin, photoaldrin and possibly a hydroperoxide. In an aq sol containing peracetic acid, aldrin was transformed to dieldrin in the dark. Oxidation of aldrin by oxygen also yielded dieldrin.

EXPOSURE LIMITS: NIOSH REL: lowest detectable limit; OSHA PEL: TWA 0.25 mg/m^3; ACGIH TLV: TWA 0.25 mg/m^3.

SYMPTOMS OF EXPOSURE: Head, dizz; nau, vomit, mal; myoclonic jerks of limbs; clonic, tonic convuls; coma; hema, azotemia.

USES: Formerly as insecticide and fumigant; manufactur and use has been discontinued in the U.S.

ALLYL ALCOHOL

SYNONYMS: AA; Allyl al; Allylic alcohol; 3-Hydroxypropene; Orvinylcarbinol; Propenol; Propenol-3; Propen-1-ol-3; 1-Propenol-3; 1-Propen-3-ol; 2-Propenol; **2-Propen-1-ol**; Propenyl alcohol; 2-Propenyl alcohol; RCRA waste number P005; UN 1098; Vinyl carbinol.

CHEMICAL DESIGNATIONS: CAS: 107-18-6; DOT: 1098; mf: C_3H_6O; fw: 58.08; RTECS: BA 5075000.

PROPERTIES: Colorless liq with a pungent, mustard-like odor. Mp: -129°, bp: 97.1°, *d*: 0.8540 @ 20/4°, fl p: 21.1° (OC), lel: 2.5%, uel: 18%, K_H: 5.00 x 10^{-6} atm·m^3/mol @ 25°, IP: 9.67 eV, log K_{oc}: 0.51 (calcd), log K_{ow}: -0.22, 0.17, S_o: misc with alcohol, chloroform, ether and petroleum ether, S_w: misc, vap d: 2.37 g/L @ 25°, 2.01 (air = 1), vp: 20 mm @ 20°.

TRANSFORMATION PRODUCTS: Will slowly polymerize over time.

EXPOSURE LIMITS: NIOSH REL: IDLH 150 ppm; OSHA PEL: TWA 2 ppm; ACGIH TLV: TWA 2 ppm, STEL 4 ppm.

SYMPTOMS OF EXPOSURE: Eye irrit; tissue damage, burns; irrit upper resp, skin; pulm edema.

USES: Manufacture of acrolein, allyl compounds, glycerol, plasticizers, resins, war gas; contact pesticide for weed seeds and certain fungi; intermediate for pharmaceuticals and other organic compounds; herbicide.

ALLYL CHLORIDE

SYNONYMS: Chlorallylene; 3-Chloroprene; 1-Chloropropene-2; 1-Chloro-2-propene; 3-Chloropropene; 3-Chloroprene-1; **3-Chloro-1-propene**; 1-Chloro-2-propylene; 3-Chloropropylene; α-Chloropropylene; 3-Chloro-1-propylene; NCI-C04615; UN 1100.

CHEMICAL DESIGNATIONS: CAS: 107-05-1; DOT: 1100; mf: C_3H_5Cl; fw: 76.53; RTECS: UC 7350000.

PROPERTIES: Colorless or yellow liq with a pungent odor. Mp: -134.5°, bp: 45°, *d*: 0.9376 @ 20/4°, fl p: -29° (OC), lel: 2.9%, uel: 11%, K_H: 1.08 x 10^{-2} atm·m^3/mol @ 25° (calcd), log K_{oc}: 1.68 (calcd), log K_{ow}: 1.79 (calcd), S_o: sol in acetone, benzene and lignoin; misc with alcohol, chloroform, ether, petroleum ether, S_w: 3.6 g/L @ 20°, vap d: 3.13 g/L @ 25°, 2.64 (air = 1), vp: 295 mm @ 20°.

TRANSFORMATION PRODUCTS: Hydrolysis under basic conditions will yield allyl alcohol (est $t_{1/2}$ = 69 days @ 25° and pH 7). Anticipated products from the reaction of allyl chloride with O_3 or hydroxyl radicals in the atmosphere are formaldehyde, formic acid, chloroacetaldehyde, chloroacetic acid and chlorinated hydroxy carbonyls.

EXPOSURE LIMITS: NIOSH REL: TWA 1 ppm 10-h, IDLH 300 ppm; OSHA PEL: TWA 1 ppm; ACGIH TLV: TWA 1 ppm, STEL 2 ppm.

SYMPTOMS OF EXPOSURE: Irrit eyes, nose, skin; pulm edema; lung, liv, kid damage; deep musc ache; deep bone ache.

USES: Preparation of epichlorohydrin, glycerol, allyl compounds; thermosetting resins for adhesives, plastics, varnishes; synthesis of pharmaceuticals, glycerol and insecticides.

ALLYL GLYCIDYL ETHER

SYNONYMS: AGE; Allyl-2,3-epoxypropyl ether; 1-Allyloxy-2,3-epoxypropane; 1,2-Epoxy-3-allyloxypropane; Glycidyl allyl ether; NCI-C56666; **[(2-Propenyl-oxy)methyl]oxirane**; UN 2219.

CHEMICAL DESIGNATIONS: CAS: 106-92-3; DOT: 2219; mf: $C_6H_{10}O_2$; fw: 114.14; RTECS: RR 0875000.

PROPERTIES: Colorless liq with a strong odor. Mp: -100°, bp: 153.9°, *d*: 0.9698 @ 20/4°, fl p: 57.2°, K_H: 3.83 x 10^{-6} atm·m^3/mol @ 20° (calcd), log K_{ow}: 0.63 (calcd), S_w: 141 g/L, vap d: 4.67 g/L @ 25°, 3.94 (air = 1), vp: 3.6 mm @ 20°.

EXPOSURE LIMITS: NIOSH REL: 15-min C 9.6 ppm, IDLH 270 ppm; OSHA PEL: C 10 ppm; ACGIH TLV: TWA 5 ppm, STEL 10 ppm.

SYMPTOMS OF EXPOSURE: Derm; eye, nose irrit; pulm irrit, edema; narco.

USES: Ingredient in epoxy resins.

4-AMINOBIPHENYL

SYNONYMS: *p*-Aminobiphenyl; 4-Aminodiphenyl; *p*-Aminodiphenyl; Anilinobenzene; Biphenylamine; 4-Biphenylamine; *p*-Biphenylamine; **[1,1′-Biphenyl]-4-amine**; Paraminodiphenyl; 4-Phenylaniline; *p*-Phenylaniline; Xenylamine.

CHEMICAL DESIGNATIONS: CAS: 92-67-1; DOT: none assigned; mf: $C_{12}H_{11}N$; fw: 169.23; RTECS: DU 8925000.

PROPERTIES: Colorless cryst which turn purple on exposure to air; floral-like odor. Mp: 50-54°, bp: 302°, *d*: 1.160 @ 20/20°, fl p: 152.8°, K_H: 3.89 x 10^{-10} atm·m^3/mol @ 25° (calcd), log K_{oc}: 2.03 (calcd), log K_{ow}: 2.78, S_o: sol in alcohol, chloroform and ether, S_w: 842 mg/L @ 20-30°, vp: 6 x 10^{-5} mm @ 20-30°.

SYMPTOMS OF EXPOSURE: Head, lethargy, dizz, dysp, atax, weak, methemoglobinemia, urinary burning, acute hemorr cystitis.

USES: Detecting sulfates; formerly used as a rubber antioxidant; cancer research.

2-AMINOPYRIDINE

SYNONYMS: Amino-2-pyridine; α-Aminopyridine; *o*-Aminopyridine; **2-Pyridinamine**; α-Pyridinamine; α-Pyridylamine; 2-Pyridylamine; UN 2671.

CHEMICAL DESIGNATIONS: CAS: 504-29-0; DOT: 2671; mf: $C_5H_6N_2$; fw: 94.12; RTECS: US 1575000.

PROPERTIES: Colorless solid or cryst with a characteristic odor. Mp: 57-58°, bp: 204° (subl), *d*: 1.073 @ 20/4° (calcd), fl p: 67.8°, pK_a: 6.86 @ 25°, log K_{ow}: -0.22, S_o: sol in acetone, alcohol, benzene and ether, S_w: > 100 wt % @ 20°.

TRANSFORMATION PRODUCTS: Releases toxic NO_x when heated to decomposition.

EXPOSURE LIMITS: NIOSH REL: IDLH 5 ppm; OSHA PEL: TWA 0.5 ppm; ACGIH TLV: TWA 0.5 ppm.

SYMPTOMS OF EXPOSURE: Head, dizz; nau; high BP; resp distress; weak; convuls; stupor.

USES: Manufacture of pharmaceuticals, especially antihistamines.

AMMONIA

SYNONYMS: Am-fol; Ammonia anhydrous; Ammonia gas; Anhydrous ammonia; Nitro-sil; R-717; Spirit of Hartshorn; UN 1005.

CHEMICAL DESIGNATIONS: CAS: 7664-41-7; DOT: 1005; mf: H_3N; fw: 17.04; RTECS: BO 0875000.

PROPERTIES: Colorless gas with a penetrating, pungent, suffocating odor. Mp: -77.8°, bp: -33.3°, *d*: 0.77 @ 0/4°, lel: 15%, uel: 28%, pK_a: 9.247 @ 25° (as NH_4^+), K_H: 2.91 x 10^{-4} atm·m^3/mol @ 20° (calcd), IP: 10.15 eV, log K_{oc}: 0.49 (calcd), log K_{ow}: 0.00, S_o: sol in chloroform, ether, methanol (16 wt % @ 25°) and ethanol (20 wt % @ 25°), S_w: 531 g/L @ 20°, vap d: 0.7714 g/L, 0.5967 (air = 1), vp: 8.7 atm @ 20°.

TRANSFORMATION PRODUCTS: Reacts with acids forming soluble ammonium salts.

EXPOSURE LIMITS: NIOSH REL: 5-min C 50 ppm, IDLH 500 ppm; OSHA PEL: TWA 50 ppm; ACGIH TLV: TWA 25 ppm, STEL 35 ppm.

SYMPTOMS OF EXPOSURE: Eye, nose, throat irrit; dysp; bronspas; chest pain; pulm edema; pink frothy sputum; skin burns; vesic.

USES: Manufacture of acrylonitrile, hydrazine hydrate, hydrogen cyanide, nitric acid, sodium carbonate, urethane, explosives, synthetic fibers, fertilizers; refrigerant; condensation catalyst; dyeing; neutralizing agent; synthetic fibers; latex preservative; fuel cells, rocket fuel; nitrocellulose; nitroparaffins; ethylenediamine, melamine; sulfite cooking liquors; developing diazo films; yeast nutrient.

n-AMYL ACETATE

SYNONYMS: Acetic acid amyl ester; **Acetic acid pentyl ester**; Amyl acetate; Amyl acetic ester; Amyl acetic ether; Banana oil; Birnenoel; Pear oil; Pent-acetate; Pent-acetate 28; 1-Pentanol acetate; 1-Pentyl acetate; *n*-Pentyl acetate; Primary amyl acetate; UN 1104.

CHEMICAL DESIGNATIONS: CAS: 628-63-7; DOT: 1104; mf: $C_7H_{14}O_2$; fw: 130.19; RTECS: AJ 1925000.

PROPERTIES: Colorless liq with a banana-like odor. Mp: -70.8°, bp: 149.25°, *d*: 0.8756 @ 20/4°, fl p: 25°, lel: 1.1%, uel: 7.5%, K_H: 3.88 x 10^{-4} atm·m^3/mol @ 25°, log K_{ow}: 2.349 (calcd), S_o: misc with alcohol and ether, S_w: 1.8 g/L @ 20°, vap d: 5.32 g/L @ 25°, 4.49 (air = 1), vp: 4 mm @ 20°.

TRANSFORMATION PRODUCTS: Anticipated hydrolysis products include acetic acid and *n*-pentyl alcohol.

EXPOSURE LIMITS: NIOSH REL: IDLH 4,000 ppm; OSHA PEL: TWA 100 ppm; ACGIH TLV: TWA 100 ppm.

SYMPTOMS OF EXPOSURE: Irrit eyes, nose; narco; derm.

USES: Solvent for lacquers and paints; leather polishes; flavoring agent; photographic film; extraction of penicillin; nail polish; printing and finishing fabrics; odorant.

sec-AMYL ACETATE

SYNONYMS: 2-Acetoxypentane; 1-Methylbutyl acetate; **2-Pentanol acetate**; 2-Pentyl acetate; *sec*-Pentyl acetate.

CHEMICAL DESIGNATIONS: CAS: 626-38-0; DOT: 1104; mf: $C_7H_{14}O_2$; fw: 130.19; RTECS: AJ 2100000.

PROPERTIES: Clear, colorless liq with a fruity odor. Bp: 134°, *d*: 0.862-0.866 @ 20/20°, fl p: 31.7°, lel: 1%, K_H: 4.87 x 10^{-4} atm·m^3/mol @ 20° (calcd), log K_{ow}: 5.26 (calcd), S_o: sol in alcohol and ether, S_w: 0.2 wt % @ 20°, vap d: 5.32 g/L @ 25°, 4.49 (air = 1), vp: 10 mm @ 35.2°.

TRANSFORMATION PRODUCTS: Anticipated hydrolysis products include acetic acid and *sec*-pentyl alcohol.

EXPOSURE LIMITS: NIOSH REL: IDLH 9,000 ppm; OSHA PEL: TWA 125 ppm; ACGIH TLV: TWA 125 ppm.

SYMPTOMS OF EXPOSURE: Irrit eyes, nose; narco; derm.

USES: Solvent for nitrocellulose and ethyl cellulose; coated paper, lacquers; cements; nail enamels, leather finishes; textile sizing and printing compounds; plastic wood.

ANILINE

SYNONYMS: Aminobenzene; Aminophen; Aniline oil; Anyvim; **Benzenamine**; Benzidam; Blue oil; C.I. 76,000; C.I. oxidation base 1; Cyanol; Krystallin; Kyanol; NCI-CO3736; Phenylamine; RCRA waste number U012; UN 1547.

CHEMICAL DESIGNATIONS: CAS: 62-53-3; DOT: 1547; mf: C_6H_7N; fw: 93.13; RTECS: BW 6650000.

PROPERTIES: Colorless to brown, oily liq with a faint ammonical odor. Mp: -6.3°, bp: 184°, *d*: 1.02173 @ 20/4°, fl p: 70°, lel: 1.3%, uel: 11%, pK_a: 4.630 @ 25°, K_H: 0.136 atm·m^3/mol @ 25° and pH 7.3, IP: 7.70 eV, log K_{oc}: 1.41, log K_{ow}: 0.781-1.09, S_o: sol in acetone, ether and lignoin; misc with alcohol, benzene and chloroform, S_w: 36.65 mg/L @ 25°, vap d: 3.81 g/L @ 25°, 3.22 (air = 1), vp: 0.3 mm @ 20°.

TRANSFORMATION PRODUCTS: Alkali or alkaline earth metals dissolve with hydrogen evolution and the formation of anilides. Irradiation of an aq soln @ 50° for 24 h resulted in a 28.5% yield of CO_2. A reversible equilibrium is quickly established when aniline covalently bonds with humates in soils forming imine linkages. These quinodal structures may oxidize to give nitrogen-substituted quinoid rings. A CO_2 yield of 46.5% was achieved when aniline adsorbed on silica gel was irradiated with light ($\lambda > 290$ nm) for 17 h. In activated sludge, 20.5% mineralized to CO_2 after 5 days. Products identified from the gas-phase reaction of O_3 with aniline in synthetic air @ 23° were nitrobenzene, formic acid, H_2O_2 and a nitrated salt having the formula: $[C_6H_5NH_3]^+NO_3^-$. Aniline degraded in pond water containing sewage sludge to catechol which further degraded to CO_2. Intermediate compounds identified in minor degradative pathways include acetanilide, phenylhydroxylamine, *cis,cis*-muconic acid, β-ketoadipic acid, levulinic acid and succinic acid. Under anaerobic conditions using a sewage inoculum, 10% of the aniline present degraded to acetanilide and 2-methylquinoline. In a 56-day experiment, ^{14}C-labelled aniline applied to soil-water suspensions under aerobic and anaerobic conditions gave $^{14}CO_2$ yields of 26.5 and 11.9%, respectively. A bacterial culture isolated from the Oconee River in North Georgia degraded aniline to the intermediate catechol. Silage samples (chopped corn plants) containing aniline were incubated in an anaerobic chamber for 2 wk @ 28°. After 3 days, aniline was biologically metabolized to formanilide, propioanilide, 3,4-dichloroaniline, 3- and 4-chloroaniline. Various microorganisms isolated from soil degraded aniline to acetanilide, 2-hydroxyacetanilide, 4-hydroxyaniline and 2 unidentified phenols.

EXPOSURE LIMITS: NIOSH REL: IDLH 100 ppm; OSHA PEL: TWA 5 ppm; ACGIH TLV: TWA 2 ppm.

SYMPTOMS OF EXPOSURE: Head; weak, dizz; cyan; atax; dysp on effort;

tacar; eye irrit.

USES: Manufacture of dyes, resins, varnishes, medicinals, perfumes, photographic chemicals, shoe blacks, chemical intermediates; solvent; vulcanizing rubber; isocyanates for urethane foams; explosives; pharmaceuticals; petroleum refining; diphenylamine; phenolics; fungicides; herbicides.

o-ANISIDINE

SYNONYMS: 2-Aminoanisole; *o*-Aminoanisole; 1-Amino-2-methoxybenzene; 2-Anisylamine; 2-Methoxy-1-aminobenzene; 2-Methoxyaniline; *o*-Methoxyaniline; **2-Methoxybenzenamine**; *o*-Methoxyphenylamine; UN 2431.

CHEMICAL DESIGNATIONS: CAS: 90-04-0; DOT: 2431; mf: C_7H_9NO; fw: 123.15; RTECS: BZ 5410000.

PROPERTIES: Colorless to pink liq with an amine-like odor. Becomes brown on exposure to air. Mp: 5°, bp: 224°, *d*: 1.0923 @ 20/4°, fl p: 118° (OC), pK_a: 4.09 @ 25°, K_H: 1.25 x 10^{-6} atm·m^3/mol @ 25° (calcd), log K_{ow}: 0.95, 1.18, S_o: sol in acetone and benzene; misc with alcohol and ether, S_w: 1.3 wt % @ 20°, vap d: 5.03 g/L @ 25°, 4.25 (air = 1), vp: < 0.1 mm @ 30°.

EXPOSURE LIMITS: NIOSH REL: IDLH 50 mg/m^3; OSHA PEL: TWA 0.5 mg/m^3; ACGIH TLV: TWA 0.1 ppm.

SYMPTOMS OF EXPOSURE: Head, dizz; cyan; RBC Heinz bodies.

USES: Manufacture of azo dyes.

p-ANISIDINE

SYNONYMS: 4-Aminoanisole; *p*-Aminoanisole; 1-Amino-4-methoxybenzene; *p*-Aminomethoxybenzene; *p*-Aminomethylphenyl ether; 4-Anisidine; *p*-Anisylamine; 4-Methoxy-1-aminobenzene; 4-Methoxyaniline; *p*-Methoxyaniline; **4-Methoxybenzenamine**; *p*-Methoxybenzenamine; 4-Methoxyphenylamine; *p*-Methoxyphenylamine.

CHEMICAL DESIGNATIONS: CAS: 104-94-9; DOT: 2431; mf: C_7H_9NO; fw: 123.15; RTECS: BZ 5450000.

PROPERTIES: Light brown solid with an amine-like odor. Mp: 57-60°, bp: 240-246°, *d*: 1.096 @ 20/4°, fl p: 30°, pK_a: 4.49 @ 25°, IP: 7.82 eV, log K_{ow}: 0.95, S_o: sol in acetone, alcohol, benzene and ether, S_w: 3.3 mg/L @ RT (calcd).

TRANSFORMATION PRODUCTS: Releases toxic NO_x when heated to decomposition.

EXPOSURE LIMITS: NIOSH REL: IDLH 50 mg/m^3; OSHA PEL: TWA 0.5 mg/m^3; ACGIH TLV: TWA 0.1 ppm.

SYMPTOMS OF EXPOSURE: Head, dizz; cyan; RBC Heinz bodies.

USES: Azo dyestuffs; chemical intermediate.

ANTHRACENE

SYNONYMS: Anthracin; Green oil; Paranaphthalene; *p*-naphthalene; Tetra olive

N2G.

CHEMICAL DESIGNATIONS: CAS: 120-12-7; DOT: none assigned; mf: $C_{14}H_{10}$; fw: 178.24; RTECS: CA 9350000.

PROPERTIES: White to yellow cryst with a bluish or violet fluorescence and a weak aromatic odor. Mp: 216.2-216.4°, bp: 339.9°, *d*: 1.24 @ 20/4°, fl p: 121.1°, lel: 0.6%, K_a > 15, K_H: 1.77-6.51 x 10^{-5} atm·m^3/mol @ 25°, IP: 7.43-7.58 eV, log K_{oc}: 4.205-4.93, log K_{ow}: 4.34-4.54, S_o (g/L): sol in alcohol (14.9), methanol (14.3), benzene (16.1), carbon disulfide (32.3), carbon tetrachloride (11.6), chloroform (11.8) and toluene (8), S_w: 43.4 μg/L @ 25°, vp: 1.95 x 10^{-4} mm @ 20°.

TRANSFORMATION PRODUCTS: Catechol is the central metabolite in the bacterial degradation of anthracene. Intermediate byproducts included 3-hydroxy-2-naphthoic acid and salicylic acid. Oxidation of anthracene adsorbed on silica gel or alumina by oxygen in the presence of UV-light yielded anthraquinone. This compound further oxidized to 1,4-dihydroxy-9,10-anthraquinone. Anthraquinone also formed by the oxidation of anthracene in diluted nitric acid or NO_x and in the dark when adsorbed on fly ash. Irradiation of anthracene (2.6 x 10^{-3} M) in cyclohexanone solns gave 9,10-anthraquinone as the principal product. A mixed bacterial community isolated from sea water foam degraded anthraquinone, a photodegradation product of anthracene, to traces of benzoic and phthalic acids. Photocatalysis of anthracene and sulfur dioxide @ -25° in various solvents yielded anthracene-9-sulfonic acid. A CO_2 yield of 16.0% was achieved when anthracene adsorbed on silica gel was irradiated with light (λ > 290 nm) for 17 h. In activated sludge, only 0.3% mineralized to CO_2 after 5 days. In a 14-day experiment, ^{14}C-labelled anthracene applied to soil-water suspensions under aerobic and anaerobic conditions gave $^{14}CO_2$ yields of 1.3 and 1.8%, respectively. The reaction of anthracene with NO_x to form 9-nitroanthracene was reported to occur in urban air from St. Louis, MO.

EXPOSURE LIMITS: No individual standards have been set, however, as a constituent in coal tar pitch volatiles, the following exposure limits have been established: NIOSH REL: 10-h TWA 0.1 mg/m^3 (cyclohexane-extractable fraction); OSHA PEL: TWA 0.2 mg/m^3 (benzene-soluble fraction); ACGIH TLV: TWA 0.2 mg/m^3 (benzene solubles).

USES: Dyes; starting material for the preparation of alizarin, phenanthrene, carbazole, 9,10-anthraquinone, 9,10-dihydroanthracene and insecticides; in calico printing; as component of smoke screens; scintillation counter crystals; organic semiconductor research; wood preservative.

ANTU

SYNONYMS: Anturat; Bantu; Chemical 109; Krysid; **1-Naphthalenylthiourea**; 1-(1-Naphthyl)-2-thiourea; α-Naphthylthiourea; *N*-1-Naphthylthiourea; α-Naphthylthiocarbamide; Rattrack.

CHEMICAL DESIGNATIONS: CAS: 86-88-4; DOT: 1651; mf: $C_{11}H_{10}N_2S$; fw: 202.27; RTECS: YT 9275000.

PROPERTIES: Colorless to gray, odorless solid. Mp: 198°, bp: decomp, *d*: 1.895

(calcd), S_o: 24.3 g/L in acetone, 86 g/L in triethylene glycol, S_w: 600 mg/L @ 20°, vp: ≈ 0 mm @ 20°.

EXPOSURE LIMITS: NIOSH REL: IDLH 100 mg/m^3; OSHA PEL: TWA 0.3 mg/m^3.

SYMPTOMS OF EXPOSURE: Vomit; dysp; cyan; coarse pulm, rales after ingestion of large doses.

USE: Rodenticide.

BENZENE

SYNONYMS: Annulene; Benxole; Benzol; Benzole; Benzolene; Bicarburet of hydrogen; Carbon oil; Coal naphtha; Coal tar naphtha; Cyclohexatriene; Mineral naphthalene; Motor benzol; NCI-C55276; Nitration benzene; Phene; Phenyl hydride; Pyrobenzol; Pyrobenzole; RCRA waste number U019; UN 1114.

CHEMICAL DESIGNATIONS: CAS: 71-43-2; DOT: 1114; mf: C_6H_6; fw: 78.11; RTECS: CY 1400000.

PROPERTIES: Clear, colorless to light yellow liq with an aromatic or gasoline-like odor. Mp: 5.533°, bp: 80.100°, *d*: 0.8765 @ 20/4°, fl p: -11°, lel: 1.3%, uel: 7.1%, pK_a: ≈ 37, K_H: 5.48 x 10^{-3} atm·m^3/mol @ 25°, IP: 9.24-9.56 eV, log K_{oc}: 1.69-2.00, log K_{ow}: 1.56-2.20, S_o: freely misc with ethanol, ether, glacial acetic acid, acetone, chloroform, carbon tetrachloride, carbon disulfide and oils, S_w: 1,780 mg/L @ 20°, vap d: 3.19 g/L @ 25°, 2.70 (air = 1), vp: 76 mm @ 20°.

TRANSFORMATION PRODUCTS: A mutant of *Pseudomonas putida* dihydroxylyzed benzene into *cis*-benzene glycol accompanied by partial dehydrogenation yielding catechol. Bacterial dioxygenases can cleave catechol at the *ortho*- and *meta*- positions to yield *cis,cis*-muconic acid and α-hydroxymuconic semialdehyde, respectively. Pure microbial cultures hydroxylated benzene to phenol and 2 unidentified phenols. Various microorganisms isolated from soil degraded aniline to acetanilide, 2-hydroxyacetanilide, 4-hydroxyaniline and 2 unidentified phenols. Muconic acid was reported to be the biooxidation product of benzene by *Nocardia corallina* V-49 using *n*-hexadecane as the substrate. Titanium dioxide suspended in an aq soln and irradiated with UV light (λ = 365 nm) converted benzene to CO_2 at a significant rate. Irradiation of benzene in an aq soln yields mucondialdehyde. Photolysis of benzene vapor (λ = 1849-2000 Å) yielded ethylene, hydrogen, methane, ethane, toluene and a polymer resembling cuprene. Other photolysis products reported under different conditions include fulvene, acetylene, substituted trienes, phenol, 2-nitrophenol, 4-nitrophenol, 2,4-dinitrophenol, 2,6-dinitrophenol, nitrobenzene, formic acid and peroxyacetyl nitrate. Under atmospheric conditions, the gas-phase reaction with hydroxyl radicals and NO_x resulted in the formation of phenol and nitrobenzene. A CO_2 yield of 40.8% was achieved when benzene adsorbed on silica gel was irradiated with light (λ > 290 nm) for 17 h. In activated sludge, 29.2% of the applied benzene mineralized to CO_2 after 5 days. In anoxic groundwater near Bemidji, MI, benzene was anaerobically biodegraded to phenol.

EXPOSURE LIMITS: NIOSH REL: TWA 0.1 ppm, 15-min C 1 ppm; OSHA

PEL: TWA 10 ppm, 10-min C 50 ppm; ACGIH TLV: TWA 10 ppm.

SYMPTOMS OF EXPOSURE: Irrit eyes, nose, resp sys; gidd; head; nau; staggered gait; ftg, anor, lass; derm; bone marrow depres; abdom pain.

USES: Manufacture of ethylbenzene (preparation of styrene monomer), dodecylbenzene (for detergents), cyclohexane (for nylon), nitrobenzene, aniline, maleic anhydride, diphenyl, benzene hexachloride, benzene sulfonic acid, phenol, dichlorobenzenes, insecticides, pesticides, fumigants, explosives, aviation fuel, flavors, perfume, medicine, dyes and many other organic chemicals; paints, coatings, plastics and resins; food processing; photographic chemicals; nylon intermediates; paint removers; rubber cement; antiknock gasoline; solvent for fats, waxes, inks, oils, paints, plastics and rubber.

BENZIDINE

SYNONYMS: Azoic diazo component 112; Benzidine base; *p*-Benzidine; 4,4′-Bianiline; *p,p′*-Bianiline; **(1,1′-Biphenyl)-4,4′-diamine**; 4,4′-Biphenyldiamine; *p,p′*-Biphenyldiamine; 4,4′-Biphenylenediamine; *p,p′*-Biphenylenediamine; C.I. 37,225; C.I. azoic diazo component 112; 4,4′-Diaminobiphenyl; *p,p′*-Diaminobiphenyl; 4,4′-Diamino-1,1′-biphenyl; 4,4′-Diaminodiphenyl; *p*-Diaminodiphenyl; *p,p′*-Diaminodiphenyl; 4,4′-Dianiline; *p,p′*-Dianiline; 4,4′-Diphenylenediamine; *p,p′*-Diphenylenediamine; Fast corinth base B; NCI-C03361; RCRA waste number U021; UN 1885.

CHEMICAL DESIGNATIONS: CAS: 92-87-5; DOT: 1885; mf: $C_{12}H_{12}N_2$; fw: 184.24; RTECS: DC 9625000.

PROPERTIES: Grayish-yellow powder or white to pale reddish cryst. Mp: 117.2°, bp: 401.7°, *d*: 1.250 @ 20/4°, pK_1 = 3.63, pK_2 = 4.70, K_H: 3.88 x 10^{-11} atm·m^3/mol @ 25° (est), log K_{oc}: 5.72, log K_{ow}: 1.34-1.81, S_o: sol in ethanol and ether (20 g/L), S_w: 500 mg/L @ 25°, vap d: 7.50 g/L @ 20°, vp: 0.83 mm @ 20° (calcd).

TRANSFORMATION PRODUCTS: Benzidine added to different soils was incubated in the dark @ 23° under a CO_2-free atmosphere. After 1 yr, 8.3-11.6% of the added benzidine degraded to CO_2 primarily by microbial metabolism and partially by hydrolysis. Tentatively identified biooxidation compounds using GC/MS include *n*-hydroxybenzidine, 3-hydroxybenzidine, 4-amino-4′-nitrobiphenyl, *N,N′*-dihydroxybenzidine, 3,3′-dihydroxybenzidine and 4,4′-dinitrobiphenyl. In the presence of H_2O_2 and acetylcholine @ pH 11 and 20°, benzidine oxidized to 4-amino-4′-nitrobiphenyl. A CO_2 yield of 40.8% was achieved when benzene adsorbed on silica gel was irradiated with light (λ > 290 nm) for 17 h. In activated sludge, < 0.1% mineralized to CO_2 after 5 days.

SYMPTOMS OF EXPOSURE: Hema, secondary anem from hemolysis, acute cystitis, acute liv disorders, derm, painful and irregular urination.

USES: Organic synthesis; manufacture of azo dyes, especially Congo Red; detection of blood stains; stain in microscopy; laboratory reagent in determining cyanide, sulfate, nicotine and some sugars; stiffening agent in rubber compounding.

BENZO[*a*]ANTHRACENE

SYNONYMS: BA; B(*a*)A; Benzanthracene; **Benz[*a*]anthracene**; 1,2-Benzanthracene; 1,2-Benz[*a*]anthracene; 2,3-Benzanthracene; Benzanthrene; 1,2-Benzanthrene; Benzoanthracene; 1,2-Benzoanthracene; Benzo[*a*]phenanthrene; Benzo[*b*]phenanthrene; 2,3-Benzophenanthrene; Naphthanthracene; RCRA waste number U018; Tetraphene.

CHEMICAL DESIGNATIONS: CAS: 56-55-3; DOT: none assigned; mf: $C_{18}H_{12}$; fw: 228.30; RTECS: CV 9275000.

PROPERTIES: Colorless leaflets or plates with a greenish-yellow fluorescence. Mp: 155-162°, bp: 400°, *d*: 1.274 @ 20/4°, pK_a: > 15, K_H: 6.6 x 10^{-7} atm·m^3/mol, IP: 7.45-8.01 eV, log K_{oc}: 6.14, log K_{ow}: 5.61-5.91, S_o: sol in ethanol, ether, acetone and benzene, S_w: 9.4-16.8 μg/L @ 25°, vp: 5 x 10^{-9} mm @ 20°.

TRANSFORMATION PRODUCTS: In an enclosed marine ecosystem containing planktonic primary production and heterotrophic benthos, the major metabolites were water soluble and could not be extracted with organic solvents. The only degradation product identified was benz[*a*]anthracene-7,12-dione. This compound also formed from the photolysis of benzo[*a*]anthracene @ 366 nm in an air-saturated, acetonitrile-water solvent. Under aerobic conditions, *Cunninghanella elegans* can degrade benzo[*a*]anthracene to 3,4-, 8,9- and 10,11-dihydrols. A strain of *Beijerinckia* oxidized benzo[*a*]anthracene producing 1-hydroxy-2-anthranoic acid as the major product. Two other metabolites identified were 2-hydroxy-3-phenanthroic acid and 3-hydroxy-2-phenanthroic acid. In a marine microcosm containing Narragansett Bay sediments and the polychaete *Mediomastis ambesita* and the bivalve *Nucula anulata*, benz[*a*]anthracene degraded to CO_2, phenols and quinones. A CO_2 yield of 25.3% was achieved when benz[*a*]anthracene adsorbed on silica gel was irradiated with light (λ > 290 nm) for 17 h. In activated sludge, < 0.1% mineralized to CO_2 after 5 days.

EXPOSURE LIMITS: No individual standards have been set, however, as a constituent in coal tar pitch volatiles, the following exposure limits have been established: NIOSH REL: 10-h TWA 0.1 mg/m^3 (cyclohexane-extractable fraction); OSHA PEL: TWA 0.2 mg/m^3 (benzene-soluble fraction); ACGIH TLV: TWA 0.2 mg/m^3 (benzene solubles).

USES: Research chemical; organic synthesis. Not manufactured commercially but is derived from industrial and experimental coal gasification operations where maximum concn detected in gas, liq and coal tar streams were 28 mg/m^3, 4.1 mg/m^3 and 18 mg/m^3, respectively.

BENZO[*b*]FLUORANTHENE

SYNONYMS: **Benz[*e*]acephenanthrylene**; 3,4-Benz[*e*]acephenanthrylene; 2,3-Benzfluoranthene; 3,4-Benzfluoranthene; Benzo[*b*]fluoranthene; Benzo[*e*]fluoranthene; 2,3-Benzofluoranthene; 3,4-Benzofluoranthene; 3,4-Benzo[*b*]fluoranthene; B(*b*)F.

CHEMICAL DESIGNATIONS: CAS: 205-99-2; DOT: none assigned; mf:

$C_{20}H_{12}$; fw: 252.32; RTECS: CU 1400000.

PROPERTIES: Solid. Mp: 163-165°, pK_a: > 15, K_H: 1.2 x 10^{-5} atm·m^3/mol @ 20-25° (calcd), log K_{oc}: 5.74, log K_{ow}: 6.57, S_o: sol in most solvents, S_w: 1.2 μg/L @ 25°, vp: 5 x 10^{-7} mm @ 20°.

EXPOSURE LIMITS: No individual standards have been set, however, as a constituent in coal tar pitch volatiles, the following exposure limits have been established: NIOSH REL: 10-h TWA 0.1 mg/m^3 (cyclohexane-extractable fraction); OSHA PEL: TWA 0.2 mg/m^3 (benzene-soluble fraction); ACGIH TLV: TWA 0.2 mg/m^3 (benzene solubles).

USE: Research chemical. Derived from industrial and experimental coal gasification operations where maximum concn detected in gas, liq and coal tar streams were 0.38 mg/m^3, 33 μg/m^3 and 3.2 mg/m^3, respectively.

BENZO[*k*]FLUORANTHENE

SYNONYMS: 8,9-Benzfluoranthene; 8,9-Benzofluoranthene; 11,12-Benzofluoranthene; 11,12-Benzo[*k*]fluoranthene; B(*k*)F; 2,3,1′,8′-Binaphthylene; Dibenzo[*b,jk*]fluorene.

CHEMICAL DESIGNATIONS: CAS: 207-08-9; DOT: none assigned; mf: $C_{20}H_{12}$; fw: 252.32; RTECS: DF 6350000.

PROPERTIES: Solid. Mp: 217°, bp: 480°, pK_a: > 15, K_H: 1.04 x 10^{-3} atm·m^3/mol (calcd), log K_{oc}: 6.64 (calcd), log K_{ow}: 6.85, S_o: sol in most solvents, S_w: 0.55 μg/L @ 25°, vp: 9.59 x 10^{-11} mm @ 25°.

EXPOSURE LIMITS: No individual standards have been set, however, as a constituent in coal tar pitch volatiles, the following exposure limits have been established: NIOSH REL: 10-h TWA 0.1 mg/m^3 (cyclohexane-extractable fraction); OSHA PEL: TWA 0.2 mg/m^3 (benzene-soluble fraction); ACGIH TLV: TWA 0.2 mg/m^3 (benzene solubles).

USE: Research chemical. Derived from industrial and experimental coal gasification operations where maximum concn detected in liq and coal tar streams were 17 μg/m^3 and 1.6 mg/m^3, respectively.

BENZOIC ACID

SYNONYMS: Benzenecarboxylic acid; Benzeneformic acid; Benzenemethanoic acid; Benzoate; Carboxybenzene; Dracylic acid; NA 9,094; Phenylcarboxylic acid; Phenylformic acid; Retarder BA; Retardex; Salvo liquid; Salvo powder; Tenn-plas.

CHEMICAL DESIGNATIONS: CAS: 65-85-0; DOT: 9094; mf: $C_7H_6O_2$; fw: 122.12; RTECS: DG 0875000.

PROPERTIES: Colorless to white needles, scales, or powder with a faint benzaldehyde-like odor. Mp: 122.13°, bp: 249.2°, *d*: 1.2659 @ 15/4°, fl p: 121°, pK_a: 4.21 @ 25°, K_H: 7.02 x 10^{-8} atm·m^3/mol (calcd), IP: 9.73 eV, log K_{oc}: 1.48-2.70, log K_{ow}: 1.81-2.03, S_o: sol in ethanol, turpentine, chloroform (222 g/L), ether (333 g/L), acetone (333 g/L), carbon tetrachloride (33 g/L), benzene (100 g/L), carbon disulfide (33 g/L), volatile and fixed oils, S_w: 2,900 mg/L @ 20°, vp: 4.5 x 10^{-3} mm @ 25°.

TRANSFORMATION PRODUCTS: Benzoic acid degrades to catechol if it is the central metabolite whereas if protocatechuic acid (3,4-dihydroxybenzoic acid) is the central metabolite, the precursor is *m*-hydroxybenzoic acid. Other compounds identified following degradation of benzoic acid to catechol include *cis,cis*-muconic acid, (+)-muconolactone, 3-oxoadipate enol lactone and 3-oxoadipate. Pure microbial cultures hydroxylated benzoic acid to 3,4-dihydroxybenzoic acid, 2- and 4-hydroxybenzoic acid. In a methanogenic enrichment culture, 91% of the added benzoic acid anaerobically biodegraded to CO_2 and methane. Titanium dioxide suspended in an aq soln and irradiated with UV light (λ = 365 nm) converted benzoic acid to CO_2 at a significant rate. An aq soln containing chlorine and irradiated with UV light (λ = 350 nm) converted benzoic acid to salicylaldehyde and unidentified chlorinated compounds. A CO_2 yield of 10.2% was achieved when benzoic acid adsorbed on silica gel was irradiated with light (λ > 290 nm) for 17 h. In activated sludge, 65.5% of the applied amount mineralized to CO_2 after 5 days.

USES: Preparation of sodium and butyl benzoates, benzoyl chloride, phenol, caprolactum and esters for perfume and flavor industry; plasticizers; manufacture of alkyl resins; preservative for food, fats and fatty oils; seasoning; tobacco; dentifrices; standard in analytical chemistry; antifungal agent; synthetic resins and coatings; pharmaceutical and cosmetic preparations; plasticizer manufacturing (to modify resins such as polyvinyl chloride, polyvinyl acetate, phenol-formaldehyde).

BENZO[*ghi*]PERYLENE

SYNONYMS: 1,12-Benzoperylene; 1,12-Benzperylene; B(*ghi*)P.

CHEMICAL DESIGNATIONS: CAS: 191-24-2; DOT: none assigned; mf: $C_{22}H_{12}$; fw: 276.34; RTECS: DI 6200500.

PROPERTIES: Solid. Mp: 276.8°, bp: > 500°, pK_a: > 15, K_H: 1.4 x 10^{-7} atm·m^3/mol @ 25° (calcd), IP: 7.24 eV, log K_{oc}: 6.89 (calcd), log K_{ow}: 7.10, S_o: sol in most solvents, S_w: 0.26 μg/L @ 25°, vp: 1.01 x 10^{-10} mm @ 25°.

EXPOSURE LIMITS: No individual standards have been set, however, as a constituent in coal tar pitch volatiles, the following exposure limits have been established: NIOSH REL: 10-h TWA 0.1 mg/m^3 (cyclohexane-extractable fraction); OSHA PEL: TWA 0.2 mg/m^3 (benzene-soluble fraction); ACGIH TLV: TWA 0.2 mg/m^3 (benzene solubles).

USE: Research chemical. Derived from industrial and experimental coal gasification operations where the maximum concn detected in coal tar streams was 2.7 mg/m^3.

BENZO[*a*]PYRENE

SYNONYMS: Benzo[*d,e,f*]chrysene; 1,2-Benzopyrene; 3,4-Benzopyrene;

6,7-Benzopyrene; Benz[*a*]pyrene; Benzo[α]pyrene; 1,2-Benzpyrene; 3,4-Benzpyrene; 3,4-Benz[*a*]pyrene; 3,4-Benzypyrene; BP; B(*a*)P; 3,4-BP; RCRA waste number U022.

CHEMICAL DESIGNATIONS: CAS: 50-32-8; DOT: none assigned; mf: $C_{20}H_{12}$; fw: 252.32; RTECS: DJ 3675000.

PROPERTIES: Odorless, yellow cryst. Mp: 181.3°, bp: 495°, *d*: 1.351, pK_a: > 15, K_H: < 2.4 x 10^{-6} atm·m^3/mol, IP: 7.23 eV, log K_{oc}: 5.60-6.29 (calcd), log K_{ow}: 5.81-6.50, S_o: sol in benzene, toluene and xylene; sparingly sol in ethanol and methanol, S_w: 3.8 μg/L @ 25°, vp: 2.4-5.6 x 10^{-9} mm @ 25°.

TRANSFORMATION PRODUCTS: Benzo[*a*]pyrene was biooxidized by *Beijerinckia* B836 to *cis*-9,10-dihydroxy-9,10-dihydrobenzo[*a*]pyrene. Under nonenzymatic conditions, this metabolite monodehydroxylated to form 9-hydroxybenzo[*a*]pyrene. Under aerobic conditions, *Cunninghanella elegans* degraded benzo[*a*]pyrene to *trans*-7,8-dihydroxy-7,8-dihydrobenzo[*a*]pyrene. Coated glass fibers exposed to air containing 100-200 ppb O_3 yielded benzo[*a*]pyrene-4,5-oxide. At 200 ppb O_3, conversion yields of 50% and 80% were observed after 1 and 4 h, respectively. Free radical oxidation and photolysis of benzo[*a*]pyrene @ 366 nm yielded the following tentatively identified products: benzo[*a*]pyrene-1,6-quinone, benzo[*a*]pyrene-3,6-quinone and benzo[*a*]pyrene-6,12-quinone. Ozonolysis to benzo[*a*]pyrene-1,6-quinone or benzo[*a*]pyrene-3,6-quinone followed by further oxidation to benzanthrone dicarboxylic anhydride was reported. In an oxygenated soln, photolysis yields a mixture of 6,12-, 1,6- and 3,6-diones. Nitration by NO_2 forms 6-nitro-, 1-nitro- and 3-nitrobenzo[*a*]pyrene. In a simulated atmosphere, direct epoxidation by O_3 led to the formation of benzo[*a*]pyrene-4,5-oxide. Benzo[*a*]pyrene reacted with benzoyl peroxide to form the 6-benzoyloxy derivative. It was reported that benzo[*a*]pyrene adsorbed on fly ash and alumina reacted with sulfur dioxide (10%) in air to form benzo[*a*]pyrene sulfonic acid. Benzo[*a*]pyrene coated on a quartz surface was subjected to O_3 and natural sunlight for 4 and 2 h, respectively. 1,6-Quinone, 3,6-quinone, 6,12-quinone of benzo[*a*]pyrene were formed in both instances. A CO_2 yield of 26.5% was achieved when benzo[*a*]pyrene adsorbed on silica gel was irradiated with light (λ > 290 nm) for 17 h. In activated sludge, < 0.1% mineralized to CO_2 after 5 days. When benzo[*a*]pyrene adsorbed from the vapor phase onto coal fly ash, silica and alumina was exposed to NO_2, no reaction occurred. However, in the presence of nitric acid, nitrated compounds were produced. Chlorination of benzo[*a*]pyrene in polluted humus poor lake water gave 11,12-dichlorobenzo[*a*]pyrene and 1,11,12-, 3,11,12-, or 3,6,11-trichlorobenzo[*a*]pyrene representing 99% of the chlorinated products formed.

EXPOSURE LIMITS: no individual standards have been set, however, as a constituent in coal tar pitch volatiles, the following exposure limits have been established: NIOSH REL: 10-h TWA 0.1 mg/m^3 (cyclohexane-extractable fraction); OSHA PEL: TWA 0.2 mg/m^3 (benzene-soluble fraction); ACGIH TLV: TWA 0.2 mg/m^3 (benzene solubles)

USE: Research chemical. Derived from industrial and experimental coal gasification operations where maximum concn detected in gas, liq and coal tar streams were 5.0 mg/m^3, 36 μg/m^3 and 3.5 mg/m^3, respectively.

BENZO[*e*]PYRENE

SYNONYMS: 1,2-Benzopyrene; 1,2-Benzpyrene; 4,5-Benzopyrene; B(*e*)P.

CHEMICAL DESIGNATIONS: CAS: 192-97-2; DOT: none assigned; mf: $C_{20}H_{12}$; fw: 252.32; RTECS: DJ 4200000.

PROPERTIES: Solid. Mp: 178-179°, *d*: 0.8769 @ 20/4°, K_H: 4.84 x 10^{-7} atm·m^3/mol @ 25° (calcd), log K_{oc}: 5.60 (calcd), log K_{ow}: 6.75 (calcd), S_w: 4.59 μg/L @ 20.0°, vp: 5.54 x 10^{-9} mm @ 25°.

USES: Chemical research.

BENZYL ALCOHOL

SYNONYMS: Benzal alcohol; Benzene carbinol; **Benzenemethanol**; Benzoyl alcohol; α-Hydroxytoluene; NCI-C06111; Phenol carbinol; Phenyl carbinol; Phenyl methanol; Phenyl methyl alcohol; α-Toluenol.

CHEMICAL DESIGNATIONS: CAS: 100-51-6; DOT: none assigned; mf: C_7H_8O; fw: 108.14; RTECS: DN 3150000.

PROPERTIES: Colorless liq with a faint, pleasant aromatic odor. Mp: -15.3 to -9°, bp: 205.3°, *d*: 1.04535 @ 20/4°, fl p: 93°, IP: 9.14 eV, log K_{oc}: 1.98 (calcd), log K_{ow}: 1.10, S_o: sol in acetone, ethanol, benzene; misc with ether and absolute alcohol, S_w: 35 g/L @ 20°, vap d: 4.42 g/L @ 25°, 3.73 (air = 1), vp: 1 mm @ 58°.

TRANSFORMATION PRODUCT: Slowly oxidizes in air to benzaldehyde.

USES: Perfumes and flavors; photographic developer for color movie films; dying nylon filament, textiles and sheet plastics; solvent for dyestuffs, cellulose, esters, casein, waxes, etc.; heat-sealing polyethylene films; bacteriostat, cosmetics, ointments, emulsions; ballpoint pen inks and stencil inks; surfactant.

BENZYL BUTYL PHTHALATE

SYNONYMS: BBP; **1,2-Benzenedicarboxylic acid butyl phenylmethyl ester**; Benzyl *n*-butyl phthalate; Butyl benzyl phthalate; *n*-Butyl benzyl phthalate; Butyl phenylmethyl 1,2-benzenedicarboxylate; NCI-C54375; Palatinol BB; Santicizer 160; Sicol 160; Unimoll BB.

CHEMICAL DESIGNATIONS: CAS: 85-68-7; DOT: none assigned; mf: $C_{19}H_{20}O_4$; fw: 312.37; RTECS: TH 9990000.

PROPERTIES: Clear oily liq. Mp: -35°, bp: 370-380°, *d*: 1.12 @ 20/4°, fl p: 110°, K_H: 1.3 x 10^{-6} atm·m^3/mol @ 25° (calcd), log K_{oc}: 1.83-2.54, log K_{ow}: 4.05-4.91, S_w: 2.82 mg/L @ 20°, vap d: 12.76 g/L @ 25°, 10.78 (air = 1), vp: 8.6 x 10^{-6} mm @ 20°.

TRANSFORMATION PRODUCTS: In anaerobic sludge diluted to 10%, benzyl butyl phthalate biodegraded to monobutyl phthalate which subsequently degraded to phthalic acid. Anticipated hydrolysis products would include mono-*n*-butyl phthalate, monobenzyl phthalate, phthalic acid, benzyl alcohol and *n*-butyl alcohol.

USES: Plasticizer used in polyvinyl chloride formulations; additive in polyvinyl acetate emulsions, ethylene glycol and ethyl cellulose.

BENZYL CHLORIDE

SYNONYMS: Benzyl chloride anhydrous; **(Chloromethyl)benzene**; Chlorophenylmethane; α-Chlorotoluene; Ω-Chlorotoluene; NCI-C06360; RCRA waste number P028; Tolyl chloride; UN 1738.

CHEMICAL DESIGNATIONS: CAS: 100-44-7; DOT: 1738; mf: C_7H_7Cl; fw: 126.59; RTECS: XS 8925000.

PROPERTIES: Colorless to pale yellow liq with a pungent, aromatic, irritating odor. Mp: -39°, bp: 179.3°, *d*: 1.1002 @ 20/4°, fl p: 60.0°, lel: 1.1%, K_H: 3.04 x 10^{-4} atm·m^3/mol @ 20° (calcd), log K_{oc}: 2.28 (calcd), log K_{ow}: 2.30, S_o: misc with alcohol, chloroform and ether, S_w: 493 mg/L @ 20°, vap d: 5.17 g/L @ 25°, 4.37 (air = 1), vp: 0.9 mm @ 20°.

TRANSFORMATION PRODUCTS: Anticipated products from the reaction of benzyl chloride with O_3 or hydroxyl radicals in the atmosphere are chloromethylphenols, benzaldehyde and Cl•. When incubated with raw sewage and raw sewage acclimated with hydrocarbons, non-chlorinated products were formed. Anticipated hydrolysis products include hydrochloric acid and benzyl alcohol (est $t_{1/2}$ = 15 h @ 25° and pH 7).

EXPOSURE LIMITS: NIOSH REL: IDLH 10 ppm, 15-min C 5 mg/m^3; OSHA PEL: TWA 1 ppm; ACGIH TLV: TWA 1 ppm.

SYMPTOMS OF EXPOSURE: Irrit eyes, nose; weak; irritable; head; skin eruption; pulm edema.

USES: Manufacture of perfumes, benzyl compounds, pharmaceutical products, resins, dyes, photographic developers; gasoline gum inhibitors; quaternary ammonium compounds; penicillin precursors; chemical intermediate.

α-BHC

SYNONYMS: Benzene hexachloride-α-isomer; α-Benzene hexachloride; ENT 9,232; α-HCH; α-Hexachloran; α-Hexachlorane; α-Hexachlorcyclohexane; α-Hexachlorocyclohexane; 1,2,3,4,5,6-Hexachloro-α-cyclohexane; **1α,2α,3β,4α,5β,6β-Hexachlorocyclohexane**; α-1,2,3,4,5,6-Hexachlorocyclohexane; α-Lindane; TBH.

CHEMICAL DESIGNATIONS: CAS: 319-84-6; DOT: none assigned; mf: $C_6H_6Cl_6$; fw: 290.83; RTECS: GV 3500000.

PROPERTIES: Brownish-to-white cryst solid with a phosgene-like odor (tech grade). Mp: 159.1°, bp: 288°, *d*: ≈ 1.87, K_H: 5.3 x 10^{-6} atm·m^3/mol @ 20° (calcd), log K_{oc}: 3.279, log K_{ow}: 3.46-3.89, S_o: sol in ethanol, benzene and chloroform, S_w: 1.63 mg/L @ 20°, vp: 2.5 x 10^{-5} mm @ 20°.

TRANSFORMATION PRODUCTS: Under aerobic conditions, indigenous microbes in contaminated soil produced pentachlorocyclohexane. However under methanogenic conditions, α-BHC was converted to chlorobenzene, 3,5-dichlorophenol and the tentatively identified compound 2,4,5-trichlorophenol. *Clostridium sphenoides* biodegraded α-BHC to δ-3,4,5,6-tetrachloro-1-cyclohexane.

USE: Not produced commercially in the U.S. and its sale is prohibited by the U.S. EPA.

β-BHC

SYNONYMS: *trans*-α-Benzenehexachloride; β-Benzenehexachloride; Benzene-*cis*-hexachloride; ENT 9,233; β-HCH; β-Hexachlorobenzene; **1α,2β,3α,4β,5α,6β-Hexachlorocyclohexane**; β-Hexachlorocyclohexane; 1,2,3,4,5,6-Hexachloro-β-cyclohexane; 1,2,3,4,5,6-Hexachloro-*trans*-cyclohexane; β-1,2,3,4,5,6-Hexachlorocyclohexane; β-Isomer; β-Lindane; TBH.

CHEMICAL DESIGNATIONS: CAS: 319-85-7; DOT: none assigned; mf: $C_6H_6Cl_6$; fw: 290.83; RTECS: GV 4375000.

PROPERTIES: Solid. Mp: 311.7°, bp: subl, *d*: 1.89 @ 19/4°, K_H: 2.3 x 10^{-7} atm·m^3/mol @ 20° (calcd), log K_{oc}: 3.322-3.553, log K_{ow}: 3.80-4.50, S_o: sol in ethanol, benzene and chloroform, S_w: 240 ppb @ 25°, vp: 2.8 x 10^{-7} mm @ 20°.

TRANSFORMATION PRODUCTS: No biodegradation of β-BHC was observed under denitrifying and sulfate-reducing conditions in a contaminated soil collected from the Netherlands.

USE: Insecticide.

δ-BHC

SYNONYMS: δ-Benzenehexachloride; ENT 9,234; δ-HCH; δ-Hexachlorocyclohexane; δ-1,2,3,4,5,6-Hexachlorocyclohexane; δ-(aeeeee)-1,2,3,4,5,6-Hexachlorocyclohexane; **1α,2α,3α,4β,5β,6β-Hexachlorocyclohexane**; 1,2,3,4,5,6-Hexachloro-δ-cyclohexane; δ-Lindane; TBH.

CHEMICAL DESIGNATIONS: CAS: 319-86-8; DOT: none assigned; mf: $C_6H_6Cl_6$; fw: 290.83; RTECS: GV 4550000.

PROPERTIES: Solid. Mp: 140.8°, bp: 60° @ 0.36 mm, *d*: ≈ 1.87, K_H: 2.5 x 10^{-7} atm·m^3/mol @ 20-25° (calcd), log K_{oc}: 3.279, log K_{ow}: 2.80, 4.14 2.80, S_o: sol in ethanol, benzene and chloroform, S_w: 21.3 ppm @ 25°, vp: 1.7 x 10^{-5} mm @ 20°.

USE: Insecticide.

BIPHENYL

SYNONYMS: Bibenzene; **1,1′-Biphenyl**; Diphenyl; Lemonene; Phenylbenzene.

CHEMICAL DESIGNATIONS: CAS: 92-52-4; DOT: none assigned; mf: $C_{12}H_{10}$; fw: 154.21; RTECS: NU 8050000.

PROPERTIES: White scales with a pleasant odor. Mp: 71°, bp: 255.9°, *d*: 0.8660 @ 20/4°, fl p: 112.8°, lel: 0.6% @ 100°, uel: 5.8% @ 155°, K_H: 1.93-4.15 x 10^{-4} atm·m^3/mol @ 25°, IP: 8.27 eV, log K_{oc}: 3.71 (calcd), log K_{ow}: 3.16-4.09, S_o: sol in alcohol, benzene and ether, S_w: 5.94-7.48 ppm @ 25°, vp: 5.84 x 10^{-4} mm @ 20.70°.

TRANSFORMATION PRODUCTS: Reported biodegradation products include 2,3-dihydro-2,3-dihydroxybiphenyl, 2,3-dihydroxybiphenyl, 2-hydroxy-6-oxo-6-phenylhexa-2,4-dienoate, 2-hydoxy-3-phenyl-6-oxohexa-2,4-dienoate, 2-oxopenta-4-enoate, phenylpyruvic acid, 2-hydroxybiphenyl, 4-hydroxybiphenyl and 4,4′-hydroxybiphenyl. Under aerobic conditions, *Beijerinckia* sp. reportedly degraded biphenyl to *cis*-2,3-dihydro-2,3-dihydroxybiphenyl. It also was reported that *Oscillatoria* sp. and *Pseudomonas putida* degraded biphenyl to 4-hydroxybiphenyl and benzoic acid, respectively. A CO_2 yield of 9.5% was achieved when biphenyl adsorbed on silica gel was irradiated with light (λ > 290 nm) for 17 h. In activated sludge, 15.2% mineralized to CO_2 after 5 days. Irradiation of biphenyl (λ > 300 nm) in the presence of nitrogen monoxide resulted in the formation of 2- and 4-nitrobiphenyl. The aq chlorination of biphenyl @ 40° over a pH range of 6.2 to 9.0 yielded *o*-chlorobiphenyl and *m*-chlorobiphenyl. In an acidic aq soln (pH = 4.5) containing bromide ions and a chlorinating agent (sodium hypochlorite), 4-bromobiphenyl formed as the major product. Minor products identified include 2-bromobiphenyl, 2,4- and 4,4′-dibromobiphenyl.

EXPOSURE LIMITS: NIOSH REL: IDLH 300 mg/m^3; OSHA PEL: TWA 0.2 ppm; ACGIH TLV: TWA 0.2 ppm.

SYMPTOMS OF EXPOSURE: Irrit throat, eyes; head; nau; ftg; numb limbs, liv damage.

USES: Heat transfer liquid; fungistat for oranges; plant disease control; manufacture of benzidine; organic synthesis.

BIS(2-CHLOROETHOXY)METHANE

SYNONYMS: BCEXM; Bis(2-chloroethyl)formal; Bis(β-chloroethyl)formal; Dichlorodiethyl formal; Dichlorodiethyl methylal; Dichloroethyl formal; Di-2-chloroethyl formal; **1,1′-[Methylenebis(oxy)]bis(2-chloroethane)**; 1,1′-[Methylenebis(oxy)]bis(2-chloroformaldehyde); Bis(β-chloroethyl)acetal ethane; Formaldehyde bis(β-chloroethylacetal); RCRA waste number U024.

CHEMICAL DESIGNATIONS: CAS: 111-91-1; DOT: 1916; mf: $C_5H_{10}Cl_2O_2$; fw: 173.04; RTECS: PA 3675000.

PROPERTIES: Colorless liq. Mp: -32.8°, bp: 218.1°, *d*: 1.2339 @ 20/20°, fl p: 110°, K_H: 3.78 x 10^{-7} atm·m^3/mol (calcd), log K_{oc}: 2.06 (calcd), log K_{ow}: 1.26 (calcd), S_w: 81 g/L @ 25° (calcd), vap d: 7.07 g/L @ 25°, 5.97 (air = 1), vp: 1 mm @ 53°.

USES: Manufacture of insecticides, polymers; degreasing solvent; intermediate for polysulfide rubber.

BIS(2-CHLOROETHYL)ETHER

SYNONYMS: Bis(β-chloroethyl)ether; Chlorex; 1-Chloro-2-(β-chloroethoxy)-ethane; Chloroethyl ether; 2-Chloroethyl ether; (β-Chloroethyl)ether; DCEE; Dichlorodiethyl ether; 2,2′-Dichlorodiethyl ether; β,β′-Dichlorodiethyl ether; Dichloroether; Dichloroethyl ether; α,α′-Dichloroethyl ether; Di(β-chloroethyl)-ether; Di(2-chloroethyl)ether; *sym*-Dichloroethyl ether; 2,2′-Dichloroethyl ether; Dichloroethyl oxide; ENT 4,504; **1,1′-Oxybis(2-chloroethane)**; RCRA waste number U025; UN 1916.

CHEMICAL DESIGNATIONS: CAS: 111-44-4; DOT: 1916; mf: $C_4H_8Cl_2O$; fw:

143.01; RTECS: KN 0875000.

PROPERTIES: Colorless liq with a strong fruity odor. Mp: -52.2 to -24.5°, bp: 178.5°, *d*: 1.2199 @ 20/4°, fl p: 55°, K_H: 1.3 x 10^{-5} atm·m^3/mol, IP: 9.85 eV, log K_{oc}: 1.15, log K_{ow}: 1.12, 1.58, S_o: sol in acetone, ethanol, benzene and ether, S_w: 10.7 g/L @ 20°, vap d: 5.84 g/L @ 25°, 4.94 (air = 1), vp: 0.71 mm @ 20°.

USES: Scouring and cleaning textiles; fumigants; processing fats, waxes, greases, cellulose esters; preparation of insecticides, butadiene, pharmaceuticals; solvent in paints, varnishes and lacquers; selective solvent for production of high-grade lubricating oils; fulling, wetting and penetrating compounds; finish removers; spotting and dry cleaning; soil fumigant; acaricide; organic synthesis.

BIS(2-CHLOROISOPROPYL)ETHER

SYNONYMS: BCIE; BCMEE; Bis(β-chloroisopropyl)ether; Bis(2-chloro-1-methylethyl)ether; 1-Chloro-2-(β-chloroisopropoxy)propane; 2-Chloroisopropyl ether; β-Chloroisopropylether; (2-Chloro-1-methylethyl)ether; Dichlorodiisopropyl ether; Dichloroisopropyl ether; 2,2′-Dichloroisopropyl ether; NCI-C50044; **2,2′-Oxybis(1-chloropropane)**; RCRA waste number U027; UN 2490.

CHEMICAL DESIGNATIONS: CAS: 108-60-1; DOT: none assigned; mf: $C_6H_{12}Cl_2O$; fw: 171.07; RTECS: KN 1750000.

PROPERTIES: Colorless liq. Mp: 96.8-101.8°, bp: 187°, *d*: 1.103 @ 20/4°, fl p: 85°, K_H: 1.1 x 10^{-4} atm·m^3/mol, log K_{oc}: 1.79, log K_{ow}: 2.58, S_o: sol in acetone, ethanol, benzene and ether, S_w: 1,700 mg/L @ 20°, vap d: 6.99 g/L @ 25°, 5.91 (air = 1), vp: 0.85 mm @ 20°.

USES: Chemical intermediate in the manufacturing of dyes, resins and pharmaceuticals; solvent and extractant for fats, waxes and greases; textile manufacturing; agent in paint and varnish removers, spotting and cleaning agents; a combatant in liver fluke infections; preparation of glycol esters in fungicidal preparations and as an insecticidal wood preservative; apparently used as a nematocide in Japan but is not registered in the U.S. for use as a pesticide.

BIS(2-ETHYLHEXYL)PHTHALATE

SYNONYMS: **1,2-Benzenedicarboxylic acid bis(2-ethylhexyl)ester**; Bioflex 81; Bioflex DOP; Bis(2-ethylhexyl)-1,2-benzenedicarboxylate; Compound 889; DAF 68; DEHP; Di(2-ethylhexyl)orthophthalate; Di(2-ethylhexyl)phthalate; Dioctyl phthalate; Di-*sec*-octyl phthalate; DOP; Ergoplast FDO; Ethylhexyl phthalate; 2-Ethylhexyl phthalate; Eviplast 80; Eviplast 81; Fleximel; Flexol DOP; Flexol plasticizer DOP; Good-rite GP 264; Hatcol DOP; Hercoflex 260; Kodaflex DOP; Mollan 0; NCI-C52733; Nuoplaz DOP; Octoil; Octyl phthalate; Palantinol AH; Phthalic acid bis(2-ethylhexyl) ester; Phthalic acid dioctyl ester; Pittsburgh PX-138; Platinol AH; Platinol DOP; RC plasticizer DOP; RCRA waste number U028; Reomol D 79P; Reomol DOP; Sicol 150; Staflex DOP; Truflex DOP; Vestinol 80; Witicizer 312.

CHEMICAL DESIGNATIONS: CAS: 117-81-7; DOT: none assigned; mf: $C_{24}H_{38}O_4$; fw: 390.57; RTECS: TI 0350000.

PROPERTIES: Colorless, oily liq with a very faint odor. Mp: -55 to -46°, bp: 385°, *d*: 0.985 @ 20/4°, fl p: 196° (OC), lel: 0.3% @ 245°, K_H: 1.1 x 10^{-5} atm·m^3/mol @ 25° (calcd), log K_{oc}: 5.0, log K_{ow}: 4.20-7.453, S_o: misc with mineral oil and hexane, S_w: 340 μg/L @ 25°, vap d: 15.96 g/L @ 25°, 13.48 (air = 1), vp: 1-3.4 x 10^{-7} mm @ 20°.

TRANSFORMATION PRODUCTS: Bis(2-ethylhexyl)phthalate degraded in both amended and unamended calcareous soils from New Mexico. After 146 days, 76 to 93% of the applied amount degraded to CO_2. No other metabolites were detected. Bis(2-ethylhexyl)phthalate hydrolyzes in water to phthalic acid and 2-ethylhexyl alcohol. Pyrolysis of bis(2-ethylhexyl)phthalate in the presence of polyvinyl chloride @ 600° for 10 min gave the following compounds: methylindene, naphthalene, 1-methylnaphthalene, 2-methylnaphthalene, biphenyl, dimethylnaphthalene, acenaphthene, fluorene, methylacenaphthene, methylfluorene, phenanthrene, anthracene, methylphenanthrene, methylanthracene, methylpyrene or fluoranthene and 17 unidentified compounds. In a 56-day experiment, ^{14}C-labelled bis(2-ethylhexyl)phthalate applied to soil-water suspensions under aerobic and anaerobic conditions gave $^{14}CO_2$ yields of 11.6 and 8.1%, respectively.

EXPOSURE LIMITS: NIOSH REL: lowest feasible limit; OSHA PEL: TWA 5 mg/m^3; ACGIH TLV: TWA 5 mg/m^3, STEL 10 mg/m^3.

SYMPTOMS OF EXPOSURE: Irrit eyes, muc memb; nau; diarr.

USES: Plasticizer; in vacuum pumps.

BROMOBENZENE

SYNONYMS: Monobromobenzene; NCI-C55492; Phenyl bromide; UN 2514.

CHEMICAL DESIGNATIONS: CAS: 108-86-1; DOT: 2514; mf: C_6H_5Br; fw: 157.01; RTECS: CY 9000000.

PROPERTIES: Mobile liq with an aromatic odor. Mp: -30.8°, bp: 156°, *d*: 1.4952 @ 20/4°, fl p: 51°, K_H: 2.40 x 10^{-3} atm·m^3/mol @ 25°, IP: 8.98 eV, 9.41 eV, log K_{oc}: 2.33 (calcd), log K_{ow}: 2.96-3.01, S_o: sol in alcohol (10.4 wt % @ 25°) and ether (71.3 wt % @ 25°); misc with benzene, chloroform and petroleum ethers, S_w: 500 mg/L @ 20°, vap d: 6.42 g/L @ 25°, 5.41 (air = 1), vp: 3.3 mm @ 20°.

TRANSFORMATION PRODUCTS: A CO_2 yield of 19.7% was achieved when bromobenzene adsorbed on silica gel was irradiated with light (λ > 290 nm) for 17 h. In activated sludge, 34.8% of the applied bromobenzene mineralized to CO_2 after 5 days. Irradiation of bromobenzene in air containing NO_x gave phenol, *p*-nitrophenol, 2,4-dinitrophenol, *p*-bromophenol, *m*-bromonitrobenzene, 3-bromo-2-nitrophenol, 3-bromo-4-nitrophenol, 3-bromo-6-nitrophenol, 2-bromo-4-nitrophenol and 2,6-dibromo-4-nitrophenol.

USES: Preparation of phenyl magnesium bromide used in organic synthesis; solvent for fats, waxes and oils; motor oil additive; crystallizing solvent; chemical intermediate.

BROMOCHLOROMETHANE

SYNONYMS: CB; CBM; Chlorobromomethane; Halon 1011; Methylene chlorobromide; Mil-B-4394-B; UN 1887.

CHEMICAL DESIGNATIONS: CAS: 74-97-5; DOT: 1887; mf: CH_2BrCl; fw: 129.39; RTECS: PA 5250000.

PROPERTIES: Clear, colorless liq. with a sweet chloroform-like odor. Mp: -86.5°, bp: 68.1°, *d*: 1.9344 @ 20/4°, K_H: 1.44 x 10^{-3} atm·m^3/mol @ 24-25° (calcd), IP: 10.77 eV, log K_{oc}: 1.43 (calcd), log K_{ow}: 1.41, S_o: sol in acetone, alcohol, benzene and ether, S_w: 0.129 M @ 25.0°, vap d: 5.29 g/L @ 25°, 4.47 (air = 1), vp: 141.07 mm @ 24.05°.

TRANSFORMATION PRODUCTS: Anticipated hydrolysis products include chloromethanol and/or bromomethanol (est $t_{1/2}$ = 44 yr @ 25° and pH 7).

EXPOSURE LIMITS: NIOSH REL: IDLH 5,000 ppm; OSHA PEL: TWA 200 ppm; ACGIH TLV: TWA 200 ppm, STEL 250 ppm.

SYMPTOMS OF EXPOSURE: Disorient, dizz; irrit eyes, throat, skin; pulm edema.

USES: Fire extinguishers; organic synthesis.

BROMODICHLOROMETHANE

SYNONYMS: BDCM; Dichlorobromomethane; NCI-C55243.

CHEMICAL DESIGNATIONS: CAS: 75-27-4; DOT: none assigned; mf: $CHBrCl_2$; fw: 163.83; RTECS: PA 5310000.

PROPERTIES: Colorless liq. Mp: -57.1°, bp: 87-90°, *d*: 1.9945 @ 20/4°, K_H: 2.12 x 10^{-4} atm·m^3/mol @ 25°, IP: 10.88 eV, log K_{oc}: 1.79, log K_{ow}: 1.88, 2.10, S_o: sol in acetone, ethanol, benzene, chloroform and ether, S_w: 4,700 mg/L @ 22°, vap d: 6.70 g/L @ 25°, 5.66 (air = 1), vp: 50 mm @ 20°.

TRANSFORMATION PRODUCTS: Anticipated hydrolysis products include dichloromethanol and/or bromochloromethanol (est $t_{1/2}$ = 137 yr @ 25° and pH 7).

USES: Component of fire extinguisher fluids; solvent for waxes, fats and resins; degreaser; flame retardant; heavy liquid for mineral and salt separations; chemical intermediate; laboratory use.

BROMOFORM

SYNONYMS: Methenyl tribromide; Methyl tribromide; NCI-C55130; RCRA waste number U225; **Tribromomethane**; UN 2515.

CHEMICAL DESIGNATIONS: CAS: 75-25-2; DOT: 2515; mf: $CHBr_3$; fw: 252.73; RTECS: PB 5600000.

PROPERTIES: Colorless liq with a chloroform-like odor. Mp: -8.3°, bp: 149.5°, *d*: 2.8899 @ 20/4°, K_H: 4.3 x 10^{-4} atm·m^3/mol @ 20°, IP: 10.51 eV, log K_{oc}: 2.06, 2.45, log K_{ow}: 2.30, 2.38, S_o: sol in ligroin; misc with benzene, chloroform, ether, petroleum ether, acetone and oils, S_w: 3,010 mg/L 20°, vap d: 10.33 g/L @ 25°,

8.72 (air = 1), vp: 4 mm @ 20°.

TRANSFORMATION PRODUCTS: The anticipated hydrolysis product is dibromomethanol ($t_{1/2}$ = 686 yr @ 25° and pH 7). To an aq soln containing 9.08 μmol bromoform was bubbled hydrogen. After 24 h, only 5% of the bromoform reacted to form methane and minor traces of ethane. In the presence of colloidal platinum catalyst, the reaction proceeded at a much faster rate forming the same end products. In an earlier study, water containing 2,000 ng/μL of bromoform and colloidal platinum catalyst was irradiated with UV light. After 20 h, about 50% of the bromoform had reacted. A duplicate experiment was performed but the concentration of bromoform was increased to 3,000 ng/μL and 0.1 g zinc was added. After 14 h, only 0.1 ng/μL bromoform remained. Presumed transformation products include methane and bromide ions. Photolysis of an aq soln containing bromoform (989 μmol) and a catalyst [Pt(colloid)/Ru(bpy)$^{2+}$/MV/EDTA] yielded the following products after 25 h (μmol detected): bromide ions (250), methylene bromide (475) and unreacted bromoform (421).

EXPOSURE LIMITS: NIOSH REL: 0.5 ppm (skin); ACGIH TLV: 0.5 ppm (skin).

SYMPTOMS OF EXPOSURE: Irrit eyes; resp sys; CNS depres; liv damage.

USES: Solvent for waxes, greases and oils; separating solids with lower densities; component of fire-resistant chemicals; geological assaying; medicine (sedative); gauge fluid; intermediate in organic synthesis.

4-BROMOPHENYL PHENYL ETHER

SYNONYMS: 4-Bromodiphenyl ether; *p*-Bromodiphenyl ether; **1-Bromo-4-phenoxybenzene**; 1-Bromo-*p*-phenoxybenzene; 4-Bromophenyl ether; *p*-Bromophenyl ether; *p*-Bromophenyl phenyl ether; Phenyl-4-bromophenyl ether; Phenyl-*p*-bromophenyl ether.

CHEMICAL DESIGNATIONS: CAS: 101-55-3; DOT: none assigned; mf: $C_{12}H_9BrO$; fw: 249.20; RTECS: none assigned.

PROPERTIES: Liq. Mp: 18.7°, bp: 305°, *d*: 1.4208 @ 20/4°, fl p: > 110°, K_H: 1.0 x 10^{-4} atm·m^3/mol, log K_{oc}: 4.94 (calcd), log K_{ow}: 5.15, S_o: sol in ether, vap d: 10.19 g/L @ 25°, 8.60 (air = 1), vp: 1.5 x 10^{-3} mm @ 20° (calcd).

USE: Research chemical.

BROMOTRIFLUOROMETHANE

SYNONYMS: Bromofluoroform; F-13B1; Freon 13-B1; Halocarbon 13B1; Halon 1301; Monobromotrifluoromethane; Refrigerant 13B1; Trifluorobromoethane; Trifluoromonobromomethane; UN 1009.

CHEMICAL DESIGNATIONS: CAS: 75-63-8; DOT: 1009; mf: $CBrF_3$; fw: 148.91; RTECS: PA 5425000.

PROPERTIES: Colorless gas with an ether-like odor. Mp: -167.8°, bp: -58 to -57°, *d*: 1.455 @ 20/4° (calcd), K_H: 5.00 x 10^{-1} atm·m^3/mol @ 25°, IP: 11.9 eV, log

K_{oc}: 2.44 (calcd), log K_{ow}: 1.54, S_o: sol in chloroform, S_w: 0.03 wt % @ 20°, vap d: 6.09 g/L @ 25°, 5.14 (air = 1), vp: 149 mm @ 20°.

EXPOSURE LIMITS: NIOSH REL: IDLH 50,000 ppm; OSHA PEL: TWA 1,000 ppm.

SYMPTOMS OF EXPOSURE: Li-head, card arrhy.

USES: Chemical intermediate; metal hardening; refrigerant; fire extinguishers.

1,3-BUTADIENE

SYNONYMS: Biethylene; Bivinyl; Butadiene; Buta-1,3-diene; α,γ-Butadiene; Divinyl erythrene; NCI-C50602; Pyrrolylene; Vinylethylene.

CHEMICAL DESIGNATIONS: CAS: 106-99-0; DOT: 1010; mf: C_4H_6; fw: 54.09; RTECS: EI 9275000.

PROPERTIES: Colorless gas with a mild aromatic odor. Mp: -108.9°, bp: -4.4°, *d*: 0.6211 @ 20/4°, fl p: -76°, lel: 2%, uel: 11.5%, K_H: 6.3 x 10^{-2} atm·m^3/mol @ 25°, IP: 9.07 eV, log K_{oc}: 2.08 (calcd), log K_{ow}: 1.99, S_o: sol in acetone, benzene, ether and alcohol (40 vol % @ RT), S_w: 735 mg/L @ 20°, vap d: 2.29 g/L @ 25°, 1.87 (air = 1), vp: 1,840 mm @ 21°.

TRANSFORMATION PRODUCTS: Will polymerize in the presence of oxygen if no inhibitor is present.

EXPOSURE LIMITS: NIOSH REL: Reduce exposure to lowest feasible limit; OSHA PEL: TWA 1,000 ppm; ACGIH TLV: TWA 10 ppm.

SYMPTOMS OF EXPOSURE: Irit eyes, nose, throat; drow, li-head; frost.

USES: Synthetic rubbers and elastomers (styrene-butadiene, polybutadiene, neoprene); organic synthesis (Diels-Alder reactions); latex paints; resins; chemical intermediate.

n-BUTANE

SYNONYMS: **Butane**; Diethyl; Methylethylmethane; UN 1011.

CHEMICAL DESIGNATIONS: CAS: 106-97-8; DOT: 1011; mf: C_4H_{10}; fw: 58.12; RTECS: EJ 4200000.

PROPERTIES: Colorless gas with a natural gas-like odor. Mp: -138.4°, bp: -0.5°, *d*: 0.5788 @ 20/4°, fl p: -138°, lel: 1.6%, uel: 8.5%, K_H: 9.30 x 10^{-1} atm·m^3/mol @ 25°, IP: 10.63 eV, log K_{ow}: 2.89, S_o: @ 17°: 150 ml/L in alcohol @ 770 mm, 25,000 ml/L in chloroform, 30,000 ml/L in ether, S_w: 61 mg/L @ 20°, vap d: 2.38 g/L @ 25°, 2.046 (air = 1), vp: 1,820 mm @ 25°.

TRANSFORMATION PRODUCTS: Major products reported from the photo-oxidation of *n*-butane with NO_x are acetaldehyde, formaldehyde and 2-butanone. Minor products included peroxyacyl nitrates and methyl, ethyl and propyl nitrates, CO and CO_2. Biacetyl, *tert*-butyl nitrate, ethanol and acetone were reported as trace products. Irradiation of *n*-butane in the presence of chlorine yielded CO, CO_2, hydroperoxides, peroxyacid and other carbonyl compounds. Nitrous acid vapor and *n*-butane in a "smog chamber" was irradiated with UV light. Major oxidation products identified included 2-butanone, acetaldehyde and *n*-butyraldehyde. Minor products included peroxyacetyl nitrate, methyl nitrate and other unidentified compounds. In the presence of methane, *Pseudomonas methanica* degraded *n*-butane to *n*-butanol, *n*-butyric acid and 2-butanone. *n*-Butane may biodegrade in 2 ways. The first is the formation of butyl hydroperoxide which decomposes to *n*-butanol followed by oxidation to butyric acid. The other pathway involves dehydrogenation yielding 1-butene which may react with water forming *n*-butanol. Complete combustion in air gives CO_2 and water.

EXPOSURE LIMITS: ACGIH TLV: TWA 800 ppm.

USES: Manufacture of synthetic rubbers, ethylene; raw material for high octane motor fuels; solvent; refrigerant; propellant in aerosols; calibrating instruments; organic synthesis.

2-BUTANONE

SYNONYMS: Butanone; Ethyl methyl ketone; Meetco; MEK; Methyl acetone; Methyl ethyl ketone; RCRA waste number U159; UN 1193; UN 1232.

CHEMICAL DESIGNATIONS: CAS: 78-93-3; DOT: 1193; mf: C_4H_8O; fw: 72.11; RTECS: EL 6475000.

PROPERTIES: Colorless liq with a sweet mint-like odor. Mp: -86.9°, bp: 79.6°, *d*: 0.8054 @ 20/4°, fl p: -9°, lel: 2%, uel: 10%, K_H: 1.05-7.0 x 10^{-5} atm·m^3/mol @ 25°, IP: 9.48 eV, log K_{oc}: 0.09 (calcd), log K_{ow}: 0.26, 0.29, S_o: misc with acetone, ethanol, benzene and ether, S_w: 294 g/L @ 20°, vap d: 2.94 g/L @ 25°, 2.49 (air = 1), vp: 71.2 mm @ 20°.

SYMPTOMS OF EXPOSURE: Irrit eyes, nose; head; dizz; vomit.

TRANSFORMATION PRODUCTS: Synthetic air containing gaseous nitrous acid and exposed to artificial sunlight (λ = 300-450 nm) photooxidized 2-butanone into peroxyacetyl nitrate and methyl nitrate. The hydroxyl-initiated photooxidation of 2-butanone in a smog chamber produced peroxyacetyl nitrate and acetaldehyde.

EXPOSURE LIMITS: NIOSH REL: IDLH 3,000 ppm, 10-h TWA 200 ppm, 15-min C 300 ppm; OSHA PEL: TWA 200 ppm; ACGIH TLV: TWA 200 ppm, STEL 300 ppm.

USES: Solvent in nitrocellulose coatings, vinyl films and "Glyptal" resins; paint removers; cements and adhesives; organic synthesis; manufacture of smokeless powders; preparation of 2-butanol, butane and amines; cleaning fluids; printing; catalyst carrier; acrylic coatings.

1-BUTENE

SYNONYMS: α-Butylene; Ethylethylene.

CHEMICAL DESIGNATIONS: CAS: 106-98-9; DOT: 1012; mf: C_4H_8; fw: 56.11; RTECS: none assigned.

PROPERTIES: Colorless liq with a weak aromatic odor. Mp: -185.3°, bp: -6.3°, *d*: 0.5951 @ 20/4°, fl p: -79°, lel: 1.6%, uel: 10.0%, K_H: 2.5 x 10^{-1} atm·m^3/mol @

25°, IP: 9.6 eV, log K_{ow}: 2.40, S_o: sol in alcohol, benzene and ether, S_w: 222 ppm @ 25°, vap d: 2.29 g/L @ 25°, 1.94 (air = 1), vp: 2,230 mm @ 25°.

TRANSFORMATION PRODUCTS: Products identified from the photoirradiation of 1-butene with NO_2 in air are epoxybutane, 2-butanone, propanal, ethanol, ethyl nitrate, CO, CO_2, methanol and nitric acid. Biooxidation of 1-butene may occur yielding 3-buten-1-ol which may further oxidize to give 3-butenoic acid. Washed cell suspensions of bacteria belonging to the genera *Mycobacterium*, *Nocardia*, *Xanthobacter* and *Pseudomonas* and growing on selected alkenes metabolized 1-butene to 1,2-epoxybutane.

USES: Polybutylenes; polymer and alkylate gasoline; intermediate for C_4 and C_5 aldehydes, alcohols, maleic acid and other organic compounds.

2-BUTOXYETHANOL

SYNONYMS: Butyl cellosolve; Butyl glycol; Butyl glycol ether; Butyl oxitol; Dowanol EB; Ektasolve; Ethylene glycol monobutyl ether; Ethylene glycol mono-*n*-butyl ether; Glycol monobutyl ether; Jeffersol EB.

CHEMICAL DESIGNATIONS: CAS: 111-76-2; DOT: 2369; mf: $C_6H_{14}O_2$; fw: 118.18; RTECS: KJ 8575000.

PROPERTIES: Colorless liq with a mild odor. Mp: -70°, bp: 171°, fl p: 60.6°, lel: 1.1%, uel: 10.6%, K_H: 2.36 x 10^{-6} atm·m^3/mol (calcd), log K_{ow}: 0.45 (calcd), S_o: sol in alcohol, ether and mineral oil, S_w: misc, *d*: 0.9015 @ 20/4°, vap d: 4.83 g/L @ 25°, 4.08 (air = 1), vp: 0.6 mm @ 20°.

EXPOSURE LIMITS: NIOSH REL: IDLH 700 ppm; OSHA PEL: TWA 50 ppm; ACGIH TLV: TWA 25 ppm.

SYMPTOMS OF EXPOSURE: Irrit eyes, nose, throat; hemolysis, hemog.

USES: Dry cleaning; solvent for nitrocellulose, cellulose acetate, resins, oil, grease, album; perfume fixative; coating compositions for paper, cloth, leather; lacquers.

n-BUTYL ACETATE

SYNONYMS: **Acetic acid butyl ester**; 1-Butanol acetate; Butyl acetate; 1-Butyl acetate; Butyl ethanoate; *n*-Butyl ethanoate; UN 1123.

CHEMICAL DESIGNATIONS: CAS: 123-86-4; DOT: 1123; mf: $C_6H_{12}O_2$; fw: 116.16; RTECS: AF 7350000.

PROPERTIES: Colorless liq with a fruity odor. Mp: -77.9°, bp: 126.5°, *d*: 0.8825 @ 20/4°, fl p: 22.2°, lel: 1.7%, uel: 7.6%, K_H: 3.3 x 10^{-4} atm·m^3/mol @ 25°, IP: 9.56 eV, 10.01 eV, log K_{ow}: 1.82, S_o: sol in benzene and most hydrocarbons; misc with alcohol and ether, S_w: 14 g/L @ 20°, vap d: 4.75 g/L @ 25°, 4.01 (air = 1), vp: 10 mm @ 20°.

TRANSFORMATION PRODUCTS: Anticipated hydrolysis products include *n*-butyl alcohol and acetic acid.

EXPOSURE LIMITS: NIOSH REL: IDLH 10,000 ppm; OSHA PEL: TWA 150 ppm; ACGIH TLV: TWA 150 ppm, STEL 200 ppm.

SYMPTOMS OF EXPOSURE: Head; drow; eyes irrit, dry; irrit resp sys.

USES: Manufacture of artificial leather, plastics, safety glass, photographic films; solvent in the production of perfumes, natural gums, synthetic resins and nitrocellulose lacquers; dehydrating agent.

sec-BUTYL ACETATE

SYNONYMS: Acetic acid 2-butoxy ester; Acetic acid *sec*-butyl ester; **Acetic acid 1-methylpropyl ester**; 2-Butanol acetate; 2-Butyl acetate; *sec*-Butyl alcohol acetate; 1-Methylpropyl acetate; UN 1123.

CHEMICAL DESIGNATIONS: CAS: 105-46-4; DOT: 1123; mf: $C_6H_{12}O_2$; fw: 116.16; RTECS: AF 7380000.

PROPERTIES: Colorless liq with a pleasant odor. Mp: -98.9°, bp: 112°, *d*: 0.8758 @ 20/4°, fl p: 16.7°, lel: 1.7%, uel: 9.8%, K_H: 1.91 x 10^{-4} atm·m^3/mol @ 20° (calcd), IP: 9.91 eV, log K_{ow}: 1.66 (calcd), S_o: sol in acetone, alcohol and ether, S_w: 0.8 wt % @ 20°, vap d: 4.75 g/L @ 25°, 4.01 (air = 1), vp: 10 mm @ 20°.

TRANSFORMATION PRODUCTS: Anticipated hydrolysis products include 2-butyl alcohol and acetic acid.

EXPOSURE LIMITS: NIOSH REL: IDLH 10,000 ppm; OSHA PEL: TWA 200 ppm; ACGIH TLV: TWA 200 ppm.

SYMPTOMS OF EXPOSURE: Irrit eyes, head, drow, dry upper resp, dry skin.

USES: Solvent for nitrocellulose lacquers; nail enamels, thinners; leather finishers.

tert-BUTYL ACETATE

SYNONYMS: Acetic acid *tert*-butyl ester; **Acetic acid 1,1-dimethylethyl ester**; *t*-Butyl acetate; Texaco lead appreciator; TLA; UN 1123.

CHEMICAL DESIGNATIONS: CAS: 540-88-5; DOT: 1123; mf: $C_6H_{12}O_2$; fw: 116.16; RTECS: AF 7400000.

PROPERTIES: Colorless liq with a fruity odor. Bp: 97-98°, *d*: 0.8665 @ 20/4°, fl p: 16.7-22.2° (est), lel: 1.5%, S_o: sol in acetic acid and acetone; misc with alcohol and ether, vap d: 4.75 g/L @ 25°, 4.01 (air = 1).

TRANSFORMATION PRODUCTS: Hydrolyzes in water to *tert*-butyl alcohol and acetic acid ($t_{1/2}$ = 140 yr @ 25° and pH 7).

EXPOSURE LIMITS: NIOSH REL: IDLH 10,000 ppm; OSHA PEL: TWA 200 ppm; ACGIH TLV: TWA 200 ppm.

SYMPTOMS OF EXPOSURE: Itch, inflamm eyes; irrit upper resp; head; narco; derm.

USES: Gasoline additive; solvent.

n-BUTYL ALCOHOL

SYNONYMS: Butan-1-ol; **1-Butanol**; *n*-Butanol; Butyl alcohol; Butyl hydroxide;

Butyric alcohol; CCS 203; 1-Hydroxybutane; Methylolpropane; NA 1,120; NBA; Propylcarbinol; Propylmethanol; RCRA waste number U031; UN 1120.

CHEMICAL DESIGNATIONS: CAS: 71-36-3; DOT: 1120; mf: $C_4H_{10}O$; fw: 74.12; RTECS: EO 1400000.

PROPERTIES: Colorless liq with a fusel oil-like odor. Mp: -89.5°, bp: 117.2°, *d*: 0.8098 @ 20/4°, fl p: 28.9°, lel: 1.4%, uel: 11.2%, K_H: 8.48 x 10^{-6} atm·m^3/mol @ 25°, IP: 10.04 eV, log K_{ow}: 0.88, 0.785, S_o: sol in acetone and benzene; misc with alcohol and ether, S_w: 77,085 mg/L @ 20°, vap d: 3.03 g/L @ 25°, 2.56 (air = 1), vp: 4.4 mm @ 20°.

TRANSFORMATION PRODUCTS: *n*-Butyl alcohol degraded rapidly, presumably by microbes, in New Mexico soils releasing CO_2. An aq soln containing chlorine and irradiated with UV light (λ = 350 nm) converted *n*-butyl alcohol into numerous chlorinated compounds which were not identified.

EXPOSURE LIMITS: NIOSH RE: IDLH 8,000 ppm; OSHA PEL: TWA 100 ppm; ACGIH TLV: C 50 ppm.

SYMPTOMS OF EXPOSURE: Irrit eyes, nose, throat; head; vert; drow; corneal inflamm, blur vision, lac, photophobia; dry cracked skin.

USES: Preparation of butyl esters; solvent for resins and coatings; hydraulic fluid; plasticizers; glycol ethers; gasoline additive.

sec-BUTYL ALCOHOL

SYNONYMS: Butanol-2; Butan-2-ol; **2-Butanol**; *sec*-Butanol; 2-Butyl alcohol; Butylene hydrate; CCS 301; Ethylmethyl carbinol; 2-Hydroxybutane; Methylethylcarbinol; SBA.

CHEMICAL DESIGNATIONS: CAS: 78-92-2; DOT: 1120; mf: $C_4H_{10}O$; fw: 74.12; RTECS: EO 1750000.

PROPERTIES: Colorless liq with a pleasant odor. Mp: -114.7°, -89°, bp: 99.5°, *d*: 0.8063 @ 20/4°, fl p: 23.9°, lel: 1.7% @ 100°, uel: 9.8% @ 100°, K_H: 1.02 x 10^{-5} atm·m^3/mol @ 25°, IP: 10 eV, log K_{ow}: 0.61, S_o: sol in acetone and benzene; misc with alcohol and ether, S_w: 201 g/L @ 20°, vap d: 3.03 g/L @ 25°, 2.56 (air = 1), vp: 12 mm @ 20°.

EXPOSURE LIMITS: NIOSH REL: IDLH 10,000 ppm; OSHA PEL: TWA 150 ppm; ACGIH TLV: TWA 100 ppm, STEL 150 ppm.

SYMPTOMS OF EXPOSURE: Eye irrit, narco, dry skin.

USES: Manufacture of flotation agents, esters (perfumes and flavors), dyestuffs, wetting agents; ingredient in industrial cleaners and paint removers; preparation of methyl ethyl ketone; solvent; organic synthesis.

tert-BUTYL ALCOHOL

SYNONYMS: *t*-Butanol; *tert*-Butanol; *t*-Butyl alcohol; *t*-Butyl hydroxide; 1,1-Dimethylethanol; **2-Methyl-2-propanol**; NCI-C55367; TBA; Trimethylcarbinol; Trimethyl methanol; UN 1120.

OSHA PEL: TWA 10 ppm; ACGIH TLV: TWA 0.5 ppm.

SYMPTOMS OF EXPOSURE: In animals: narco, inco, weak, cyan, pulm, eye irrit, paralysis.

USES: Chemical intermediate; solvent.

CAMPHOR

SYNONYMS: 2-Bornanone; 2-Camphanone; Camphor-natural; Formosa camphor; Gum camphor; Japan camphor; 2-Keto-1,7,7-trimethylnorcamphane; Laurel camphor; Matricaria camphor; Norcamphor; 2-Oxobornane; Synthetic camphor; **1,7,7-Trimethylbicyclo[2.2.1]heptan-2-one**; UN 2717.

CHEMICAL DESIGNATIONS: CAS: 76-22-2; DOT: 2717; mf: $C_{10}H_{16}O$; fw: 152.24; RTECS: EX 1225000.

PROPERTIES: Colorless solid with a fragrant, penetrating odor. Mp: 174-179°, bp: 204° (subl), *d*: 0.990 @ 25/4°, fl p: 65.6°, lel: 0.6%, uel: 3.5%, K_H: 3.00 x 10^{-5} atm·m^3/mol @ 20° (calcd), log K_{ow}: 2.42 (calcd), S_o @ 25° (g/L): acetone (2,500), alcohol (1,000), benzene (2,500), chloroform (2,000), ether (1,000), glacial acetic acid (2,500), oil of turpentine (667); sol in aniline, carbon disulfide, decalin, methylhexalin, nitrobenzene, petroleum ether, tetralin, higher alcohols, in fixed and volatile oils, S_w: ≈ 1.25 g/L @ 25°, vp: 0.18 mm @ 20°.

EXPOSURE LIMITS: NIOSH REL: IDLH 200 mg/m^3; OSHA PEL: TWA 2 ppm.

SYMPTOMS OF EXPOSURE: Irrit eyes, skin, muc memb; nau, vomit, diarr; head; dizz; irrational, epileptiform convuls.

USES: Plasticizer for cellulose esters and ethers; manufacture of incense, celluloid and other plastics; in lacquers, explosives and embalming fluids; pyrotechnics; moth repellent; preservative in pharmaceuticals and cosmetics; odorant/flavorant in household, pharmaceutical and industrial products; tooth powders.

CARBARYL

SYNONYMS: Arylam; Carbamine; Carbatox; Carbatox 60; Carbatox 75; Carpolin; Carylderm; Cekubaryl; Crag sevin; Denapon; Devicarb; Dicarbam; ENT 23,969; Experimental insecticide 7744; Gamonil; Germain's; Hexavin; Karbaspray; Karbatox; Karbosep; Methylcarbamate 1-naphthalenol; Methylcarbamate-1-naphthol; Methyl carbamic acid 1-naphthyl ester; *N*-Methyl-1-naphthylcarbamate; *N*-Methyl-α-naphthylcarbamate; *N*-Methyl-α-naphthylurethan; NA 2,757; NAC; **1-Naphthalenol methylcarbamate**; 1-Naphthol-*N*-methylcarbamate; 1-Naphthyl methylcarbamate; 1-Naphthyl-*N*-methylcarbamate; α-Naphthyl-*N*-methylcarbamate; OMS-29; Panam; Ravyon; Rylam; Seffein; Septene; Sevimol; Sevin; Sok; Tercyl; Toxan; Tricarnam; UC 7744; Union Carbide 7,744.

CHEMICAL DESIGNATIONS: CAS: 63-25-2; DOT: 2757; mf: $C_{12}H_{11}NO_2$; fw: 201.22; RTECS: FC 5950000.

PROPERTIES: Colorless solid or white cryst. Mp: 142.2°, *d*: 1.232 @ 20/20°, fl p: 195°, K_H: 1.27 x 10^{-5} atm·m^3/mol @ 20° (calcd), log K_{oc}: 2.02-2.59, log K_{ow}: 2.36-2.56, S_o: moderately sol in acetone, cyclohexanone, dimethylformamide and isophorone, S_w: 104 mg/L @ 20°, vp: 6.578 x 10^{-6} atm @ 25°.

TRANSFORMATION PRODUCTS: Ozonation of carbaryl in water yielded 1-naphthol, naphthoquinone, phthalic anhydride, *N*-formylcarbamate of 1-naphthol, naphthoquinones and acidic compounds. Photodegrades to 2-hydroxy-1,4-naphthoquinone. In aq solns, carbaryl hydrolyzes to 1-naphthol, methylamine and CO_2. Releases toxic NO_x when heated to decomposition. Fourteen soil fungi metabolized methyl-^{14}C-labelled carbaryl via hydroxylation to 1-naphthyl *N*-hydroxy methylcarbamate, 4-hydroxy-1-naphthyl methylcarbamate and 5-hydroxy-1-naphthyl methylcarbamate. When ^{14}C-carbonyl-labeled carbaryl (200 ppm) was added to 5 different soils and incubated @ 25° for 32 days, evolution of ^{14}C-CO_2 varied from 2.2-37.4%.

EXPOSURE LIMITS: NIOSH REL: 10-h TWA 5 mg/m^3, IDLH 600 mg/m^3; OSHA PEL: TWA 5 mg/m^3; ACGIH TLV: TWA 5 mg/m^3.

SYMPTOMS OF EXPOSURE: Miosis, blur, tear; nasal discharge; salv; sweat; abdom cramps, nau, vomit, diarr; tremor; cyan; convuls; skin irrit.

USE: Contact insecticide.

CARBOFURAN

SYNONYMS: Bay 70143; **2,3-Dihydro-2,2-dimethyl-7-benzofuranol methyl carbamate**; 2,2-Dimethyl-7-coumaranyl *N*-methyl carbamate; 2,2-Dimethyl-2,3-dihydro-7-benzofuranyl-*N*-methyl carbamate; ENT 27,164; Furadan; Methyl carbamic acid 2,3-dihydro-2,2-dimethyl-7-benzofuranyl ester; NIA 10242.

CHEMICAL DESIGNATIONS: CAS: 1563-66-2; DOT: none assigned; mf: $C_{12}H_{15}NO$; fw: 221.26; RTECS: none assigned.

PROPERTIES: White cryst solid. Mp: 150-153°, bp: decomposes > 150°, *d*: 1.18 @ 20/20°, K_H: 3.88 x 10^{-8} atm·m^3/mol @ 30-33° (calcd), log K_{oc}: 1.98-2.20, log K_{ow}: 1.60, 1.63, S_o (g/L): methylene chloride (> 200), 2-propanol (20-50), S_w: 320 mg/L @ 20°, vp: 2 x 10^{-5} mm @ 33°.

TRANSFORMATION PRODUCTS: Carbofuran phenol is formed from the hydrolysis of carbofuran @ pH 7.0. Carbofuran phenol was also found to be the major biodegradation product by *Azospirillum lipoferum* and *Streptomyces* spp. isolated from a flooded alluvial soil. In soils, microorganisms degradated carbofuran to carbofuran phenol, 3-hydroxycarbofuran, 3-ketocarbofuran and 3-ketocarbofuran. Under *in vitro* conditions, 15 of 20 soil fungi degraded carbofuran to one or more of the following compounds: 3-hydroxycarbofuran, 3-ketocarbofuran, carbofuran phenol and 3-hydroxyphenol. 2,3-Dihydro-2,2-dimethylbenzofuran-4,7-diol and 2,3-dihydro-3-keto-2,2-dimethylbenzofuran-7-yl carbamate were formed when carbofuran dissolved in water was irradiated by sunlight for 5 days. Releases toxic NO_x when heated to decomposition.

EXPOSURE LIMITS: OSHA PEL: TWA 0.1 mg/m^3; ACGIH TLV: TWA 0.1 mg/m^3.

USES: Systematic insecticide, nematocide and acaricide.

CARBON DISULFIDE

SYNONYMS: Carbon bisulfide; Carbon bisulphide; Carbon disulphide; Carbon sulfide; Carbon sulphide; Dithiocarbonic anhydride; NCI-C04591; RCRA waste number P022; Sulphocarbonic anhydride; UN 1131; Weeviltox.

CHEMICAL DESIGNATIONS: CAS: 75-15-0; DOT: 1131; mf: CS_2; fw: 76.13; RTECS: FF 6650000.

PROPERTIES: Clear, colorless to pale yellow liq; ethereal odor when pure; tech and reagent grades have foul odors. Mp: -111.5°, bp: 46.2°, *d*: 1.2632 @ 20/4°, fl p: -30°, lel: 1.3%, uel: 50%, K_H: 1.33 x 10^{-2} atm·m^3/mol (calcd), IP: 10.06 eV, 10.080 eV, log K_{oc}: 2.38-2.55 (calcd), log K_{ow}: 1.84, 2.16 (calcd), S_o: sol in ethanol, chloroform and ether, S_w: 2.3 g/L @ 22°, vap d: 3.11 g/L @ 25°, 2.63 (air = 1), vp: 297.5 mm @ 20°.

TRANSFORMATION PRODUCTS: In alkaline solns, carbon disulfide hydrolyzes to CO_2 and hydrogen disulfide. In an aq alkaline soln containing H_2O_2, dithiopercarbonate, sulfide, elemental sulfur and polysulfides may be expected. In an aq, alkaline soln (pH ≥ 8), carbon disulfide reacted with H_2O_2 forming sulfate and carbonate ions. However, when the pH is lowered to 7 and 7.4, colloidal sulfur is formed. Burns with a blue flame releasing CO_2 and sulfur dioxide. Oxidation of carbon disulfide in the troposphere produces carbonyl sulfide. The atmospheric half-lives of carbon disulfide and carbonyl sulfide were estimated to be ≈ 2 yr and 13 days, respectively.

EXPOSURE LIMITS: NIOSH REL: 10-h TWA 1 ppm, 15-min C 10 ppm, IDLH 500 ppm; OSHA PEL: TWA 20 ppm, C 30 ppm, 30-min 100 ppm; ACGIH TLV: TWA 10 ppm.

SYMPTOMS OF EXPOSURE: Dizz, head, poor sleep, ftg, ner; anor, low-wgt; psychosis; polyneur; Parkinson-like; ocular changes; CVS; GI; burns, derm.

USES: Manufacture of viscose rayon, cellophane, flotation agents, ammonium salts, carbon tetrachloride, carbanilide, paints, enamels, paint removers, varnishes, tallow, textiles, rocket fuel; soil disinfectants; electronic vacuum tubes; herbicides; grain fumigants; solvent for fats, resins, phosphorus, sulfur, bromine, iodine and rubber; petroleum and coal tar refining.

CARBON TETRACHLORIDE

SYNONYMS: Benzinoform; Carbona; Carbon chloride; Carbon tet; ENT 4,705; Fasciolin; Flukoids; Freon 10; Halon 104; Methane tetrachloride; Necatorina; Necatorine; Perchloromethane; R 10; RCRA waste number U211; Tetrachloormetaan; Tetrachlorocarbon; **Tetrachloromethane**; Tetrafinol; Tetraform; Tetrasol; UN 1846; Univerm; Vermoestricid.

CHEMICAL DESIGNATIONS: CAS: 56-23-5; DOT: 1846; mf: CCl_4; fw: 153.82; RTECS: FG 4900000.

PROPERTIES: Clear, colorless, heavy, liq with a strong sweetish, ether-like odor. Mp: -22.96°, bp: 76.5°, *d*: 1.5940 @ 20/4°, K_H: 3.02 x 10^{-2} atm·m^3/mol @ 25°, IP: 11.47 eV, log K_{oc}: 2.35-2.64, log K_{ow}: 2.73, 2.83, S_o: misc with ethanol, benzene, chloroform, ether, carbon disulfide, petroleum ether, solvent naphtha and volatile oils, S_w: 800 mg/L @ 20°, vap d: 6.29 g/L @ 25°, 5.31 (air = 1), vp: 90 mm @ 20°.

TRANSFORMATION PRODUCTS: Under laboratory conditions, carbon tetrachloride partially hydrolyzed to chloroform and CO_2. Complete hydrolysis yielded CO_2 and hydrochloric acid ($t_{1/2}$ = 7,000 and 40.5 yr @ 25° and pH 7). Chloroform was also formed by microbial degradation of carbon tetrachloride using denitrifying bacteria. An anaerobic species of *Clostridium* biodegraded carbon tetrachloride by reductive dechlorination yielding trichloromethane (chloroform), dichloromethane and unidentified products. Anticipated products from the reaction of carbon tetrachloride with O_3 or hydroxyl radicals in the atmosphere are phosgene and Cl•. Phosgene is hydrolyzed readily to hydrogen chloride and CO_2.

EXPOSURE LIMITS: NIOSH REL: 1-h C 2 ppm; OSHA PEL: TWA 10 ppm, C 25 ppm, 5-min/4-h 200 ppm peak; ACGIH TLV: TWA 5 ppm.

SYMPTOMS OF EXPOSURE: CNS depres; nau, vomit; liv, kid damage; skin irrit.

USES: Preparation of refrigerants, aerosols and propellants; metal degreasing; agricultural fumigant; chlorinating unsaturated organic compounds; production of semiconductors; solvent for fats, oils, rubber, etc; dry cleaning operations; industrial extractant; spot remover; fire extinguisher manufacturing; preparation of dichlorodifluoromethane; veterinary medicine; organic synthesis.

CHLORDANE

SYNONYMS: 1,068; Aspon-chlordane; Belt; CD-68; Chlordan; γ-Chlordan; Chloridan; Chlorindan; Chlor kil; Chlorodane; Chlortox; Corodane; Cortilan-neu; Dichlorochlordene; Dowklor; ENT 9,932; ENT 25,552-X; HCS 3,260; Kypchlor; M 140; M 410; NA 2,762; NCI-C00099; Niran; Octachlor; 1,2,4,5,6,7,8,8-Octachlor-2,3,3a,4,7,7a-hexahydro-4,7-methanoindane; Octachlorodihydrodicyclopentadiene; 1,2,4,5,6,7,8,8-Octachloro-2,3,3a,4,7,7a-hexahydro-4,7-methanoindene; **1,2,4,5,6,7,8,8-Octachloro-2,3,3a,4,7,7a-hexahydro-4,7-methano-1*H*-indene**; 1,2,4,5,6,7,8,8-Octachloro-3a,4,7,7a-hexahydro-4,7-methyleneindane; Octachloro-4,7-methanohydroindane; Octachloro-4,7-methanotetrahydroindane; 1,2,4,5,6,7,8,8-Octachloro-4,7-methano-3a,4,7,7a-tetrahydroindane; 1,2,4,5,6,7,8,8-Octachloro-3a,4,7,7a-tetrahydro-4,7-methanoindan; 1,2,4,5,6,7,8,8-Octachloro-3a,4,7,7a-tetrahydro-4,7-methanoindane; Octaklor; Octaterr; Orthoklor; RCRA waste number U036; SD 5,532; Shell SD-5532; Synklor; Tat chlor 4; Topichlor 20; Topiclor; Topiclor 20; Toxichlor; Velsicol 1,068.

CHEMICAL DESIGNATIONS: CAS: 57-74-9; DOT: 2762; mf: $C_{10}H_6Cl_8$; fw: 409.78; RTECS: PB 9800000.

PROPERTIES: Colorless to amber to yellowish-brown viscous liq with an aromatic, slight pungent odor similar to chlorine. Mp: < 25°, bp: 175° @ 2 mm, *d*: 1.59-1.63 @ 20/4°, lel: 0.7% (kerosene soln), uel: 5% (kerosene soln), K_H: 4.8 x 10^{-5} atm·m^3/mol @ 25°, log K_{oc}: 4.58-5.57, log K_{ow}: 6.00, S_o: misc with aliphatic and aromatic solvents, S_w: 56 ppb @ 25°, vap d: 16.75 g/L @ 25°, 14.15 (air = 1), vp:

10^{-6} mm @ 20°.

TRANSFORMATION PRODUCTS: In an alkaline medium or solvent, carrier, diluent or emulsifier having an alkaline reaction, chlorine will be released. Technical grade chlordane passed over a 5% platinum catalyst @ 200° resulted in the formation of tetrahydrodicyclopentadiene.

EXPOSURE LIMITS: NIOSH REL: IDLH 500 mg/m^3; OSHA PEL: TWA 0.5 mg/m^3; ACGIH TLV: TWA 0.5 mg/m^3, STEL 2 mg/m^3.

SYMPTOMS OF EXPOSURE: Blur vision; conf; atax; delirium; cough; abdom pain, nau, vomit, diarr; irrity; tremor, convuls, anuria.

USES: Insecticide and fumigant.

cis-CHLORDANE

SYNONYMS: α-Chlordane; β-Chlordane; **1,2,4,5,6,7,8,8-Octachloro-2,3,3a,4,7,7a-hexahydro-4,7-methano-1*H*-indene**; α-1,2,4,5,6,7,8,8-Octachloro-3a,4,7,7a-tetrahydro-4,7-methanoindan.

CHEMICAL DESIGNATIONS: CAS: 5103-74-2; DOT: none assigned; mf: $C_{10}H_6Cl_8$; fw: 409.78; RTECS: PC 01750000.

PROPERTIES: Solid. Mp: 107.0-108.8°, bp: 175° (isomeric mixture), K_H: 8.75 x 10^{-4} atm·m^3/mol @ 23°, log K_{oc}: 5.40-6.00, log K_{ow}: 5.93 (calcd), S_o: misc with aliphatic and aromatic solvents, S_w: 51 μg/L @ 20-25°, vp: 8.3 x 10^{-5} mm @ 23° (calcd).

TRANSFORMATION PRODUCTS: In an alkaline medium or solvent, carrier, diluent or emulsifier having an alkaline reaction, chlorine will be released. Irradiation of *cis*-chlordane by a 450-W high-pressure mercury lamp gave photo-*cis*-chlordane.

SYMPTOMS OF EXPOSURE: Blur vision; conf; atax; delirium; cough; abdom pain, nau, vomit, diarr; irrity; tremor, convuls, anuria.

USE: Insecticide.

trans-CHLORDANE

SYNONYMS: α-Chlordan; *cis*-Chlordan; α-Chlordane; α(*cis*)-Chlordane; γ-Chlordane; 1,2,4,5,6,7,8,8-Octachloro-3a,4,7,7a-tetrahydro-4,7-methanoindan.

CHEMICAL DESIGNATIONS: CAS: 5103-71-9; DOT: none assigned; mf: $C_{10}H_6Cl_8$; fw: 409.78; RTECS: PB 9705000.

PROPERTIES: Solid. Mp: 103.0-105.0°, bp: 175° (isomeric mixture), K_H: 1.34 x 10^{-3} atm·m^3/mol @ 23°, log K_{oc}: 5.48, 6.00, log K_{ow}: 8.69, 9.65 (calcd), S_o: misc with aliphatic and aromatic solvents, vp: 9.75 x 10^{-6} mm @ 30° (est).

TRANSFORMATION PRODUCTS: In an alkaline medium or solvent, carrier, diluent or emulsifier having an alkaline reaction, chlorine will be released. Irradiation of *trans*-chlordane by a 450-W high-pressure mercury lamp gave photo-*trans*-chlordane.

SYMPTOMS OF EXPOSURE: Blur vision; conf; atax; delirium; cough; abdom

pain, nau, vomit, diarr; irrity; tremor, convuls, anuria.
USE: Insecticide.

CHLOROACETALDEHYDE

SYNONYMS: 2-Chloroacetaldehyde; Chloroacetaldehyde monomer; 2-Chloroethanal; 2-Chloro-1-ethanal; Monochloroacetaldehyde; RCRA waste number P023; UN 2232.

CHEMICAL DESIGNATIONS: CAS: 107-20-0; DOT: 2232; mf: C_2H_3ClO; fw: 78.50; RTECS: AB 2450000.

PROPERTIES: Colorless liq with an irritating, acrid odor. Mp: -16.1° (40% soln), bp: 85-85.5° @ 748 mm, *d*: 1.236 @ 20/4°, fl p: 53°, IP: 10.48 eV, S_o: sol in ether, acetone and methanol, S_w: > 50 wt %, vap d: 3.21 g/L @ 25°, 2.71 (air = 1), vp: 100 mm @ 20°.

TRANSFORMATION PRODUCTS: Polymerizes on standing.

EXPOSURE LIMITS: NIOSH REL: IDLH 250 ppm; OSHA PEL: C 1 ppm.

SYMPTOMS OF EXPOSURE: Irrit skin, eyes, muc memb; skin burns; sys damage; pulm edema; skin, resp sens.

USES: Removing barks from trees; manufacture of 2-aminothiazole; chemical intermediate; organic synthesis.

α-CHLOROACETOPHENONE

SYNONYMS: CAF; CAP; Chemical mace; 2-Chloroacetophenone; Ω-Chloroacetophenone; Chloromethyl phenyl ketone; **2-Chloro-1-phenylethanone**; CN; Mace; NCI-C55107; Phenacyl chloride; Phenyl chloromethyl ketone; Tear gas; UN 1697.

CHEMICAL DESIGNATIONS: CAS: 532-27-4; DOT: 1697; mf: C_8H_7ClO; fw: 154.60; RTECS: AM 6300000.

PROPERTIES: Colorless solid with a sharp, penetrating odor. Mp: 56.5°, bp: 247°, *d*: 1.324 @ 15/4°, fl p: 117.8°, IP: 9.5 eV, 9.44 eV, S_o: sol in acetone, alcohol, benzene and ether, S_w: misc, vp: 0.004-0.012 mm @ 20°.

TRANSFORMATION PRODUCTS: Releases toxic chloride fumes when heated to decomposition.

EXPOSURE LIMITS: NIOSH REL: IDLH 100 mg/m^3; OSHA PEL: TWA 0.05 ppm.

SYMPTOMS OF EXPOSURE: Irrit skin, eyes, resp sys; pulm edema.

USES: Riot control agent.

4-CHLOROANILINE

SYNONYMS: 1-Amino-4-chlorobenzene; 1-Amino-*p*-chlorobenzene; 4-Aminochlorobenzene; *p*-Aminochlorobenzene; 4-Chloraniline; *p*-Chloraniline; *p*-Chloroaniline; **4-Chlorobenzamine**; *p*-Chlorobenzamine; 4-Chlorophenylamine;

p-Chlorophenylamine; NCI-C02039; RCRA waste number P024; UN 2018; UN 2019.

CHEMICAL DESIGNATIONS: CAS: 106-47-8; DOT: 2018 (solid), 2019 (liq); mf: C_6H_6ClN; fw: 127.57; RTECS: BX 0700000.

PROPERTIES: Yellowish-white solid with a mild, sweetish odor. Mp: 70°, 72.5°, bp: 232°, *d*: 1.429 @ 19/4°, fl p: > 1,205°, pK_a: 3.98, 4.15, K_H: 1.07×10^{-5} atm·m^3/mol @ 25° (calcd), log K_{oc}: 1.98-3.13, log K_{ow}: 1.83-2.78, S_o: sol in ethanol and ether, freely sol in acetone and carbon disulfide, S_w: 4.7 g/L @ 25°, vp: 1.5×10^{-2} mm @ 20°.

TRANSFORMATION PRODUCTS: Under artificial sunlight, river water containing 2-5 ppm 4-chloroaniline photodegraded to 4-aminophenol and unidentified polymers. In an anaerobic medium, the bacteria of the *Paracoccus* sp. converted 4-chloroaniline to 1,3-bis(*p*-chlorophenyl)triazene and 4-chloroacetanilide with product yields of 80% and 5%, respectively. In a field experiment, 4-chloroaniline-^{14}C was applied to a soil at a depth of 10 cm. After 20 wk, 32.4% of the applied amount was recovered in soil. Metabolites identified include 4-chloroformanilide, 4-chloroacetanilide, 4-chloronitrobenzene, 4-chloronitrosobenzene, 4,4′-dichloroazoxybenzene and 4,4′-dichloroazobenzene. Photooxidation of 4-chloroaniline in air-saturated water using UV light (λ > 290 nm) produced 4-chloronitrobenzene and 4-chloronitrosobenzene. About 6 h later, 4-chloroaniline reacted completely leaving dark purple condensation products. 4-Chloroaniline covalently bonds with humates in soils to form quinoidal structures followed by oxidation to yield a nitrogen-substituted quinoid ring. Catechol, a humic acid monomer, reacted with 4-chloroaniline yielding 4,5-bis(4-chlorophenylamino)-3,5-cyclohexadiene-1,2-dione. A CO_2 yield of 27.7% was achieved when 4-chloroaniline adsorbed on silica gel was irradiated with light (λ > 290 nm) for 17 h. In activated sludge, 22.7% mineralized to CO_2 after 5 days. In a 56-day experiment, ^{14}C-labelled 4-chloroaniline applied to soil-water suspensions under aerobic and anaerobic conditions gave $^{14}CO_2$ yields of 3.0 and 2.3%, respectively. Silage samples (chopped corn plants) containing 4-chloroaniline were incubated in an anaerobic chamber for 2 wk @ 28°. After 3 days, 4-chloroaniline was biologically metabolized to 4-chloroacetanilide and another compound tentatively identified as 4-chloroformanilide.

USES: Dye intermediate; pharmaceuticals; agricultural chemicals.

CHLOROBENZENE

SYNONYMS: Benzene chloride; Chlorbenzene; Chlorbenzol; Chlorobenzol; MCB; Monochlorbenzene; Monochlorobenzene; NCI-C54886; Phenyl chloride; RCRA waste number U037; UN 1134.

CHEMICAL DESIGNATIONS: CAS: 108-90-7; DOT: 1134; mf: C_6H_5Cl; fw: 112.56; RTECS: CZ 0175000.

PROPERTIES: Clear, colorless, liq with a sweet almond odor. Mp: -45.6°, bp: 131.5°, *d*: 1.1058 @ 20/4°, fl p: 28°, lel: 1.3%, uel: 7.1%, K_H: 3.93×10^{-3} atm·m^3/mol @ 25°, IP: 9.07 eV, 9.14 eV, log K_{oc}: 1.92, 2.52, log K_{ow}: 2.71-2.98,

S_o: sol in ethanol, benzene, chloroform, ether and carbon tetrachloride, S_w: 500 mg/L @ 20°, vap d: 4.60 g/L @ 25°, 3.88 (air = 1), vp: 9 mm @ 20°.

TRANSFORMATION PRODUCTS: Under artificial sunlight, chlorobenzene (2-5 ppm) in river water degraded to phenol and chlorophenol. However, in a 1% acetonitrile/distilled water soln, 28% photolyzed to phenol, chloride ion and acetanilide with product yields of 55, 112 and 2%, respectively. Titanium dioxide suspended in an aq soln and irradiated with UV light (λ = 365 nm) converted chlorobenzene to CO_2 at a significant rate. In a similar experiment, irradiation of chlorobenzene in a titanium dioxide aq suspension yielded 3 monochlorophenols, chlorohydroquinone and hydroxyhydroquinone. Photooxidation of chlorobenzene in air containing NO in a Pyrex glass vessel and a quartz vessel gave *m*-chloronitrobenzene, 2-chloro-6-nitrophenol, 2-chloro-4-nitrophenol, 4-chloro-2-nitrophenol, *p*-nitrophenol, 3-chloro-4-nitrophenol, 3-chloro-6-nitrophenol and 3-chloro-2-nitrophenol. Anticipated products from the reaction of chlorobenzene with O_3 or hydroxyl radicals in air are chlorophenols and ring cleavage compounds. A CO_2 yield of 18.5% was achieved when chlorobenzene adsorbed on silica gel was irradiated with light (λ > 290 nm) for 17 h. In activated sludge, 31.5% mineralized to CO_2 after 5 days. A mixed culture of soil bacteria or a *Pseudomonas* sp. transformed chlorobenzene to chlorophenol. Pure microbial cultures isolated from soil hydroxylated chlorobenzene to 2- and 4-hydroxychlorobenzene. The sunlight irradiation of chlorobenzene (20 g) in a borosilicate glass-stoppered Erlenmeyer flask for 28 days yielded 1,060 ppm monochlorobiphenyl. In the absence of O_2, chlorobenzene reacted with Fenton's reagent forming chlorophenols, dichlorobiphenyls and phenolic polymers as major intermediates. With O_2, chlorobenzoquinone, chlorinated and nonchlorinated diols formed.

EXPOSURE LIMITS: NIOSH REL: IDLH 2,400 ppm; OSHA PEL: TWA 75 ppm; ACGIH TLV: TWA 75 ppm.

SYMPTOMS OF EXPOSURE: Irrit skin, eyes, nose; drow, inco; liv damage.

USES: Preparation of phenol, *p*-chlorophenol, chloronitrobenzene, aniline, *o*-, *m*- and *p*-nitrochlorobenzenes; solvent carrier for methylene diisocyanate; solvent; insecticide, pesticide and dyestuffs intermediate; heat transfer agent.

o-CHLOROBENZYLIDENEMALONONITRILE

SYNONYMS: 2-Chlorobenzalmalononitrile; *o*-Chlorobenzalmalononitrile; 2-Chlorobenzylidenemalononitrile; 2-Chlorobnm; **[(2-Chlorophenyl)methylene]propanedinitrile**; CS; *β*,*β*-Dicyano-*o*-chlorostyrene; NCI-C55118; OCBM; USAF KF-11.

CHEMICAL DESIGNATIONS: CAS: 2698-41-1; DOT: none assigned; mf: $C_{10}H_5ClN_2$; fw: 188.61; RTECS: OO 3675000.

PROPERTIES: White powder with a pepper-like odor. Mp: 95-96°, bp: 310-315°, *d*: 1.472 (calcd), lel: 0.025 g/L (dust), S_o: sol in acetone, benzene, dioxane, ethyl acetate and methylene chloride, vp: 3.4 x 10^{-5} mm @ 20°.

TRANSFORMATION PRODUCTS: Hydrolyzes in water to *o*-chlorobenzaldehyde and malononitrile.

SYMPTOMS OF EXPOSURE: Pain, burn eyes, lac, conj, eryt eyelids, blepharo spas; irrit throat, cough, chest const; head; eryt vesic skin.

EXPOSURE LIMITS: NIOSH REL: IDLH 2 mg/m^3; OSHA PEL: TWA 0.05 mg/m^3; ACGIH TLV: C 0.05 ppm.

USES: Riot control agent.

p-CHLORO-*m*-CRESOL

SYNONYMS: Aptal; Baktol; Baktolan; Candaseptic; *p*-Chlor-*m*-cresol; Chlorocresol; 4-Chlorocresol; *p*-Chlorocresol; 4-Chloro-*m*-cresol; 6-Chloro-*m*-cresol; 4-Chloro-1-hydroxy-3-methylbenzene; 2-Chlorohydroxytoluene; 2-Chloro-5-hydroxytoluene; 4-Chloro-3-hydroxytoluene; 6-Chloro-3-hydroxytoluene; *p*-Chlorometacresol; **4-Chloro-3-methylphenol**; *p*-Chloro-3-methylphenol; 3-Methyl-4-chlorophenol; Ottafact; Parmetol; Parol; PCMC; Peritonan; Preventol CMK; Raschit; Raschit K; Rasenanicon; RCRA waste number U039.

CHEMICAL DESIGNATIONS: CAS: 59-50-7; DOT: none assigned; mf: C_7H_7ClO; fw: 142.59; RTECS: GO 7100000.

PROPERTIES: Colorless, white, or pinkish cryst with a slight phenolic odor. On exposure to air it slowly becomes light brown. Mp: 63-68°, bp: 235°, pK_a: 9.549 @ 25°, K_H: 1.78 x 10^{-6} atm·m^3/mol (calcd), log K_{oc}: 2.89 (calcd), log K_{ow}: 2.95, 3.10, S_o: sol in ethanol, ether, benzene, chloroform, acetone, petroleum ether, fixed oils, terpenes and aq alkaline solns, S_w: 3.85 g/L @ 25°, vp: 5 x 10^{-2} mm @ 20° (est).

USES: External germicide; preservative for gums, glues, paints, inks, textile and leather products; topical antiseptic (veterinarian).

CHLOROETHANE

SYNONYMS: Aethylis; Aethylis chloridum; Anodynon; Chelen; Chlorethyl; Chloridum; Chloryl; Chloryl anesthetic; Ether chloratus; Ether hydrochloric; Ether muriatic; Ethyl chloride; Hydrochloric ether; Kelene; Monochlorethane; Monochloroethane; Muriatic ether; Narcotile; NCI-C06224; UN 1037.

CHEMICAL DESIGNATIONS: CAS: 75-00-3; DOT: 1037; mf: C_2H_5Cl; fw: 64.52; RTECS: KH 7525000.

PROPERTIES: Colorless liq with an ethereal odor. Mp: -136.4°, bp: 12.3°, *d*: 0.8978 @ 20/4°, fl p: -50°, lel: 3.16%, uel: 14%, K_H: 8.5 x 10^{-3} atm·m^3/mol @ 25°, IP: 10.97 eV, log K_{oc}: 0.51 (calcd), log K_{ow}: 1.43, S_o: sol in ethanol and ether, S_w: 5.74 g/L @ 20°, vap d: 2.76 kg/m^3 @ 20°, vp: 1,011 mm @ 20°.

TRANSFORMATION PRODUCTS: Hydrolyzes in water forming ethanol ($t_{1/2}$ = 38 days @ 25° and pH 7). In the atmosphere, formyl chloride is the initial photooxidation product which reacts with mositure forming hydrochloric acid and CO. Burns with a smoky, greenish flame releasing hydrogen chloride.

EXPOSURE LIMITS: NIOSH REL: IDLH 20,000 ppm; OSHA PEL: TWA 1,000 ppm; ACGIH TLV: TWA 1,000 ppm.

SYMPTOMS OF EXPOSURE: Inco, inebriate; abdom cramps; card arrhy, card arrest, liv, kid damage.

USES: Intermediate for tetraethyl lead and ethyl cellulose; anesthetic; organic synthesis; alkylating agent; refrigeration; analytical reagent; solvent for phosphorus, sulfur, fats, oils, resins and waxes; insecticides.

2-CHLOROETHYL VINYL ETHER

SYNONYMS: 2-Chlorethyl vinyl ether; **(2-Chloroethoxy)ethene**; RCRA waste number U042; Vinyl 2-chloroethyl ether; Vinyl β-chloroethyl ether.

CHEMICAL DESIGNATIONS: CAS: 110-75-8; DOT: none assigned; mf: C_4H_7ClO; fw: 106.55; RTECS: KN 6300000.

PROPERTIES: Colorless liq. Mp: -70.3°, bp: 108°, *d*: 1.0475 @ 20/4°, fl p: 16°, K_H: 2.5 x 10^{-4} atm·m^3/mol, log K_{oc}: 0.82, log K_{ow}: 1.28 (calcd), S_o: sol in ethanol and ether, S_w: 15 g/L @ 20°, vap d: 4.36 g/L @ 25°, 3.68 (air = 1), vp: 26.75 mm @ 20°.

TRANSFORMATION PRODUCTS: Chlorination of 2-chloroethyl vinyl ether to α-chloroethyl ethyl ether or β-chloroethyl ethyl ether may occur in water treatment facilities. The *alpha* compound is very unstable in water and decomposes almost as fast as it is formed. Though stable in sodium hydroxide solns, in dilute acid solns hydrolysis will yield acetaldehyde and chlorohydrin.

USES: Anesthetics, sedatives and cellulose ethers; copolymer of 95% ethyl acrylate with 5% 2-chloroethyl vinyl ether is used to produce an acrylic elastomer.

CHLOROFORM

SYNONYMS: Formyl trichloride; Freon 20; Methane trichloride; Methenyl chloride; Methenyl trichloride; Methyl trichloride; NCI-C02686; R 20; R 20 (refrigerant); RCRA waste number U044; TCM; Trichloroform; **Trichloromethane**; UN 1888.

CHEMICAL DESIGNATIONS: CAS: 67-66-3; DOT: 1888; mf: $CHCl_3$; fw: 119.38; RTECS: FS 9100000.

PROPERTIES: Clear, water-white, volatile liq with a strong, sweet, ethereal odor. Mp: -63.5°, bp: 61.7°, *d*: 1.4890 @ 20/4°, K_H: 3.39 x 10^{-3} atm·m^3/mol @ 25°, IP: 11.42 eV, log K_{oc}: 1.64-1.94, log K_{ow}: 1.90-1.97, S_o: sol in acetone; misc with ethanol, ether, benzene and ligroin, S_w: 8 g/L @ 20°, vap d: 4.88 g/L @ 25°, 4.12 (air = 1), vp: 160 mm @ 20°.

TRANSFORMATION PRODUCTS: An anaerobic species of *Clostridium* biodegraded chloroform (a metabolite of carbon tetrachloride) by reductive dechlorination yielding methylene chloride and unidentified products. Complete mineralization was reported when distilled deionized water containing trichloromethane (avg concn = 118 ppm) and 0.1 wt % titanium dioxide as a catalyst was irradiated with UV light. Mineralization products included CO_2 and hydrochloric acid. In a similar experiment, titanium dioxide suspended in an aq soln and irradiated with UV light (λ = 365 nm) converted chloroform to CO_2 at a significant rate. Intermediate compounds were not identified. Anticipated hydrolysis product is dichloromethanol and hydrochloric acid ($t_{1/2}$ = 3,500 and

1,849.6 yr @ 25° and pH 7). An aq soln containing 300 ng/μL chloroform and colloidal platinum catalyst was irradiated with UV light. After 15 h, only 10 ng/μL chloroform remained. A duplicate experiment was performed but 0.1 g zinc was added to the system. After ≈ 2 h, 10 ng/μL chloroform remained and 210 ng/μL methane was produced. Photolysis of an aq soln containing chloroform (314 μmol) and the catalyst [Pt(colloid)/Ru(bpy)$^{2+}$/MV/EDTA] yielded the following products after 15 h (μmol detected): chloride ions (852), methane (265), ethylene (0.05), ethane (0.52) and unreacted chloroform (10.5). In the troposphere, photolysis of chloroform via hydroxyl radicals may yield formyl chloride, CO, hydrogen chloride and phosgene as the principal products. Phosgene is hydrolyzed readily to hydrogen chloride and CO_2.

EXPOSURE LIMITS: NIOSH REL: 1-h C 2 ppm; OSHA PEL: C 50 ppm; ACGIH TLV: TWA 10 ppm.

SYMPTOMS OF EXPOSURE: Dizz, mental dullness; nau; head; ftg; anes; hepatomegaly; eye, skin irrit.

USES: Fluorocarbon refrigerants and plastics; solvent for natural products; analytical chemistry; soil fumigant; insecticides; preparation of chlorodifluoromethane, methyl fluoride, salicylaldehyde; cleaning electronic circuit boards; anesthetics.

2-CHLORONAPHTHALENE

SYNONYMS: β-Chloronaphthalene; RCRA waste number U047.

CHEMICAL DESIGNATIONS: CAS: 91-58-7; DOT: none assigned; mf: $C_{10}H_7Cl$; fw: 162.62; RTECS: QJ 2275000.

PROPERTIES: Monoclinic plates or leaflets. Mp: 61°, bp: 256°, *d*: 1.2656 @ 16/4°, K_H: 6.12 x 10^{-4} atm·m^3/mol (calcd), log K_{oc}: 3.93 (calcd), log K_{ow}: 4.07, S_o: sol in ethanol, benzene, ether, chloroform and carbon disulfide, S_w: 6.74 mg/L @ 25° (calcd), vp: 1.7 x 10^{-2} mm @ 25° (calcd).

TRANSFORMATION PRODUCTS: Reported biodegradation products include 8-chloro-1,2-dihydro-1,2-dihydroxynaphthalene and 3-chlorosalicylic acid.

USES: Chlorinated naphthalenes were formerly used in the production of electric condensers, insulating electric condensers, electric cables and wires; additive for high pressure lubricants.

p-CHLORONITROBENZENE

SYNONYMS: **1-Chloro-4-nitrobenzene**; 4-Chloronitrobenzene; 4-Chloro-1-nitrobenzene; 4-Nitrochlorobenzene; *p*-Nitrochlorobenzene; PCNB; PNCB; UN 1578.

CHEMICAL DESIGNATIONS: CAS: 100-00-5; DOT: 1578; mf: $C_6H_4ClNO_2$; fw: 157.56; RTECS: CZ 1050000.

PROPERTIES: Yellow solid with a sweet odor. Mp: 83.6°, bp: 242°, *d*: 1.520 @ 18/4°, fl p: 127.2°, K_H: < 6.91 x 10^{-3} atm·m^3/mol @ 20° (calcd), IP: 9.99 eV, log K_{oc}: 2.68 (calcd), log K_{ow}: 2.39, 2.41, S_o: sol in acetone and alcohol, S_w: 0.003 wt %

@ 20°, vp: < 1 mm @ 20°.

TRANSFORMATION PRODUCTS: An aq soln containing *p*-chloronitrobenzene and a titanium dioxide suspension was irradiated with UV light (λ > 290 nm). 2-Chloro-5-nitrophenol was identified as a minor degradation product. Continued irradiation caused further degradation to yield CO_2, water, hydrochloric and nitric acids. Irradiation in air and nitrogen gave 4-chloro-2-nitrophenol and *p*-chlorophenol, respectively. Under aerobic conditions, the yeast *Rhodosporidium* sp. metabolized 4-chloronitrobenzene to 4-chloroacetanilide and 4-chloro-2-hydroxyacetanilide as final major metabolites. Intermediate compounds identified include 4-chloronitrosobenzene, 4-chlorophenylhydroxylamine and 4-chloroaniline. Under continuous flow conditions involving feeding, aeration, settling and reflux, a mixture of *p*-chloronitrobenzene and 2,4-dinitrochlorobenzene was reduced 61-70% after 8-13 days by *Arthrobacter simplex*, a microorganism isolated from industrial waste. A similar experiment was conducted using 2 aeration columns. One column contained *A. simplex*, the other a mixture of *A. simplex* and microorganisms isolated from soil (*Streptomyces coelicolor*, *Fusarium* sp. probably *aquaeductum* and *Trichoderma viride*). After 10 days, 89.5-91% of the nitrocompounds was reduced. *p*-Chloronitrobenzene was reduced to *p*-chloroaniline and 6 unidentified compounds.

EXPOSURE LIMITS: NIOSH REL: IDLH 1,000 ppm; OSHA PEL: TWA 1 mg/m^3.

SYMPTOMS OF EXPOSURE: Anoxia, unpleasant taste; mild anem; dizz, weak; nau, vomit; dysp.

USES: Intermediate for dyes; rubber and agricultural chemicals; manufacture of *p*-nitrophenol.

1-CHLORO-1-NITROPROPANE

SYNONYMS: Chloronitropropane; Korax; Lanstan.

CHEMICAL DESIGNATIONS: CAS: 600-25-9; DOT: none assigned; mf: $C_3H_6ClNO_2$; fw: 123.54; RTECS: TX 5075000.

PROPERTIES: Colorless liq with an unpleasant odor. Mp: < 25°, bp: 141-143°, *d*: 1.209 @ 20/20°, fl p: 62.2°, K_H: 1.57 x 10^{-1} atm·m^3/mol @ 20-25° (calcd), log K_{oc}: 3.34 (calcd), log K_{ow}: 4.25 (calcd), S_o: sol in alcohol and ether, S_w: 6 mg/L @ 20°, vap d: 5.05 g/L @ 25°, 4.26 (air = 1), vp: 5.8 mm @ 25°.

EXPOSURE LIMITS: NIOSH REL: IDLH 2,000 ppm; OSHA PEL: TWA 20 ppm; ACGIH TLV: TWA 2 ppm.

SYMPTOMS OF EXPOSURE: In animals: irrit lungs, eyes; liv, kid, heart, blood vessel damage.

USE: Fungicide.

2-CHLOROPHENOL

SYNONYMS: 1-Chloro-2-hydroxybenzene; 1-Chloro-*o*-hydroxybenzene; 2-

Chlorohydroxybenzene; ***o*-Chlorophenol**; 1-Hydroxy-2-chlorobenzene; 1-Hydroxy-*o*-chlorobenzene; 2-Hydroxychlorobenzene; RCRA waste number U048; UN 2020; UN 2021.

CHEMICAL DESIGNATIONS: CAS: 95-57-8; DOT: 2020 (liq); 2021 (solid); mf: C_6H_5ClO; fw: 128.56; RTECS: SK 2625000.

PROPERTIES: Pale amber liq with a slight phenolic odor. Mp: 9.0°, bp: 174.9°, *d*: 1.2634 @ 20/4°, fl p: 64°, pK_a: 8.48 @ 25°, K_H: 5.6 x 10^{-7} atm·m^3/mol @ 25° (est), IP: 9.28 eV, log K_{oc}: 1.71-3.69, log K_{ow}: 2.15-2.19, S_o: sol in ethanol, benzene and ether, S_w: 28.5 g/L @ 20°, vap d: 5.25 g/L @ 25°, 4.44 (air = 1), vp: 2.25 mm @ 25°.

TRANSFORMATION PRODUCTS: Monochlorophenols exposed to sunlight (UV radiation) produced catechol and other hydroxybenzenes. Titanium dioxide suspended in an aq soln and irradiated with UV light (λ = 365 nm) converted 2-chlorophenol to CO_2 at a significant rate. In a similar experiment, irradiation of an aq soln containing 2-chlorophenol and titanium dioxide with UV light (λ > 340 nm) resulted in the formation of chlorohydroquinone and trace amounts of catechol. Hydroxylation of both of these compounds forms the intermediate hydroxyhydroquinone which degrades quickly to unidentified carboxylic acids and carbonyl compounds. Irradiation of an aq soln @ 296 nm @ pH 8-13 yielded different products. Photolysis at a pH nearly equal to the dissociation constant yielded pyrocatechol. At an elevated pH, 2-chlorophenol is almost completely ionized, photolysis yielded cyclopentadienic acid. Irradiation of an aq soln @ 296 nm containing H_2O_2 converted 2-chlorophenol to catechol and 2-chlorohydroquinone. In the dark, NO (10^{-3} vol %) reacted with 2-chlorophenol forming 4-nitro-2-chlorophenol and 6-nitro-2-chlorophenol with yields of 36 and 30%, respectively. Wet oxidation of 2-chlorophenol @ 320° yielded formic and acetic acids. Wet oxidation of 2-chlorophenol at an elevated pressure and temperature gave the following products: acetone, acetaldehyde, formic, acetic, maleic, oxalic, muconic and succinic acids. Chloroperoxidase, a fungal enzyme isolated from *Caldariomyces fumago*, reacted with 2-chlorophenol yielding traces of 2,4,6-trichlorophenol, 2,4- and 2,6-dichlorophenols.

USES: Component of disinfectant formulations; chemical intermediate for phenolic resins; solvent for polyester fibers, antiseptic (veterinarian); preparation of 4-nitroso-2-methylphenol and other compounds.

4-CHLOROPHENYL PHENYL ETHER

SYNONYMS: 4-Chlorodiphenyl ether; *p*-Chlorodiphenyl ether; **1-Chloro-4-phenoxybenzene**; 1-Chloro-*p*-phenoxybenzene; 4-Chlorophenyl ether; *p*-Chlorophenyl ether; *p*-Chlorophenyl phenyl ether; Monochlorodiphenyl oxide.

CHEMICAL DESIGNATIONS: CAS: 7005-72-3; DOT: none assigned; mf: $C_{12}H_9ClO$; fw: 204.66; RTECS: none assigned.

PROPERTIES: Liq. Mp: -8°, bp: 284-285°, *d*: 1.193 @ 20/4°, fl p: 110°, K_H: 2.2 x 10^{-4} atm·m^3/mol, log K_{oc}: 3.60 (calcd), log K_{ow}: 4.08, S_w: 3.3 mg/L @ 25°, vap d: 8.36 g/L @ 25°, 7.06 (air = 1), vp: 2.7 x 10^{-3} mm @ 25° (calcd).

TRANSFORMATION PRODUCTS: In a methanolic soln irradiated with UV light (λ > 290 nm), dechlorination of 4-chlorophenyl phenyl ether resulted in the formation of diphenyl ether. Photolysis of an aq soln containing 10% acetonitrile with UV light (λ = 230-400 nm) yielded 4-hydroxybiphenyl ether and chloride ion.

USE: Research chemical.

CHLOROPICRIN

SYNONYMS: Acquinite; Chlor-o-pic; Dolochlor; G 25; Larvacide 100; Microlysin; NA 1,583; NCI-C00533; Nitrochloroform; Nitrotrichloromethane; Pic-clor; Picfume; Picride; Profume A; PS; S 1; **Trichloronitromethane**; Tri-clor; UN 1580.

CHEMICAL DESIGNATIONS: CAS: 76-06-2; DOT: 1580; mf: CCl_3NO_2; fw: 164.38; RTECS: PB 6300000.

PROPERTIES: Colorless, oily liq with a sharp, penetrating odor. Mp: -64.5°, bp: 111.8°, *d*: 1.6558 @ 20/4°, fl p: detonates, K_H: 8.4 x 10^{-2} atm·m^3/mol, log K_{oc}: 0.82 (calcd), log K_{ow}: 1.03, S_o: sol in acetic acid and acetone; misc with benzene, carbon disulfide and ethanol, S_w: 2,270 mg/L @ 0°, vap d: 6.72 g/L @ 25°, 5.67 (air = 1), vp: 16.9 mm @ 20°.

TRANSFORMATION PRODUCTS: Releases very toxic fumes of chlorides and NO_x when heated to decomposition. Photodegrades under simulated atmospheric conditions to phosgene and nitrosyl chloride. Photolysis of nitrosyl chloride yields chlorine and NO.

EXPOSURE LIMITS: NIOSH REL: IDLH 4 ppm; OSHA PEL: TWA 0.1 mg/m^3.

SYMPTOMS OF EXPOSURE: Eye irrit, lac; cough, pulm edema; nau, vomit; skin irrit.

USES: Disinfecting cereals and grains; fumigant and soil insecticide; dyestuffs; fungicide; rat exterminator; organic synthesis; war gas.

CHLOROPRENE

SYNONYMS: Chlorobutadiene; 2-Chlorobutadiene-1,3; 2-Chlorobuta-1,3-diene; **2-Chloro-1,3-butadiene**; β-Chloroprene; Neoprene; UN 1991.

CHEMICAL DESIGNATIONS: CAS: 126-99-8; DOT: 1991; mf: C_4H_5Cl; fw: 88.54; RTECS: EI 9625000.

PROPERTIES: Colorless liq with an ethereal odor. Mp: -130°, bp: 59.4°, *d*: 0.9583 @ 20/4°, fl p: -20°, lel: 4%, uel: 20%, K_H: 3.20 x 10^{-2} atm·m^3/mol (calcd), IP: 8.83 eV, S_o: sol in acetone, benzene and ether, vap d: 3.62 g/L @ 25°, 3.06 (air = 1), vp: 179 mm @ 20°.

TRANSFORMATION PRODUCTS: Anticipated products from the reaction of chloroprene with O_3 or hydroxyl radicals in the atmosphere are formaldehyde, 2-chloroacrolein, OHCCHO, ClCOCHO, $H_2CCHCClO$, chlorohydroxy acids and aldehydes.

EXPOSURE LIMITS: NIOSH REL: 15-min C 1 ppm; OSHA PEL: TWA 25 ppm; ACGIH TLV: TWA 10 ppm.

SYMPTOMS OF EXPOSURE: Irrit eyes, resp sys; ner, irrity; derm, alopecia.

USE: Manufacture of neoprene.

CHLORPYRIFOS

SYNONYMS: Brodan; Chlorpyrifos-ethyl; Detmol U.A.; *O,O*-Diethyl-*O*-3,5,6-trichloro-2-pyridyl phosphorothioate; Dowco-179; Dursban; Dursban F; ENT 27,311; Eradex; Lorsban; NA 2,783; OMS-0971; Phosphorothionic acid *O,O*-diethyl *O*-(3,5,6-trichloro-2-pyridyl)ester; Pyrinex.

CHEMICAL DESIGNATIONS: CAS: 2921-88-2; DOT: none assigned; mf: $C_9H_{11}Cl_3NO_3PS$; fw: 350.59; RTECS: TF 6300000.

PROPERTIES: White granular cryst with a mild mercaptan odor. Mp: 41.5-43.5°, bp: ≈ 160° (decomp), *d*: 1.398 @ 43.5/4°, K_H: 4.16 x 10^{-6} atm·m^3/mol @ 25°, log K_{oc}: 3.77-4.13, log K_{ow}: 4.80-5.267, S_o (kg/kg): acetone (6.5), benzene (7.9), chloroform (6.3), S_w: 2 ppm @ 25°, vp: 1.87 x 10^{-5} mm @ 25°.

TRANSFORMATION PRODUCTS: Hydrolyzes in water forming 3,5,6-trichloro-2-pyridinol, *O*-ethyl *O*-hydrogen-*O*-(3,5,6-trichloro-2-pyridyl)phosphorothioate and *O,O*-dihydrogen-*O*-(3,5,6-trichloro-2-pyridyl)phosphorothioate. Reported half-lives in buffered distilled water @ 25° @ pH 8.1, 6.9 and 4.7 are 22.8, 35.3 and 62.7 days, respectively. 3,5,6-Trichloro-2-pyridinol was also reported to form by the photolysis of chlorpyrifos in water. Continued photolysis yielded chloride ions, CO_2, ammonia and possibly polyhydroxychloropyridines. The following photolytic half-lives in water at north 40° latitude were reported: 31 days during midsummer at a depth of 10^{-3} cm; 345 days during midwinter at a depth of 10^{-3} cm; 43 days at a depth of 1 m; 2.7 yr during midsummer at a depth of 1 m in river water.

EXPOSURE LIMITS: ACGIH TLV: TWA 0.2 mg/m^3, STEL 0.6 mg/m^3.

USE: Insecticide.

CHRYSENE

SYNONYMS: Benz[*a*]phenanthrene; Benzo[*a*]phenanthrene; Benzo[α]phenanthrene; 1,2-Benzophenanthrene; 1,2-Benzphenanthrene; 1,2-Dibenzonaphthalene; 1,2,5,6-Dibenzonaphthalene; RCRA waste number U050.

CHEMICAL DESIGNATIONS: CAS: 218-01-9; DOT: none assigned; mf: $C_{18}H_{12}$; fw: 228.30; RTECS: GC 0700000.

PROPERTIES: Orthorhombic plates exhibiting strong fluorescence under UV light. Mp: 258.2°, bp: 448°, *d*: 1.274 @ 20/4°, pK_a: > 15, K_H: 7.26 x 10^{-20} atm·m^3/mol (calcd), IP: 7.85 eV, log K_{oc}: 5.39 (calcd), log K_{ow}: 5.60-5.91, S_o: sol in ether, acetic acid and ethanol, S_w: 1.8-6 μg/L @ 25°, vp: 6.3 x 10^{-7} mm @ 20°.

TRANSFORMATION PRODUCTS: Based on structurally related compounds, chrysene may undergo photolysis to yield quinones and/or hydroxy derivatives.

EXPOSURE LIMITS: No individual standards have been set, however, as a constituent in coal tar pitch volatiles, the following exposure limits have been established: NIOSH REL: 10-h TWA 0.1 mg/m^3 (cyclohexane-extractable fraction); OSHA PEL: TWA 0.2 mg/m^3 (benzene-soluble fraction); ACGIH TLV: TWA 0.2 mg/m^3 (benzene solubles).

USE: Organic synthesis. Derived from industrial and experimental coal gasification operations where maximum concn detected in gas, liq and coal tar streams were 7.3, 0.16 and 8.6 mg/m^3, respectively.

CROTONALDEHYDE

SYNONYMS: 2-Butenal; *trans*-2-Butenal; **(*E*)-2-Butenal**; Crotenaldehyde; Crotonal; (*E*)-Crotonaldehyde; Crotonic aldehyde; **1,2-Ethanediol dipropanoate**; Ethylene glycol dipropionate; Ethylene dipropionate; Ethylene propionate; β-Methylacrolein; NCI-C56279; Propylene aldehyde; RCRA waste number U053; Topanel.

CHEMICAL DESIGNATIONS: CAS: 123-73-9; DOT: 1143; mf: C_4H_6O; fw: 70.09; RTECS: GP 9625000.

PROPERTIES: Clear, colorless to straw-colored liq with a pungent, irritating, suffocating odor. Mp: -74°, -76.5°, bp: 99°, 104-105°, *d*: 0.8477 @ 20.5/4°, fl p: 12.8°, lel: 2.1%, uel: 15.5%, K_H: 1.96 x 10^{-5} atm·m^3/mol, IP: 9.73 eV, S_o: sol in acetone and ether; misc with alcohol, benzene, gasoline, kerosene, solvent naphtha and toluene, S_w: 155 g/L, vap d: 2.41 (air = 1), 2.86 g/L @ 25°, 2.41 (air = 1), vp: 19 mm @ 20°.

TRANSFORMATION PRODUCTS: Slowly oxidizes in air to crotonic acid.

EXPOSURE LIMITS: NIOSH REL: IDLH 400 ppm; OSHA PEL: TWA 2 ppm; ACGIH TLV: 2 ppm.

SYMPTOMS OF EXPOSURE: Irrit eyes, resp sys, skin.

USES: Preparation of *n*-butyl alcohol, butyraldehyde, 2-ethylhexyl alcohol, quinaldine; chemical warfare; insecticides; leather tanning; alcohol denaturant; solvent; warning agent in fuel gases; purification of lubricating oils; organic synthesis.

CYCLOHEPTANE

SYNONYMS: Heptamethylene; Suberane; UN 2241.

CHEMICAL DESIGNATIONS: CAS: 291-64-5; DOT: 2241; mf: C_7H_{14}; fw: 98.19; RTECS: GU 3140000.

PROPERTIES: Oily liq. Mp: -12°, bp: 118.5°, *d*: 0.8098 @ 20/4°, fl p: 6°, lel: 1.1%, uel: 6.7%, log K_{ow}: 2.64 (calcd), S_o: sol in alcohol, benzene, chloroform, ether and lignoin, S_w: 30 ppm @ 25°, vap d: 4.01 g/L @ 25°, 3.39 (air = 1).

TRANSFORMATION PRODUCTS: Cycloheptane may be oxidized by microbes to cycloheptanol which may further oxidize to give cycloheptanone.

USES: Organic synthesis; gasoline component.

CYCLOHEXANE

SYNONYMS: Benzene hexahydride; Hexahydrobenzene; Hexamethylene; Hexanaphthene; RCRA waste number U056; UN 1145.

CHEMICAL DESIGNATIONS: CAS: 110-82-7; DOT: 1145; mf: C_6H_{12}; fw: 84.16; RTECS: GU 6300000.

PROPERTIES: Colorless liq with a sweet odor. Mp: 6.5°, bp: 80.7°, *d*: 0.7785 @ 20/4°, fl p: -18°, lel: 1.3%, uel: 8.4%, pK_a: ≈ 45, K_H: 1.94 x 10^{-1} atm·m^3/mol @ 25°, IP: 9.88 eV, 11.00 eV, log K_{ow}: 3.44, S_o: sol in lignoin and methanol (570 g/L @ 25°); misc with acetone, benzene, carbon tetrachloride, ethanol and ethyl ether, S_w: 55 mg/L @ 20°, vap d: 3.44 g/L @ 25°, 2.91 (air = 1), vp: 77 mm @ 20°.

TRANSFORMATION PRODUCTS: Microbial degradation products reported include cyclohexanol, 1-oxa-2-oxocycloheptane, 6-hydroxyheptanoate, 6-oxohexanoate, adipic acid, acetyl-CoA and succinyl-CoA and cyclohexanone.

EXPOSURE LIMITS: NIOSH REL: IDLH 10,000 ppm; OSHA PEL: TWA 300 ppm; ACGIH TLV: TWA 300 ppm.

SYMPTOMS OF EXPOSURE: Irrit eyes, resp sys; drowsy; derm; narco, coma; dizz; nau.

USES: Manufacture of nylon; solvent for cellulose ethers, fats, oils, waxes, resins, bitumens, crude rubber; paint and varnish removers; extracting essential oils; glass substitutes; solid fuels; fungicides; gasoline component; organic synthesis.

CYCLOHEXANOL

SYNONYMS: Adronal; Anol; Cyclohexyl alcohol; Hexahydrophenol; Hexalin; Hydralin; Hydrophenol; Hydroxycyclohexane; Naxol.

CHEMICAL DESIGNATIONS: CAS: 108-93-0; DOT: none assigned; mf: $C_6H_{12}O$; fw: 100.16; RTECS: GV 7875000.

PROPERTIES: Colorless, viscous liq with a camphor-like odor. Mp: 25.1°, bp: 161.1°, *d*: 0.9624 @ 20/4°, fl p: 67.8°, lel: 2.4%, K_H: 5.74 x 10^{-6} atm·m^3/mol @ 25°, IP: 10 eV, log K_{ow}: 1.23, S_o: sol in acetone and ether; misc with aromatic hydrocarbons, ethanol, ethyl acetate, linseed oil and petroleum solvents, S_w: 36 g/L @ 20°, vap d: 4.09 g/L @ 25°, 3.46 (air = 1), vp: 1 mm @ 20°.

TRANSFORMATION PRODUCTS: Reported biodegradation products include cyclohexanone, 2-hydroxyhexanone, 1-oxa-2-oxocycloheptane, 6-hydroxyheptanonate, 6-oxohexanoate and adipate.

EXPOSURE LIMITS: NIOSH REL: IDLH 3,500 ppm; OSHA PEL: 50 ppm; ACGIH TLV: 50 ppm.

SYMPTOMS OF EXPOSURE: Irrit eyes, nose, throat; narco; irrit skin.

USES: In paint and varnish removers; solvent for lacquers and resins; manufacture of adipic acid, benzene, cyclohexene, bromocyclohexane, chlorocyclohexane, cyclohexanone, nitrocyclohexane and solid fuel for camp stoves; fungicidal formulations; polishes; plasticizers; soap manufacturing; emulsified products; blending agent; recrystallizing steroids; germicides; plastics; organic synthesis.

CYCLOHEXANONE

SYNONYMS: Anone; Cyclohexyl ketone; Hexanon; Hytrol O; Ketohexamethylene; Nadone; NCI-C55005; Pimelic ketone; Pimelin ketone; RCRA waste number U057; Sextone; UN 1915.

CHEMICAL DESIGNATIONS: CAS: 108-94-1; DOT: 1915; mf: $C_6H_{10}O$; fw: 98.14; RTECS: GW 1050000.

PROPERTIES: Colorless to pale yellow liq with a peppermint-like odor. Mp: -47 to -16.4°, bp: 155.6°, *d*: 0.9478 @ 20/4°, fl p: 43.9°, lel: 1.1% @ 100°, uel: 9.4%, K_H: 1.2 x 10^{-5} atm·m^3/mol @ 25°, IP: 9.14 eV, log K_{ow}: 0.81, S_o: sol in acetone, alcohol, benzene, chloroform and ether, S_w: 23 g/L @ 20°, vap d: 4.01 g/L @ 25°, 3.39 (air = 1), vp: 4 mm @ 20°.

EXPOSURE LIMITS: NIOSH REL: 10-h TWA 25 ppm, IDLH 5,000 ppm; OSHA PEL: TWA 50 ppm; ACGIH TLV: TWA 25 ppm.

SYMPTOMS OF EXPOSURE: Irrit eyes, muc memb; CNS, narco; skin.

USES: Solvent for cellulose acetate, crude rubber, natural resins, nitrocellulose, vinyl resins, waxes, fats, oils, shellac, pesticides; preparation of adipic acid and caprolactum; wood stains; paint and varnish removers; degreasing of metals; spot remover; lube oil additive; leveling agent in dyeing and delustering silk.

CYCLOHEXENE

SYNONYMS: Benzene tetrahydride; Tetrahydrobenzene; 1,2,3,4-Tetrahydrobenzene; UN 2256.

CHEMICAL DESIGNATIONS: CAS: 110-83-8; DOT: 2256; mf: C_6H_{10}; fw: 82.15; RTECS: GW 2500000.

PROPERTIES: Colorless liq with a sweet odor. Mp: -103.5°, bp: 83°, *d*: 0.8102 @ 20/4°, fl p: -12.2°, lel: 1%, uel: 5% @ 100°, K_H: 4.6 x 10^{-2} atm·m^3/mol @ 25°, IP: 8.72-9.20 eV, log K_{ow}: 2.86, S_o: sol in acetone, alcohol, benzene and ether, S_w: 213 ppm @ 25°, vap d: 3.36 g/L @ 25°, 2.84 (air = 1), vp: 67 mm @ 20°.

TRANSFORMATION PRODUCTS: Cyclohexene may biodegrade to cyclohexanone. Gaseous products formed from the reaction of cyclohexene with O_3 were (% yield): formic acid (12), CO (18), CO_2 (42), ethylene (1) and valeraldehyde (17). Cyclohexene reacts with chlorine dioxide in water forming 2-cyclohexen-1-one.

EXPOSURE LIMITS: NIOSH REL: IDLH 10,000 ppm; OSHA PEL: TWA 300 ppm; ACGIH TLV: TWA 300 ppm.

SYMPTOMS OF EXPOSURE: Irrit skin, eyes, resp sys; drow.

USES: Manufacture of adipic acid, hexahydrobenzoic acid, maleic acid; catalyst solvent; oil extraction; organic synthesis.

CYCLOPENTADIENE

SYNONYMS: **1,3-Cyclopentadiene**; Pentole; R-pentine; Pyropentylene.

CHEMICAL DESIGNATIONS: CAS: 542-92-7; DOT: none assigned; mf: C_5H_6;

fw: 66.10; RTECS: GY 1000000.

PROPERTIES: Colorless liq with a turpentine-like odor. Mp: -97.2°, -85°, bp: 40.0°, *d*: 0.8021 @ 20/4°, fl p: < 32.2°, pK_a: 15, IP: 8.97 eV, log K_{ow}: 2.34 (calcd), S_o: misc with acetone, benzene, carbon tetrachloride and ether; sol in acetic acid, aniline and carbon disulfide, S_w: 0.0103 mol/L @ RT, vap d: 2.70 g/L @ 25°, 2.28 (air = 1).

TRANSFORMATION PRODUCTS: Polymerizes to dicyclopentadiene on standing. Cyclopentadiene may be oxidized by microbes to cyclopentanone.

EXPOSURE LIMITS: NIOSH REL: IDLH 2,000 ppm; OSHA PEL: TWA 75 ppm; ACGIH TLV: 75 ppm.

SYMPTOMS OF EXPOSURE: Irrit eyes, nose.

USES: Manufacture of resins; chlorinated insecticides; organic synthesis (Diels-Alder reaction).

CYCLOPENTANE

SYNONYMS: Pentamethylene; UN 1146.

CHEMICAL DESIGNATIONS: CAS: 287-92-3; DOT: 1146; mf: C_5H_{10}; fw: 70.13; RTECS: GY 2390000.

PROPERTIES: Colorless mobile liq. Mp: -93.9°, bp: 49.2°, *d*: 0.7457 @ 20/4°, fl p: -7°, lel: 1.5%, pK_a: ≈ 44, K_H: 1.86 x 10^{-1} atm·m^3/mol @ 25°, IP: 10.53 eV, log K_{ow}: 3.00, S_o: sol in acetone and benzene; misc with alcohol, ether and other hydrocarbon solvents, S_w: 156 ppm @ 25°, vap d: 2.87 g/L @ 25°, 2.42 (air = 1), vp: 400 mm @ 31.0°.

TRANSFORMATION PRODUCTS: Cyclopentane may be oxidized by microbes to cyclopentanol which may further oxidize to give cyclopentanone.

EXPOSURE LIMITS: ACGIH TLV: TWA 600 ppm.

USES: Solvent for cellulose ethers; azeotropic distillation agent; motor fuel; organic synthesis.

CYCLOPENTENE

SYNONYM: UN 2246.

CHEMICAL DESIGNATIONS: CAS: 142-29-0; DOT: 2246; mf: C_5H_8; fw: 68.12; RTECS: GY 5950000.

PROPERTIES: Colorless liq. Mp: -135°, bp: 44.2°, *d*: 0.7720 @ 20/4°, fl p: -34°, K_H: 6.3 x 10^{-2} atm·m^3/mol @ 25°, IP: 9.01 eV, log K_{ow}: 2.45 (calcd), S_o: sol in alcohol, benzene, ether and petroleum, S_w: 535 ppm @ 25°, vap d: 2.78 g/L @ 25°, 2.35 (air = 1).

TRANSFORMATION PRODUCTS: Cyclopentene may be oxidized by microbes to cyclopentanol which may further oxidize to cycloheptanone. Gaseous products formed from the reaction of cyclopentene with O_3 were (% yield): formic acid (11), CO (35), CO_2 (42), ethylene (12), formaldehyde (13) and butyraldehyde (11). Particulate products identified include succinic acid, glutaraldehyde, 5-

oxopentanoic acid and glutaric acid.

USES: Cross-linking agent; organic synthesis.

2,4-D

SYNONYMS: Agrotect; Amidox; Amoxone; Aqua-kleen; BH 2,4-D; Brush-rhap; B-Selektonon; Chipco turf herbicide 'D'; Chloroxone; Crop rider; Crotilin; D 50; 2,4-D acid; Dacamine; Debroussaillant 600; Decamine; Ded-weed; Ded-weed LV-69; Desormone; Dichlorophenoxyacetic acid; **(2,4-Dichlorophenoxy)acetic acid**; Dicopur; Dicotox; Dinoxol; DMA-4; Dormone; Emulsamine BK; Emulsamine E-3; ENT 8,538; Envert 171; Envert DT; Esteron; Esteron 76 BE; Esteron 44 weed killer; Esteron 99; Esteron 99 concentrate; Esteron brush killer; Esterone four; Estone; Farmco; Fernesta; Fernimine; Fernoxone; Ferxone; Foredex 75; Formula 40; Hedonal; Herbidal; Ipaner; Krotiline; Lawn-keep; Macrondray; Miracle; Monosan; Moxone; NA 2,765; Netagrone; Netagrone 600; NSC 423; Pennamine; Pennamine D; Phenox; Pielik; Planotox; Plantgard; RCRA waste number U240; Rhodia; Salvo; Spritz-hormin/2,4-D; Spritz-hormit/2,4-D; Super D weedone; Transamine; Tributon; Trinoxol; U 46; U-5,043; U 46DP; Vergemaster; Verton; Verton D; Verton 2D; Vertron 2D; Vidon 638; Visko-rhap; Visko-rhap drift herbicides; Visko-rhap low volatile 4L; Weedar; Weddar-64; Weddatul; Weed-b-gon; Weedez wonder bar; Weedone; Weedone LV4; Weed-rhap; Weed tox; Weedtrol.

CHEMICAL DESIGNATIONS: CAS: 94-75-7; DOT: 2765; mf: $C_8H_6Cl_2O_3$; fw: 221.04; RTECS: AG 6825000.

PROPERTIES: Odorless, white to pale yellow prismatic cryst. Mp: 140-141°, bp: 160° @ 0.4 mm, *d*: 1.416 @ 25/4°, pK_a: 2.73, 2.90, K_H: 1.95 x 10^{-2} atm·m^3/mol @ 20° (calcd), log K_{oc}: 1.68-2.73, log K_{ow}: 1.47-4.88, S_o: @ 25°: carbon tetrachloride (1 g/L), ethyl ether (270 g/L), acetone (850 g/L) and ethyl alcohol (1,300 g/L), S_w: 890 ppm @ 25°, vp: 0.0047 mm @ 20°.

TRANSFORMATION PRODUCTS: Reported metabolic products in bean and soybean plants include 4-*O*-β-glucosides of 4-hydroxy-2,5-dichlorophenoxyacetic acid, 4-hydroxy-2,3-dichlorophenoxyacetic acid, *N*-(2,4-dichlorophenoxyacetyl)-L-aspartic acid and *N*-(2,4-dichlorophenocyacetyl)-L-glutamic acid. Metabolites identified in cereals and strawberries include 1-*O*-(2,4-dichlorophenoxyacetyl)-β-D-glucose and 2,4-dichlorophenol, respectively. When 2,4-D was applied to resistant grasses, 3-(2,4-dichlorophenoxy)propionic acid formed. 2,4-D degraded in anaerobic sewage sludge to 4-chlorophenol. In moist soils, 2,4-D degraded to 2,4-dichlorophenol, 2,4-dichloroanisole and CO_2. Photolysis of 2,4-D in distilled water using mercury arc lamps (λ = 254 nm) or by natural sunlight yielded 2,4-dichlorophenol, 4-chlorocatechol, 2-hydroxy-4-chlorophenoxyacetic acid, 1,2,4-benzenetriol and polymeric humic acids. In a helium pressurized reactor containing ammonium nitrate and polyphosphoric acid at temperatures of 121 and 232°, 2,4-D was oxidized to CO_2, water and hydrochloric acid. CO_2, chloride, aldehydes, oxalic and glycolic acids, were reported as ozonation products of 2,4-D in water @ pH 8. In a 5-day experiment, ^{14}C-labelled 2,4-D applied to soil-water suspensions under aerobic and anaerobic conditions gave $^{14}CO_2$ yields of 0.5 and 0.7%, respectively. Reacts with alkalies, metals and amines forming water soluble salts.

EXPOSURE LIMITS: NIOSH REL: IDLH 500 mg/m^3; OSHA PEL: TWA 10 mg/m^3; ACGIH TLV: 10 mg/m^3.

SYMPTOMS OF EXPOSURE: Weak stupor, hyporflexia, musc twitch, convuls, derm.

USES: Herbicide; weed killer and defoliant.

p,p'-DDD

SYNONYMS: 1,1-Bis(4-chlorophenyl)-2,2-dichloroethane; 1,1-Bis(*p*-chlorophenyl)-2,2-dichloroethane; 2,2-Bis(4-chlorophenyl)-1,1-dichloroethane; 2,2-Bis(*p*-chlorophenyl)-1,1-dichloroethane; DDD; 4,4'-DDD; 1,1-Dichloro-2,2-bis(*p*-chlorophenyl)ethane; 1,1-Dichloro-2,2-di(4-chlorophenyl)ethane; 1,1-Dichloro-2,2-di(*p*-chlorophenyl)ethane; Dichlorodiphenyldichloroethane; 4,4'-Dichlorodiphenyldichloroethane; *p,p'*-Dichlorodiphenyldichloroethane; **1,1'-(2,2-Dichloroethylidene)bis[4-chlorobenzene]**; Dilene; ENT 4,225; ME-1,700; NA 2,761; NCI-C00475; RCRA waste number U060; Rhothane; Rhothane D-3; Rothane; TDE; 4,4'-TDE; *p,p'*-TDE; Tetrachlorodiphenylethane.

CHEMICAL DESIGNATIONS: CAS: 72-54-8; DOT: 2761; mf: $C_{14}H_{10}Cl_4$; fw: 320.05; RTECS: KI 0700000.

PROPERTIES: White cryst solid. Mp: 109-112°, bp: 193°, *d*: 1.476 @ 20/4°, K_H: 2.16 x 10^{-5} atm·m^3/mol (calcd), log K_{oc}: 5.38, log K_{ow}: 5.061-6.217, S_w: 20-90 μg/L @ 25°, vap d: 17.2 ng/L @ 30°, vp: 4.68 x 10^{-6} mm @ 25°.

TRANSFORMATION PRODUCTS: It was reported that *p,p'*-DDD, a major biodegradation product of *p,p'*-DDT, was degraded by *Aerobacter aerogenes* under aerobic conditions to yield 1-chloro-2,2-bis(*p*-chlorophenyl)ethylene, 1-chloro-2,2-bis(*p*-chlorophenyl)ethane and 1,1-bis(*p*-chlorophenyl)ethylene. Under anaerobic conditions, however, 4 additional compounds were identified: bis(*p*-chlorophenyl)acetic acid, *p,p'*-dichlorodiphenylmethane, *p,p'*-dichlorobenzhydrol and *p,p'*-dichlorobenzophenone. Under reducing conditions, indigenous microbes in Lake Michigan sediments degraded DDD to 2,2-bis(*p*-chlorophenyl)ethane and 2,2-bis(*p*-chlorophenyl)ethanol. Incubation of *p,p'*-DDD with hematin and ammonia gave 4,4'-dichlorobenzophenone, 1-chloro-2,2-bis(*p*-chlorophenyl)-ethylene and bis(*p*-chlorophenyl)acetic acid methyl ester.

USES: Dusts, emulsions and wettable powders for contact control of leaf rollers and other insects on vegetables and tobacco.

p,p'-DDE

SYNONYMS: 2,2-Bis(4-chlorophenyl)-1,1-dichloroethene; 2,2-Bis(*p*-chlorophenyl)-1,1-dichloroethene; 1,1-Bis(4-chlorophenyl)-2,2-dichloroethylene; 1,1-Bis(*p*-chlorophenyl)-2,2-dichloroethylene; DDE; 4,4'-DDE; DDT dehydrochloride; 1,1-Dichloro-2,2-bis(*p*-chlorophenyl)ethylene; Dichlorodiphenyldichloroethylene;

p,p'-Dichlorodiphenyldichloroethylene; **1,1'-(Dichloroethenylidene)bis(4-chlorobenzene)**; NCI-C00555.

CHEMICAL DESIGNATIONS: CAS: 72-55-9; DOT: 2761; mf: $C_{14}H_8Cl_4$; fw: 319.03; RTECS: KV 9450000.

PROPERTIES: Solid. Mp: 88-90°, K_H: 1.22 x 10^{-3} atm·m^3/mol @ 23°, log K_{oc}: 5.386, 6.00, log K_{ow}: 5.69-6.956, S_o: sol in fats and most solvents, S_w: 40 ppb @ 20°, vap d: 109 ng/L @ 30°, vp: 1.57 x 10^{-5} mm @ 25°.

TRANSFORMATION PRODUCTS: May degrade to bis(chlorophenyl)acetic acid in water or oxidize to *p,p'*-dichlorobenzophenone using UV light as a catalyst. In an air-saturated distilled water medium irradiated with monochromic light (λ = 313 nm), *p,p'*-DDE degraded to *p,p'*-dichlorobenzophenone, 1,1-bis(*p*-chlorophenyl)-2-chloroethylene (DDMU) and 1-(4-chlorophenyl)-1-(2,4-dichlorophenyl)-2-chloroethylene (*o*-chloro DDMU). Identical photoproducts also were observed using tap water containing Mississippi River sediments.

USES: Military product; chemical research.

p,p'-DDT

SYNONYMS: Agritan; Anofex; Arkotine; Azotox; 2,2-Bis(4-chlorophenyl)-1,1,1-trichloroethane; 2,2-Bis(*p*-chlorophenyl)-1,1,1-trichloroethane; α,α-Bis(*p*-chlorophenyl)-β,β,β-trichloroethane; 1,1-Bis(*p*-chlorophenyl)-2,2,2-trichloroethane; Bosan Supra; Bovidermol; Chlorophenothan; Chlorophenothane; Chlorophenotoxum; Citox; Clofenotane; DDT; 4,4'-DDT; Dedelo; Deoval; Detox; Detoxan; Dibovan; Dichlorodiphenyltrichloroethane; *p,p'*-Dichlorodiphenyltrichloroethane; 4,4'-Dichlorodiphenyltrichloroethane; Dicophane; Didigam; Didimac; Diphenyltrichloroethane; Dodat; Dykol; ENT 1,506; Estonate; Genitox; Gesafid; Gesapon; Gesarex; Gesarol; Guesapon; Gyron; Havero-extra; Ivoran; Ixodex; Kopsol; Mutoxin; NCI-C00464; Neocid; Parachlorocidum; PEB1; Pentachlorin; Pentech; PPzeidan; RCRA waste number U061; Rukseam; Santobane; Trichlorobis(4-chlorophenyl)ethane; Trichlorobis(*p*-chlorophenyl)ethane; 1,1,1-Trichloro-2,2-bis(*p*-chlorophenyl)ethane; 1,1,1-Trichloro-2,2-di(4-chlorophenyl)ethane; 1,1,1-Trichloro-2,2-di(*p*-chlorophenyl)ethane; **1,1'-(2,2,2-Trichloroethylidene)bis[4-chlorobenzene]**; Zeidane; Zerdane.

CHEMICAL DESIGNATIONS: CAS: 50-29-3; DOT: none assigned; mf: $C_{14}H_9Cl_5$; fw: 354.49; RTECS: KJ 3325000.

PROPERTIES: Odorless to slightly fragrant colorless cryst or white powder. Mp: 108.5°, bp: 185°, 260°, *d*: 1.56 @ 15/4°, fl p: 72.2-77.2°, K_H: 1.29 x 10^{-5} atm·m^3/mol @ 23°, log K_{oc}: 5.146-6.26, log K_{ow}: 4.89-6.914, S_o: sol in cyclohexane, morpholine, pyridine, dioxane, ether, acetone (580 g/L), benzene (780 g/L), benzyl benzoate (420 g/L), carbon tetrachloride (450 g/L), chlorobenzene (740 g/L), cyclohexanone (1,160 g/L), ethyl ether (280 g/L), gasoline (100 g/L), isopropanol (30 g/L), kerosene (80-100 g/L), morpholine (750 g/L), peanut oil (110 g/L), pine oil (100-160 g/L), tetralin (610 g/L), tributyl phosphate (500 g/L), S_w: 1.2-5.5 μg/L @ 25°, vap d: 13.6 ng/L @ 30°, vp: 1.7 x 10^{-10} mm @ 20°.

TRANSFORMATION PRODUCTS: In soils under anaerobic conditions, *p,p'*-DDT is rapidly converted to *p,p'*-DDD and very slowly to *p,p'*-DDE under aerobic conditions. Other reported degradation products under aerobic and anaerobic conditions by various microbes include 1,1'-bis(*p*-chlorophenyl)-2-chloroethane, 1,1'-bis(*p*-chlorophenyl)-2-hydroxyethane and *p*-chlorophenyl acetic acid. It was also reported that *p,p'*-DDE formed by hydrolyzing *p,p'*-DDT. In 1 day, *p,p'*-DDT reacted rapidly with reduced hematin forming *p,p'*-DDD and unidentified products. The white rot fungus *Phanerochaete chrysosporium* degraded *p,p'*-DDT yielding the following metabolites: 1,1-dichloro-2,2-bis(4-chlorophenyl)ethane (*p,p'*-DDD), 2,2,2-trichloro-1,1-bis(4-chlorophenyl)ethanol (dicofol), 2,2-dichloro-1,1-bis(4-chlorophenyl)ethanol and 4,4'-dichlorobenzophenone. Mineralization of *p,p'*-DDT by the white rot fungi *Pleurotus ostreatus*, *Phellinus weirri* and *Polyporus versicolor* was also demonstrated. *Aerobacter aerogenes* degraded *p,p'*-DDT under aerobic conditions to *p,p'*-DDD, *p,p'*-DDE, 1-chloro-2,2-bis(*p*-chlorophenyl)ethylene, 1-chloro-2,2-bis(*p*-chlorophenyl)ethane and 1,1-bis(*p*-chlorophenyl)ethylene. Under anaerobic conditions the same organism produced 4 additional compounds. These were bis(*p*-chlorophenyl)acetic acid, *p,p'*-dichlorodiphenylmethane, *p,p'*-dichlorobenzhydrol and *p,p'*-dichlorobenzophenone. Other degradation products of *p,p'*-DDT under aerobic and anaerobic conditions in soils by various cultures not previously mentioned include 1,1-bis(*p*-chlorophenyl)-2,2,2-trichloroethanol (Kelthane) and *p*-chlorobenzoic acid. Under aerobic conditions, the amoeba *Acanthamoeba castellanii* (Neff strain ATCC 30.010) degraded *p,p'*-DDT to *p,p'*-DDE, *p,p'*-DDD and dibenzophenone (DBP). Incubation of *p,p'*-DDT with hematin and ammonia gave *p,p'*-DDD, *p,p'*-DDE, bis(*p*-chlorophenyl)acetonitrile, 1-chloro-2,2-bis(*p*-chlorophenyl)ethylene, 4,4'-dichlorobenzophenone and bis(*p*-chlorophenyl)acetic acid, methyl ester. Thirty-five microorganisms isolated from marine sediment and marine water samples taken from Hawaii and Houston, TX were capable of degrading *p,p'*-DDT. *p,p'*-DDD was identified as the major metabolite. Minor transformation products included 2,2-bis(*p*-chlorophenyl)ethanol (DDOH), 2,2-bis(*p*-chlorophenyl)ethane (DDNS) and *p,p'*-DDE. Photolysis of *p,p'*-DDT in nitrogen-sparged methanol solvent by UV light (λ = 260 nm) produced DDD and 1,1-bis(*p*-chlorophenyl)-2-chloro-ethylene (DDMU). But photolysis of *p,p*-'DDT @ 280 nm in an oxygenated methanol soln yielded a complex mixture containing the methyl ester of 2,2-bis(*p*-chlorophenyl)acetic acid. *p,p'*-DDT in an aq soln containing suspended titanium dioxide as a catalyst and irradiated with UV light (λ > 340 nm) formed chloride ions. Based on the amount of chloride ions generated, CO_2 and hydrochloric acid were reported as the end products. In a 42-day experiment, ^{14}C-labelled *p,p'*-DDT applied to soil-water suspensions under aerobic and anaerobic conditions gave $^{14}CO_2$ yields of 0.8 and 0.7%, respectively.

EXPOSURE LIMITS: NIOSH REL: lowest detectable limit; OSHA PEL: TWA 1 mg/m^3; ACGIH TLV: TWA 1 mg/m^3.

SYMPTOMS OF EXPOSURE: Pares tongue, lips, face; tremor; appre, diss, conf, mal; head; convuls; paresis hands; vomit; irrit eyes, skin.

USES: Use as an insecticide is now prohibited; chemical research; nonsystemic stomach and contact insecticide.

DECAHYDRONAPHTHALENE

SYNONYMS: Bicyclo[4.4.0]decane; Dec; Decalin; Decalin solvent; Dekalin; Dekalin; Naphthalane; Naphthane; Perhydronaphthalene; UN 1147.

CHEMICAL DESIGNATIONS: CAS: 91-17-8; DOT: none assigned; mf: $C_{10}H_{18}$; fw: 138.25; RTECS: QJ 3150000.

PROPERTIES: Colorless liq with a methanol-like odor. Mp: -43° (*cis*), -30.4° (*trans*), bp: 195.6° (*cis*), 187.2° (*trans*), *d*: 0.8965 @ 20/4° (*cis*), 0.8699 @ 20/4° (*trans*), fl p: 58° (commercial mixture), lel: 0.7% @ 100°, uel: 4.9% @ 100°, K_H: 39.2 atm·m^3/mol @ 25° (calcd), log K_{ow}: 4.00 (calcd), S_o: sol in acetone, alcohol, benzene, ether and chloroform; misc with propanol, isopropanol and most ketones and ethers, S_w: 0.889 ppm @ 25°, vap d: 5.65 g/L @ 25°, 4.77 (air = 1), vp: 195.77 and 187.27 mm @ 25° for *cis*- and *trans*- isomers, respectively.

USES: Solvent for naphthalene, waxes, fats, oils, resins, rubbers; motor fuel and lubricants; cleaning machinery; substitute for turpentine; shoe-creams; stain remover.

n-DECANE

SYNONYMS: **Decane**; Decyl hydride; UN 2247.

CHEMICAL DESIGNATIONS: CAS: 124-18-5; DOT: 2247; mf: $C_{10}H_{22}$; fw: 142.28; RTECS: HD 6550000.

PROPERTIES: Colorless liq. Mp: -29.7°, bp: 174.1°, *d*: 0.730 @ 20/4°, fl p: 46.1°, lel: 0.8%, uel: 5.4%, K_H: 1.87 x 10^{-1} atm·m^3/mol @ 25° (calcd), log K_{ow}: 6.69, S_o: sol in alcohol and ether, S_w: 0.022 ppm @ 25°, vap d: 5.82 g/L @ 25°, 4.91 (air = 1), vp: 1.35 mm @ 25°.

TRANSFORMATION PRODUCTS: *n*-Decane may biodegrade in 2 ways. The first is the formation of decyl hydroperoxide which decomposes to *n*-decanol followed by oxidation to decanoic acid. The other pathway involves dehydrogenation to 1-decene which may react with water giving *n*-decanol.

USES: Solvent; standardized hydrocarbon; manufacturing paraffin products; jet fuel research; paper processing industry; rubber industry; organic synthesis.

DIACETONE ALCOHOL

SYNONYMS: DAA; Diacetone; Diacetonyl alcohol; Diketone alcohol; Dimethylacetonylcarbinol; 4-Hydroxy-2-keto-4-methylpentane; 4-Hydroxy-4-methylpentanone-2; 4-Hydroxy-4-methylpentan-2-one; **4-Hydroxy-4-methyl-2-pentanone**; 2-Methyl-2-pentanol-4-one; Pyranton; Pyranton A; UN 1148.

CHEMICAL DESIGNATIONS: CAS: 123-42-2; DOT: 1148; mf: $C_6H_{12}O_2$; fw: 116.16; RTECS: SA 9100000.

PROPERTIES: Colorless liq with a mild, pleasant odor. Mp: -44°, bp: 164°, 167.9°, *d*: 0.9387 @ 20/4°, fl p: 13° (OC), lel: 1.8%, uel: 6.9%, S_o: sol in alcohol and ether, S_w: misc, vap d: 4.75 g/L @ 25°, 4.01 (air = 1), vp: 0.8 mm @ 20°.

EXPOSURE LIMITS: NIOSH REL: 10-h 50 ppm, IDLH 2,100 ppm; OSHA PEL: TWA 50 ppm.

SYMPTOMS OF EXPOSURE: Irrit eyes, nose, throat; corneal tissure damage; narco; skin irrit.

USES: Solvent for celluloid, cellulose acetate, fats, oils, waxes, nitrocellulose and resins; wood preservatives; rayon and artificial leather; imitation gold leaf; extraction of resins and waxes; antifreeze mixtures; laboratory reagent; preservative for animal tissue; dyeing mixtures; stripping agent for textiles.

DIBENZ[*a,h*]ANTHRACENE

SYNONYMS: 1,2:5,6-Benzanthracene; DBA; 1,2,5,6-DBA; DB[*a,h*]A; 1,2:5,6-Dibenzanthracene; 1,2:5,6-Dibenz[*a*]anthracene; 1,2:5,6-Dibenzoanthracene; Dibenzo[*a,h*]anthracene; RCRA waste number U063.

CHEMICAL DESIGNATIONS: CAS: 53-70-3; DOT: none assigned; mf: $C_{22}H_{14}$; fw: 278.36; RTECS: HN 2625000.

PROPERTIES: Monoclinic or orthorhombic cryst or leaflets. Mp: 269-270°, bp: 524°, *d*: 1.282, pK_a: > 15, K_H: 7.33 x 10^{-9} atm·m^3/mol @ 20-25° (calcd), IP: 7.28 eV, log K_{oc}: 6.22, log K_{ow}: 5.97-6.58, S_o: sol in most solvents including petroleum ether, benzene, toluene, xylene and oils; slightly sol in alcohol and ether, S_w: 0.5 μg/L @ 25°, vp: ≈ 10^{-10} mm @ 20°.

TRANSFORMATION PRODUCTS: A CO_2 yield of 45.3% was achieved when dibenz[*a,h*]anthracene adsorbed on silica gel was irradiated with light (λ > 290 nm) for 17 h. In activated sludge, < 0.1% of the applied benzene mineralized to CO_2 after 5 days.

EXPOSURE LIMITS: No individual standards have been set, however, as a constituent in coal tar pitch volatiles, the following exposure limits have been established: NIOSH REL: 10-h TWA 0.1 mg/m^3 (cyclohexane-extractable fraction); OSHA PEL: TWA 0.2 mg/m^3 (benzene-soluble fraction); ACGIH TLV: TWA 0.2 mg/m^3 (benzene solubles).

USE: Research chemical. Though not produced commercially in the U.S., dibenz[*a,h*]anthracene is derived from industrial and experimental coal gasification operations where maximum concn detected in gas and coal tar streams were 6.1 μg/m^3 and 3.4 mg/m^3, respectively.

DIBENZOFURAN

SYNONYMS: Biphenylene oxide; Diphenylene oxide.

CHEMICAL DESIGNATIONS: CAS: 132-64-9; DOT: none assigned; mf: $C_{12}H_8O$; fw: 168.20; RTECS: none assigned.

PROPERTIES: Solid. Mp: 86-87°, bp: 287°, *d*: 1.0886 @ 99/4°, IP: 8.59 eV, log K_{oc}: 3.91-4.10 (calcd), log K_{ow}: 4.12-4.31, S_o: sol in acetic acid, acetone, ethanol and ether, S_w: 10 mg/L @ 25°, vp: 0.35 Pa @ 25°.

TRANSFORMATION PRODUCT: Chlorination of dibenzofuran in tap water may yield chlorodibenzofuran.

USE: Research chemical. Derived from industrial and experimental coal gasification operations where the maximum concn detected in gas tar streams was 12 mg/m^3.

1,4-DIBROMOBENZENE

SYNONYM: *p*-Dibromobenzene.

CHEMICAL DESIGNATIONS: CAS: 106-37-6; DOT: 2711mf: $C_6H_4Br_2$; fw: 235.91; RTECS: CZ 1791000.

PROPERTIES: Colorless liq with a pleasant, aromatic odor. Mp: 87.3°, bp: 218-219°, *d*: 1.9767 @ 25/4°, K_H: 5.0 x 10^{-4} atm·m^3/mol @ 25°, log K_{oc}: 3.20 (calcd), log K_{ow}: 3.79, S_o: misc with acetone, alcohol, benzene, carbon tetrachloride, ether and *n*-heptane, S_w: 16.5 mg/L @ 25°, vp: 0.161 mm @ 25°.

USES: Solvent for oils; ore flotation; motor fuels; organic synthesis.

DIBROMOCHLOROMETHANE

SYNONYMS: Chlorodibromomethane; CDBM; NCI-C55254.

CHEMICAL DESIGNATIONS: CAS: 124-48-1; DOT: none assigned; mf: $CHBr_2Cl$; fw: 208.28; RTECS: PA 6360000.

PROPERTIES: Clear, colorless, heavy liq. Mp: -23 to -21°, bp: 118-122°, *d*: 2.451 @ 20/4°, K_H: 8.7 x 10^{-4} atm·m^3/mol @ 20°, IP: 10.59 eV, log K_{oc}: 1.92, log K_{ow}: 2.08, S_o: misc with oils, dichloropropane and isopropanol, S_w: 4 g/L @ 20°, vap d: 8.51 g/L @ 25°, 7.19 (air = 1), vp: 76 mm @ 20°.

TRANSFORMATION PRODUCTS: Anticipated hydrolysis products include dibromomethanol and/or bromochloromethanol ($t_{1/2}$ = 274 yr @ 25° and pH 7). To an aq soln containing 18.8 μmol bromodichloromethane was bubbled hydrogen. After 24 h, only 18% of the bromodichloromethane reacted forming methane and minor traces of ethane. In the presence of colloidal platinum catalyst, the reaction proceeded at a much faster rate forming the same end products. In an earlier study, water containing 2,000 ng/μL of dibromochloromethane and colloidal platinum catalyst was irradiated with UV light. After 20 h, dibromochloromethane degraded to 80 ng/μL bromochloromethane, 22 ng/μL methyl chloride and 1,050 ng/μL methane. A duplicate experiment was performed but 1 g zinc was added. After about 1 h, total degradation was achieved. Presumed transformation products include methane, bromide and chloride ions.

USES: Manufacture of fire extinguishing agents, propellants, refrigerants and pesticides; organic synthesis.

1,2-DIBROMO-3-CHLOROPROPANE

SYNONYMS: BBC 12; 1-Chloro-2,3-dibromopropane; 3-Chloro-1,2-dibromopropane; DBCP; Dibromochloropropane; Fumagon; Fumazone; Fumazone 86; Fumazone 86E; NCI-C00500; Nemabrom; Nemafume; Nemagon; Nemagon 20; Nemagon 20G; Nemagon 90; Nemagon 206; Nemagon soil fumigant; Nemanax; Nemapaz; Nemaset; Nematocide; Nematox; Nemazon; OS 1,987; Oxy DBCP; RCRA waste number U066; SD-1897; UN 2872.

CHEMICAL DESIGNATIONS: CAS: 96-12-8; DOT: 2872; mf: $C_3H_5Br_2Cl$; fw: 236.36; RTECS: TX 8750000.

PROPERTIES: Yellow to brown liq with a pungent odor. Mp: 6.1°, bp: 196°, *d*: 2.05 @ 20/4°, fl p: 76.7° (OC), K_H: 2.49 x 10^{-4} atm·m^3/mol @ 20° (calcd), log K_{oc}: 2.11, log K_{ow}: 2.63 (calcd), S_o: misc with oils, dichloropropane and isopropanol, S_w: 1,270 ppm, vap d: 9.66 g/L @ 25°, 8.16 (air = 1), vp: 0.8 mm @ 20°.

TRANSFORMATION PRODUCTS: Soil water cultures converted 1,2-dibromo-3-chloropropane to *n*-propanol, bromide and chloride ions. Precursors to the alcohol formation include allyl chloride and allyl alcohol. Hydrolysis of 1,2-dibromo-3-chloropropane yielded the intermediates 2-bromo-3-chloropropene and 2,3-dibromopropene. Both were readily converted to 2-bromoallyl alcohol.

EXPOSURE LIMITS: NIOSH REL: 10-h TWA 10 ppm; OSHA PEL: TWA 1 ppb.

USES: Soil fumigant, nematocide; pesticide; organic synthesis.

DIBROMODIFLUOROMETHANE

SYNONYMS: Difluorodibromomethane; Freon 12-B2; Halon 1202; UN 1941.

CHEMICAL DESIGNATIONS: CAS: 75-61-6; DOT: 1941; mf: CBr_2F_2; fw: 209.82; RTECS: PA 7525000.

PROPERTIES: Colorless liq or gas with a characteristic odor. Mp: -147.2°, bp: 24.5°, *d*: 2.297 @ 20°, IP: 11.07 eV, S_o: sol in acetone, alcohol, benzene and ether, vap d: 8.58 g/L @ 25°, 7.24 (air = 1), vp: 688 mm @ 20° (calcd).

EXPOSURE LIMITS: NIOSH REL: IDLH 2,500 ppm; OSHA PEL: TWA 100 ppm.

SYMPTOMS OF EXPOSURE: Frost; irrit nose, throat; drow; unconscious; liv damage.

USES: Synthesis of dyes; quaternary ammonium compounds; pharmaceuticals; fire-extinguishing agent.

DI-*n*-BUTYL PHTHALATE

SYNONYMS: 1,2-Benzenedicarboxylate; **1,2-Benzenedicarboxylic acid dibutyl ester**; *o*-Benzenedicarboxylic acid dibutyl ester; Benzene-*o*-dicarboxylic acid di-*n*-butyl ester; Butyl phthalate; *n*-Butyl phthalate; Celluflex DPB; DBP; Dibutyl-1,2-benzenedicarboxylate; Dibutyl phthalate; Elaol; Hexaplas M/B; Palatinol C; Phthalic acid dibutyl ester; Polycizer DBP; PX 104; RCRA waste number U069; Staflex DBP; Witicizer 300.

CHEMICAL DESIGNATIONS: CAS: 84-74-2; DOT: 9095; mf: $C_{16}H_{22}O_4$; fw:

278.35; RTECS: TI 0875000.

PROPERTIES: Colorless oily liq with a mild aromatic odor. Mp: -35°, bp: 340°, *d*: 1.046 @ 20/4°, fl p: 157°, lel: 0.5% @ 235°, uel: 2.5% (calcd), K_H: 6.3 x 10^{-5} atm·m^3/mol, log K_{oc}: 3.14, log K_{ow}: 4.31-4.79, S_o: sol in ethanol, benzene, ether; very sol in acetone, S_w: 10.1 mg/L @ 20°, vap d: 11.38 g/L @ 25°, 9.61 (air = 1), vp: 10^{-5} mm @ 25°.

TRANSFORMATION PRODUCTS: Under aerobic conditions using a fresh-water hydrosoil, mono-*n*-butyl phthalate and phthalic acid were produced. Under anaerobic conditions, phthalic acid was not present. In anaerobic sludge, di-*n*-butyl phthalate degraded as follows: monobutyl phthalate to phthalic acid to protocatechuic acid followed by ring cleavage and mineralization. A variety of microorganisms were capable of degrading of di-*n*-butyl phthalate in the following degradation scheme: di-*n*-butyl phthalate to mono-*n*-butyl phthalate to phthalic acid to 3,4-dihydroxybenzoic acid and other unidentified products. Di-*n*-butyl phthalate was degraded to benzoic acid by tomatoe cell suspension cultures (*Lycopericon lycopersicum*). An aq soln containing titanium dioxide and subjected to UV radiation (λ > 290 nm) produced hydroxyphthalates and dihydroxy-phthalates as intermediates. Hydrolyzes in water to phthalic acid and *n*-butyl alcohol. Pyrolysis of di-*n*-butyl phthalate in presence of polyvinyl chloride @ 600° gave the following compounds: indene, methylindene, naphthalene, 1-methyl-naphthalene, 2-methylnaphthalene, biphenyl, dimethylnaphthalene, acenaphthene, fluorene, methylacenaphthene, methylfluorene and 6 unidentified compounds.

EXPOSURE LIMITS: NIOSH REL: IDLH 9,300 mg/m^3; OSHA PEL: TWA 5 mg/m^3; ACGIH TLV: TWA 5 mg/m^3.

SYMPTOMS OF EXPOSURE: Irrit nasal passages, upper resp, stomach; light sensitivity.

USES: Manufacture of plasticizers, insect repellents, printing inks, paper coatings, explosives, adhesives, safety glass; organic synthesis.

1,2-DICHLOROBENZENE

SYNONYMS: Chloroben; Chloroden; Cloroben; DCB; 1,2-DCB; *o*-DCB; 1,2-Dichlorbenzene; *o*-Dichlorbenzene; 1,2-Dichlorbenzol; *o*-Dichlorbenzol; *o*-Dichlorobenzene; 1,2-Dichlorobenzol; *o*-Dichlorobenzol; Dilantin DB; Dilatin DB; Dizene; Dowtherm E; NCI-C54944; ODB; ODCB; Orthodichlorobenzene; Orthodichlorobenzol; RCRA waste number U070; Special termite fluid; Termitkil; UN 1591.

CHEMICAL DESIGNATIONS: CAS: 95-50-1; DOT: 1591; mf: $C_6H_4Cl_2$; fw: 147.00; RTECS: CZ 4500000.

PROPERTIES: Clear, colorless liq with a pleasant aromatic odor. Mp: -17.5°, bp: 180.5°, *d*: 1.3064 @ 20/4°, fl p: 66°, lel: 2.2%, uel: 9.2%, K_H: 1.2 x 10^{-3} atm·m^3/mol @ 20°, IP: 9.06 eV, log K_{oc}: 2.26-3.23, log K_{ow}: 3.34-3.433, S_o: sol in acetone, carbon tetrachloride and ligroin; misc with alcohol, ether and benzene, S_w: 100 mg/L @ 20°, vap d: 6.01 g/L @ 25°, 5.07 (air = 1), vp: 1.282 mm @ 25°.

TRANSFORMATION PRODUCTS: *Pseudomonas* sp. isolated from sewage samples produced 3,4-dichloro-*cis*-1,2-dihydroxycyclohexa-3,5-diene. Subsequent degradation of this metabolite yielded 3,4-dichlorocatechol which underwent ring cleavage to form 2,3-dichloro-*cis,cis*-muconate followed by hydrolysis to form 5-chloromaleylacetic acid. Titanium dioxide suspended in an aq soln and irradiated with UV light (λ = 365 nm) converted 1,2-dichlorobenzene to CO_2 at a significant rate. Anticipated products from the reaction of 1,2-dichlorobenzene with O_3 or hydroxyl radicals in the atmosphere are chlorinated phenols, ring cleavage products and nitro compounds. The sunlight irradiation of 1,2-dichlorobenzene (20 g) in a 100 ml borosilicate glass-stoppered Erlenmeyer flask for 56 days yielded 2,270 ppm 2,3′,4′-trichlorobiphenyl.

EXPOSURE LIMITS: NIOSH REL: IDLH 1,700 ppm; OSHA PEL: C 50 ppm; ACGIH TLV: C 50 ppm.

SYMPTOMS OF EXPOSURE: Irrit eyes, nose; liv, kid damage; skin blister.

USES: Preparation of 3,4-dichloroaniline; solvent for a wide variety of organic compounds and for oxides of nonferrous metals; solvent carrier in products of toluene diisocyanate; intermediate for dyes; fumigant and insecticide; degreasing hides and wool; metal polishes; industrial air control; disinfectant; heat transfer medium.

1,3-DICHLOROBENZENE

SYNONYMS: 1,3-DCB; *m*-DCB; 1,3-Dichlorbenzene; *m*-Dichlorbenzene; 1,3-Dichlorbenzol; *m*-Dichlorbenzol; *m*-Dichlorobenzene; 1,3-Dichlorobenzol; *m*-Dichlorobenzol; RCRA waste number U071; UN 1591.

CHEMICAL DESIGNATIONS: CAS: 541-73-1; DOT: none assigned; mf: $C_6H_4Cl_2$; fw: 147.00; RTECS: CZ 4499000.

PROPERTIES: Colorless liq. Mp: -25 to -22°, bp: 173°, *d*: 1.2881 @ 20/4°, fl p: 63°, lel: 2.02% (est), uel: 9.2% (est), K_H: 1.8 x 10^{-3} atm·m^3/mol @ 20°, IP: 9.12 eV, log K_{oc}: 3.23, log K_{ow}: 3.38-3.72, S_o: sol in ethanol, acetone, ether, benzene, carbon tetrachloride and ligroin, S_w: 69 mg/L @ 22°, vap d: 6.01 g/L @ 25°, 5.07 (air = 1), vp: 1.9 mm @ 25°.

TRANSFORMATION PRODUCTS: Anticipated products from the reaction of 1,3-dichlorobenzene with O_3 or hydroxyl radicals in the atmosphere are chlorinated phenols, ring cleavage products and nitro compounds. The sunlight irradiation of 1,3-dichlorobenzene (20 g) in a 100 ml borosilicate glass-stoppered Erlenmeyer flask for 56 days yielded 520 ppm trichlorobiphenyl.

USES: Fumigant and insecticide; organic synthesis.

1,4-DICHLOROBENZENE

SYNONYMS: 4-Chlorophenyl chloride; *p*-Chlorophenyl chloride; 1,4-DCB; 4-DCB; *p*-DCB; Di-chloricide; 4-Dichlorobenzene; *p*-Dichlorobenzene; 4-Dichlorobenzol; *p*-Dichlorobenzol; Evola; NCI-C54955; Paracide; Para crystals; Paradi; Paradichlorobenzene; Paradichlorobenzol; Paradow; Paramoth; Paranuggetts;

Parazene; Parodi; PDB; PDCB; Persia-Perazol; RCRA waste number U072; Santochlor; UN 1592.

CHEMICAL DESIGNATIONS: CAS: 106-46-7; DOT: 1592; mf: $C_6H_4Cl_2$; fw: 147.00; RTECS: CZ 4550000.

PROPERTIES: White cryst with a penetrating mothball-like odor. Mp: 53.1°, bp: 174.4°, *d*: 1.2475 @ 20/4°, fl p: 65.6°, lel: 2.5%, K_H: 1.5 x 10^{-3} atm·m^3/mol @ 20°, IP: 8.94 eV, 9.07 eV, log K_{oc}: 2.20 (log K_{om}), log K_{ow}: 3.355-3.62, S_o: sol in ethanol, acetone, ether, benzene, carbon tetrachloride, ligroin, carbon disulfide and chloroform, S_w: 49 mg/L @ 22°, vp: 0.7 mm @ 25°.

TRANSFORMATION PRODUCTS: Under artificial sunlight, river water containing 2-5 ppm of 1,4-dichlorobenzene photodegraded to chlorophenol and phenol. Anticipated products from the reaction of 1,4-dichlorobenzene with O_3 or hydroxyl radicals in the atmosphere are chlorinated phenols, ring cleavage products and nitro compounds. A CO_2 yield of 5.1% was achieved when 1,4-dichlorobenzene adsorbed on silica gel was irradiated with light (λ > 290 nm) for 17 h. In activated sludge, < 0.1% mineralized to CO_2 after 5 days. Irradiation of 1,4-dichlorophenol in air containing NO_x gave 2,5-dichloro-6-phenol (major product), 2,5-dichloronitrobenzene, 2,5-dichlorophenol and 2,5-dichloro-4-nitrophenol. The sunlight irradiation of 1,4-dichlorobenzene (20 g) in a 100 ml borosilicate glass-stoppered Erlenmeyer flask for 56 days yielded 1,860 ppm 4,2′,5′-trichlorobiphenyl.

EXPOSURE LIMITS: NIOSH REL: IDLH 1,000 ppm; OSHA PEL: TWA 75 ppm; ACGIH TLV: TWA 75 ppm, STEL 110 ppm.

SYMPTOMS OF EXPOSURE: Head; eye irrit, swell periobital; profuse rhinitis, anor, nau, vomit; low-wgt; jaun, cirr.

USES: Moth repellent; general insecticide, fumigant and germicide; space odorant; manufacture of 2,5-dichloroaniline and dyes; pharmacy; disinfectant and chemical intermediate.

3,3′-DICHLOROBENZIDINE

SYNONYMS: C.I. 23,060; Curithane C126; DCB; 4,4′-Diamino-3,3′-dichlorobiphenyl; 4,4′-Diamino-3,3′-dichlorodiphenyl; Dichlorobenzidine; Dichlorobenzidine base; *m,m*′-Dichlorobenzidine; **3,3′-Dichloro-1,1′-(biphenyl)-4,4′-diamine**; 3,3′-Dichlorobiphenyl-4,4′-diamine; 3,3′-Dichloro-4,4′-biphenyldiamine; 3,3′-Dichloro-4,4′-diaminobiphenyl; 3,3′-Dichloro-4,4′-diamino-1,1-biphenyl; RCRA waste number U073.

CHEMICAL DESIGNATIONS: CAS: 91-94-1; DOT: none assigned; mf: $C_{12}H_{10}Cl_2N_2$; fw: 253.13; RTECS: DD 0525000.

PROPERTIES: Colorless to grayish-purple cryst with a mild odor. Mp: 132-133°, fl p: ≥ 200 °C, pK_a: < 4, K_H: 4.5 x 10^{-8} atm·m^3/mol @ 25° (est), log K_{oc}: 3.30 (calcd), log K_{ow}: 3.51, S_o: sol in ethanol, benzene and glacial acetic acid, S_w: 3.11 ppm @ 25°, vp: 4.2 x 10^{-9} mm @ 25° (est).

TRANSFORMATION PRODUCTS: An aq soln subjected to UV radiation caused a rapid degradation to monochlorobenzidine, benzidine and several unidentified chromophores. A CO_2 yield of 41.2% was achieved when 3,3′-dichlorobenzidine adsorbed on silica gel was irradiated with light ($\lambda > 290$ nm) for 17 h. In activated sludge, 2.7% mineralized to CO_2 after 5 days.

SYMPTOMS OF EXPOSURE: Allergic skin reaction, sens; head; dizz; caustic burns; frequent urination, dys; hema; GI upsets; upper resp infection; derm.

USES: Intermediate for azo dyes and pigments; curing agent for isocyanate-terminated polymers and resins; rubber and plastic compounding ingredient; formerly used as chemical intermediate for direct red 61 dye.

DICHLORODIFLUOROMETHANE

SYNONYMS: Algofrene type 2; Arcton 6; Difluorodichloromethane; Electro-CF 12; Eskimon 12; F 12; FC 12; Fluorocarbon 12; Freon 12; Freon F-12; Frigen 12; Genetron 12; Halon; Halon 122; Isceon 122; Isotron 2; Kaiser chemicals 12; Ledon 12; Propellant 12; R 12; RCRA waste number U075; Refrigerant 12; Ucon 12; Ucon 12/halocarbon 12; UN 1028.

CHEMICAL DESIGNATIONS: CAS: 75-71-8; DOT: 1028; mf: CCl_2F_2; fw: 120.91; RTECS: PA 8200000.

PROPERTIES: Colorless liq or gas with an ethereal odor. Mp: -158°, bp: -29.8°, *d*: 1.35 @ 15/4°, K_H: 0.425 atm·m^3/mol @ 25°, IP: 11.97 eV, 12.31 eV, log K_{oc}: 2.56 (calcd), log K_{ow}: 2.16, S_o: sol in acetic acid, acetone, chloroform, ether and ethanol, S_w: 280 mg/L @ 25°, vap d: 4.94 g/L @ 25°, 4.17 (air = 1), vp: 4,250 mm @ 20°.

EXPOSURE LIMITS: NIOSH REL: IDLH 50,000 ppm; OSHA PEL: TWA 1,000 ppm; ACGIH TLV: TWA 1,000 ppm.

SYMPTOMS OF EXPOSURE: Dizz, tremors, unconscious, card arrhy, card arrest.

USES: Refrigerant; aerosol propellant; blowing agent; low temperature solvent; chilling cocktail glasses; freezing of foods by direct contact; leak-detecting agent.

1,3-DICHLORO-5,5-DIMETHYLHYDANTOIN

SYNONYMS: Dactin; DCA; DDH; **1,3-Dichloro-5,5-dimethyl-2,4-imidazolidinedione**; Dichlorodimethylhydantoin; Halane; NCI-C03054; Omchlor.

CHEMICAL DESIGNATIONS: CAS: 118-52-5; DOT: none assigned; mf: $C_5H_6Cl_2N_2O_2$; fw: 197.03; RTECS: MU 0700000.

PROPERTIES: White solid with a chlorine-like odor. Mp: 132°, subl @ 100°, *d*: 1.5 @ 20/20°, fl p: 174.4°, S_o @ 25° (wt %): benzene (9.2), carbon tetrachloride (12.5), chloroform (14), 1,2-dichloroethane (32.0), methylene chloride (30.0), 1,1,2,2-tetrachloroethane (17.0), S_w: 0.21 wt % @ 25°.

TRANSFORMATION PRODUCTS: Reacts with water (pH = 7.0) releasing hypochlorous acid. At pH 9, nitrogen chloride is formed.

EXPOSURE LIMITS: NIOSH REL: IDLH 5 mg/m^3; OSHA PEL: TWA 0.2 mg/m^3.

SYMPTOMS OF EXPOSURE: Irrit eyes, muc memb, resp sys.

USES: Chlorinating agent; industrial deodorant, disinfectant; intermediate for amino acids, drugs and insecticides; polymerization catalyst; stabilizer for vinyl chloride polymers; household laundry bleach; water treatment; organic synthesis.

1,1-DICHLOROETHANE

SYNONYMS: Chlorinated hydrochloric ether; 1,1-Dichlorethane; *asym*-Dichloroethane; Ethylidene chloride; Ethylidene dichloride; 1,1-Ethylidene dichloride; NCI-C04535; RCRA waste number U076; UN 2362.

CHEMICAL DESIGNATIONS: CAS: 75-34-3; DOT: 2362; mf: $C_2H_4Cl_2$; fw: 98.96; RTECS: KI 0175000.

PROPERTIES: Colorless liq with a chloroform-like odor. Mp: -97.4°, bp: 57.3°, *d*: 1.1757 @ 20/4°, fl p: -6°, lel: 5.6%, uel: 16%, K_H: 5.45 x 10^{-3} atm·m^3/mol @ 25°, log K_{oc}: 1.48, log K_{ow}: 1.78, 1.79, S_o: misc with ethanol, S_w: 5,500 mg/L @ 20°, vap d: 4.04 g/L @ 25°, 3.42 (air = 1), vp: 234 mm @ 25°.

TRANSFORMATION PRODUCTS: Under anoxic conditions, indigenous microbes in uncontaminated sediments produced vinyl chloride. Titanium dioxide suspended in an aq soln and irradiated with UV light (λ = 365 nm) converted dichloroethane to CO_2 at a significant rate. A glass bulb containing air and 1,1-dichloroethane degraded outdoors to CO_2 and hydrochloric acid ($t_{1/2}$ = 17 wk). The initial photodissociation product of 1,1-dichloroethane was reported to be chloroacetyl chloride. This compound is readily hydrolyzed to hydrochloric acid and chloroacetic acid. Hydrolysis of 1,1-dichloroethane under alkaline conditions yielded vinyl chloride ($t_{1/2}$ = 61.3 yr @ 25° and pH 7).

EXPOSURE LIMITS: NIOSH REL: IDLH 4,000 ppm; OSHA PEL: TWA 100 ppm; ACGIH TLV: TWA 200 ppm, STEL 250 ppm.

SYMPTOMS OF EXPOSURE: CNS depres; skin irrit; drow; unconscious; liv, kid damage.

USES: Extraction solvent; insecticide and fumigant; preparation of vinyl chloride; paint, varnish and finish removers; degreasing and drying metal parts; ore flotation; solvent for plastics, oils and fats; chemical intermediate for 1,1,1-trichloroethane; in rubber cementing, fabric spreading and fire extinguishers; formerly used as an anesthetic; organic synthesis.

1,2-DICHLOROETHANE

SYNONYMS: 1,2-Bichloroethane; Borer sol; Brocide; 1,2-DCA; 1,2-DCE; Destruxol borer-sol; Dichloremulsion; 1,2-Dichlorethane; Dichlormulsion; α,β-Dichloroethane; *sym*-Dichloroethane; Dichloroethylene; Dutch liquid; Dutch oil; EDC; ENT 1,656; Ethane dichloride; Ethene dichloride; Ethylene chloride; Ethylene dichloride; 1,2-Ethylene dichloride; Freon 150; Glycol dichloride; NCI-C00511; RCRA waste number U077; UN 1184.

CHEMICAL DESIGNATIONS: CAS: 107-06-2; DOT: 2362; mf: $C_2H_4Cl_2$; fw:

98.96; RTECS: KI 0525000.

PROPERTIES: Colorless liq with a pleasant odor. Mp: -35.3°, bp: 83.5°, *d*: 1.2351 @ 20/4°, fl p: 13° (OC), lel: 6.2%, uel: 16%, K_H: 1.31 x 10^{-3} atm·m^3/mol @ 25°, IP: 9.64 eV, 11.04 eV, log K_{oc}: 1.15, 1.279, log K_{ow}: 1.45, 1.48, S_o: misc with ethanol, chloroform and ether, S_w: 8.69 g/L @ 20°, vap d: 4.04 g/L @ 25°, 3.42 (air = 1), vp: 64 mm @ 20°.

TRANSFORMATION PRODUCTS: Titanium dioxide suspended in an aq soln and irradiated with UV light (λ = 365 nm) converted dichloroethane to CO_2 at a significant rate. Anticipated products from the reaction of 1,2-dichloroethane with O_3 or hydroxyl radicals in the atmosphere are chloroacetaldehyde, chloroacetyl chloride, formaldehyde and ClHCHO. Hydrolysis of 1,2-dichloroethane under alkaline and neutral conditions yielded vinyl chloride and ethylene glycol, respectively ($t_{1/2}$ = 72.0 yr @ 25° and pH 7). *Methanococcus thermolithotrophicus*, *Methanococcus deltae* and *Methanobacterium thermoautotrophicum* metabolized 1,2-dichloroethane releasing methane and ethylene.

EXPOSURE LIMITS: NIOSH REL: 10-h TWA 1 ppm, 15-min C 2 ppm; OSHA PEL: TWA 50 ppm, C 100 ppm, 5-min/3-h peak 200 ppm; ACGIH TLV: TWA 10 ppm.

SYMPTOMS OF EXPOSURE: CND depres; nau, vomit; derm; irrit eyes, corneal opac.

USES: Vinyl chloride solvent; lead scavenger in antiknock unleaded gasoline; paint, varnish and finish remover; metal degreasers; soap and scouring compounds; wetting and penetrating agents; preparation of ethylenediamine; ore flotation; tobacco flavoring; soil and foodstuff fumigant.

1,1-DICHLOROETHYLENE

SYNONYMS: 1,1-DCE; **1,1-Dichloroethene**; *asym*-Dichloroethylene; NCI-C54262; RCRA waste number U078; Sconatex; VC; VDC; Vinylidene chloride; Vinylidene chloride (II); Vinylidene dichloride; Vinylidine chloride.

CHEMICAL DESIGNATIONS: CAS: 75-35-4; DOT: 1303; mf: $C_2H_2Cl_2$; fw: 96.94; RTECS: KV 9275000.

PROPERTIES: Colorless liq with a mild, sweet odor resembling chloroform. Mp: -122.1°, bp: 32°, *d*: 1.218 @ 20/4°, fl p: -15°, lel: 6.5%, uel: 15.5%, K_H: 1.5 x 10^{-2} atm·m^3/mol @ 25°, IP: 9.81 eV, log K_{oc}: 1.81, log K_{ow}: 1.48, 2.13, S_o: slightly sol in ethanol, ether, acetone, benzene and chloroform, S_w: 273-6,400 mg/L @ 25°, vap d: 3.96 g/L @ 25°, 3.35 (air = 1), vp: 591 mm @ 25°.

TRANSFORMATION PRODUCTS: In a methanogenic aquifer material, 1,1-dichloroethylene biodegraded to vinyl chloride. Under anoxic conditions, indigenous microbes in uncontaminated sediments degraded 1,1-dichloroethylene to vinyl chloride. Photooxidation of 1,1-dichloroethylene in the presence of NO_2 and air yielded phosgene, chloroacetyl chloride, formic acid, hydrochloric acid, CO, formaldehyde and O_3. Above 0° in the presence of oxygen or other catalysts, 1,1-dichloroethylene will polymerize to a plastic. The alkaline hydrolysis of 1,1-dichloroethylene yielded chloroacetylene ($t_{1/2}$ 1.2 x 10^8 yr @ 25° and pH 7).

EXPOSURE LIMITS: ACGIH TLV: TWA 5 ppm, STEL 20 ppm.

SYMPTOMS OF EXPOSURE: Irrit eyes, resp sys; CNS depres.

USES: Synthetic fibers and adhesives; chemical intermediate in vinylidene fluoride synthesis; comonomer for food packaging, coating resins and modacrylic fibers.

trans-1,2-DICHLOROETHYLENE

SYNONYMS: Acetylene dichloride; *trans*-Acetylene dichloride; 1,2-Dichloroethene; **(*E*)-1,2-Dichloroethene**; *trans*-Dichloroethylene; 1,2-*trans*-Dichloroethene; 1,2-*trans*-Dichloroethylene; *sym*-Dichloroethylene; Dioform.

CHEMICAL DESIGNATIONS: CAS: 156-60-5; DOT: 1150 (isomeric mixture); mf: $C_2H_2Cl_2$; fw: 96.94; RTECS: KV 9400000.

PROPERTIES: Colorless liq with a sweet pleasant odor. Mp: -50°, bp: 47.5°, *d*: 1.2565 @ 20/4°, fl p: 2°, lel: 9.7%, uel: 12.8%, K_H: 5.32 x 10^{-3} atm·m^3/mol, IP: 9.64 eV, log K_{oc}: 1.77, log K_{ow}: 2.06, 2.09, S_o: misc with acetone, ethanol, ether; very sol in benzene and chloroform, S_w: 6.26 g/L @ 25°, vap d: 3.96 g/L @ 25°, 3.35 (air = 1), vp: 265 mm @ 20°.

TRANSFORMATION PRODUCTS: In a methanogenic aquifer material, *trans*-1,2-dichloroethylene biodegraded to vinyl chloride. Under anoxic conditions *trans*-1,2-dichloroethylene, when subjected to indigenous microbes in uncontaminated sediments, degraded to vinyl chloride. CO, formic and hydrochloric acids were reported to be photooxidation products. Slowly decomposes in the presence of air, light and moisture releasing hydrogen chloride ($t_{1/2}$ = 2.1 x 10^{10} yr @ 25° and pH 7).

EXPOSURE LIMITS: NIOSH REL: IDLH 4,000 ppm (isomeric mixture); ACGIH TLV: TWA 200 ppm (isomeric mixture).

SYMPTOMS OF EXPOSURE: Irrit eyes, resp sys; CNS depres.

USES: A mixture of *cis*- and *trans*- isomers is used as a solvent for fats, phenols, camphor, etc.; ingredient in perfumes; low temperature solvent for sensitive substances such as caffeine; refrigerant; organic synthesis.

DICHLOROFLUOROMETHANE

SYNONYMS: Algofrene type 5; Arcton 7; Dichloromonofluoromethane; Fluorocarbon 21; Fluorodichloromethane; Freon 21; Genetron 21; Halon 21; Refrigerant 21; UN 1029.

CHEMICAL DESIGNATIONS: CAS: 75-43-4; DOT: 1029; mf: $CHCl_2F$; fw: 120.91; RTECS: PA 8400000.

PROPERTIES: Colorless liq or gas with an ether-like odor. Mp: -158°, -135°, bp: -29.8°, 8.9°, *d*: 1.75 @ -115/4°, K_H: ≈ 2.42 x 10^{-2} atm·m^3/mol @ 20-30° (calcd), IP: 12.39 eV, log K_{oc}: 1.57 (calcd), log K_{ow}: 1.55, S_o: sol in acetic acid, alcohol and ether, S_w: 1 wt % @ 20°, vap d: 4.94 g/L @ 25°, 4.17 (air = 1), vp: 2 atm @ 28.4°.

EXPOSURE LIMITS: NIOSH REL: IDLH 50,000 ppm; OSHA PEL: TWA 1,000 ppm; ACGIH TLV: TWA 10 ppm.

SYMPTOMS OF EXPOSURE: Asphyxia, card arrhy, card arrest.

USES: Fire extinguishers; solvent; refrigerant.

sym-DICHLOROMETHYL ETHER

SYNONYMS: BCME; Bis(chloromethyl)ether; Bis-cme; Chloro(chloromethoxy)methane; Chloromethyl ether; Dimethyl-1,1′-dichloroether; *sym*-Dichlorodimethyl ether; Dichloromethyl ether; **Oxybis(chloromethane)**; RCRA waste number P016; UN 2249.

CHEMICAL DESIGNATIONS: CAS: 542-88-1; DOT: none assigned; mf: $C_2H_4Cl_2O$; fw: 114.96; RTECS: KN 1575000.

PROPERTIES: Colorless liq with a suffocating odor. Mp: -41.5°, bp: 104°, *d*: 1.315 @ 20/4°, fl p: < 19°, S_o: sol in alcohol, ether and benzene, S_w: decomp, vap d: 4.70 g/L @ 25°, 3.97 (air = 1).

TRANSFORMATION PRODUCTS: Reacts with water forming hydrochloric acid and formaldehyde. Anticipated products from the reaction of *sym*-dichloromethyl ether with O_3 or hydroxyl radicals in the atmosphere, excluding the decomposition products formaldehyde and hydrochloric acid, are chloromethyl formate and formyl chloride.

EXPOSURE LIMITS: ACGIH TLV: TWA 0.001 ppm.

SYMPTOMS OF EXPOSURE: Irrit eyes, skin, muc memb of resp sys; pulm congestion, edema; corn damage, nec; reduced pulm function, cough, dysp, wheez; hemoptysis, bronchial secretions.

USE: Chloromethylating agent.

2,4-DICHLOROPHENOL

SYNONYMS: 3-Chloro-4-hydroxychlorobenzene; DCP; 2,4-DCP; 2,4-Dichlorohydroxybenzene; 4,6-Dichlorohydroxybenzene; NCI-C55345; RCRA waste number U081.

CHEMICAL DESIGNATIONS: CAS: 120-83-2; DOT: none assigned; mf: $C_6H_4Cl_2O$; fw: 163.00; RTECS: SK 8575000.

PROPERTIES: Colorless to yellow cryst with a sweet, musty or medicinal odor. Mp: 45°, bp: 210°, 216°, *d*: 1.40 @ 15/4°, fl p: 93.3° (OC), pK_a: 7.85 @ 25°, K_H: 3.23 x 10^{-6} atm·m^3/mol @ 25° (est), log K_{oc}: 2.10–3.60, log K_{ow}: 3.06–3.23, S_o: sol in ethanol, benzene, ether, chloroform and carbon tetrachloride, S_w: 4,600 mg/L @ 20°, vp: 8.9 x 10^{-2} mm @ 25°.

TRANSFORMATION PRODUCTS: In distilled water, photolysis occurs at a slower rate than in estuarine waters containing humic substances. Photolysis products identified in distilled water were the 3 isomers of chlorocyclopentadienic acid. In a similar experiment, titanium dioxide suspended in an aq soln and irradiated with UV light (λ = 365 nm) converted 2,4-dichlorophenol to CO_2 at a significant rate. An aq soln containing H_2O_2 and irradiated by UV light (λ = 296 nm) converted 2,4-dichlorophenol to chlorohydroquinone and 1,4-dihydroquinone. In freshwater lake sediments, anaerobic reductive dechlorination produced 4-chlorophenol. A CO_2 yield of 50.4% was achieved when 2,4-dichlorophenol adsorbed on silica gel was irradiated with light (λ > 290 nm) for 17 h. In activated sludge, 2.8% mineralized to CO_2 after 5 days. Chloroperoxidase, a fungal enzyme isolated from *Caldariomyces fumago*, converted 9-12% of 2,4-dichlorophenol to 2,4,6-trichlorophenol.

USES: A chemical intermediate in the manufacture of the pesticide 2,4-dichlorophenoxyacetic acid (2,4-D) and other compounds for use as germicides, antiseptics and seed disinfectants.

1,2-DICHLOROPROPANE

SYNONYMS: α,β-Dichloropropane; ENT 15,406; NCI-C55141; Propylene chloride; Propylene dichloride; α,β-Propylene dichloride; RCRA waste number U083.

CHEMICAL DESIGNATIONS: CAS: 78-87-5; DOT: 1279; mf: $C_3H_6Cl_2$; fw: 112.99; RTECS: TX 9625000.

PROPERTIES: Colorless liq with a sweet, chloroform-like odor. Mp: -100.4°, -70°, bp: 96.4°, *d*: 1.560 @ 20/4°, fl p: 15.6°, lel: 3.4%, uel: 14.5%, K_H: 2.94 x 10^{-3} atm·m^3/mol @ 25°, IP: 10.87 eV, log K_{oc}: 1.431, 1.71, log K_{ow}: 2.28, S_o: misc with organic solvents, S_w: 2,700 mg/L @ 20°, vap d: 4.62 g/L @ 25°, 3.90 (air = 1), vp: 42 mm @ 20°.

TRANSFORMATION PRODUCTS: Hydrolysis in distilled water @ 25° produced 1-chloro-2-propanol and hydrochloric acid ($t_{1/2}$ = 8,613 days). Distilled water irradiated with UV light (λ = 290 nm) yielded the following photolysis products: 2-chloro-1-propanol, allyl chloride, allyl alcohol and acetone ($t_{1/2}$ in distilled water = 50 min but $t_{1/2}$ in distilled water containing H_2O_2 = 30 min). Ozonolysis yielded CO_2 at low O_3 concentrations.

EXPOSURE LIMITS: NIOSH REL: IDLH 2,000 ppm; OSHA PEL: TWA 75 ppm; ACGIH TLV: TWA 75 ppm, STEL 110 ppm.

SYMPTOMS OF EXPOSURE: Eye irrit, drow, li-head, irrit skin.

USES: Preparation of tetrachloroethylene and carbon tetrachloride; lead scavenger for antiknock fluids; metal cleanser; soil fumigant for nematodes; solvent for oils, fats, gums, waxes and resins; spotting agent.

cis-1,3-DICHLOROPROPYLENE

SYNONYMS: *cis*-1,3-Dichloropropene; *cis*-1,3-Dichloro-1-propene; (*Z*)-1,3-Dichloropropene; **(*Z*)-1,3-Dichloro-1-propene**; 1,3-Dichloroprop-1-ene; *cis*-1,3-Dichloro-1-propylene.

CHEMICAL DESIGNATIONS: CAS: 10061-01-5; DOT: 2047 (isomeric mixture); mf: $C_3H_4Cl_2$; fw: 110.97; RTECS: UC 8325000.

PROPERTIES: Colorless liq with a chloroform-like odor. Mp: -84° (isomeric mixture), bp: 104.3°, *d*: 1.224 @ 20/4°, fl p: 35° (isomeric mixture), lel: 5.3% (isomeric mixture), uel: 14.5% (isomeric mixture), K_H: 3.55 x 10^{-3} atm·m^3/mol, log K_{oc}: 1.36, 1.68, log K_{ow}: 1.41, S_o: sol in benzene, chloroform and ether, S_w: 2,700 mg/L @ 25°, vap d: 4.54 g/L @ 25°, 3.83 (air = 1), vp: 25 mm @ 20°.

TRANSFORMATION PRODUCTS: Hydrolysis in distilled water @ 25° produced 2-chloro-3-propenol and hydrochloric acid ($t_{1/2}$ = 1 day). Hydrolyzes in wet soil forming *cis*-3-chloroallyl alcohol. *cis*-1,3-Dichloropropylene was reported to hydrolyze to 3-chloro-2-propen-1-ol and can be biologically oxidized to 3-chloropropenoic acid which is further oxidized to formylacetic acid. Decarboxylation of this compound yields CO_2. Chloroacetaldehyde, formyl chloride and chloroacetic acid were formed from the ozonation of dichloropropylene @ ≈ 23° and ≈ 730 mm. Chloroacetaldehyde and formyl chloride also formed from the reaction of dichloropropylene with hydroxyl radicals.

USES: The isomeric mixture is used as a soil fumigant and a nematocide.

trans-1,3-DICHLOROPROPYLENE

SYNONYMS: (*E*)-1,3-Dichloropropene; *trans*-1,3-Dichloropropene; **(*E*)-1,3-Dichloro-1-propene**; *trans*-1,3-Dichloro-1-propene; 1,3-Dichloroprop-1-ene; *trans*-1,3-Dichloro-1-propylene.

CHEMICAL DESIGNATIONS: CAS: 10061-02-6; DOT: 2047 (isomeric mixture); mf: $C_3H_4Cl_2$; fw: 110.97; RTECS: UC 8320000.

PROPERTIES: Liq with a chloroform-like odor. Mp: -84° (isomeric mixture), bp: 77° @ 757 mm, 112.1°, *d*: 1.1818 @ 20/4°, fl p: 5.3° (isomeric mixture), lel: 5.3% (isomeric mixture), uel: 14.5% (isomeric mixture), K_H: 3.55 x 10^{-3} atm·m^3/mol, log K_{oc}: 1.415, 1.68, log K_{ow}: 1.41, S_o: sol in benzene, chloroform and ether, S_w: 2,800 mg/L @ 25°, vap d: 4.54 g/L @ 25°, 3.83 (air = 1), vp: 25 mm @ 20°.

TRANSFORMATION PRODUCTS: Hydrolysis in distilled water @ 25° produced 2-chloro-3-propenol and hydrochloric acid ($t_{1/2}$ = 2 days). Hydrolysis in wet soil resulted in the formation of *trans*-3-chloroallyl alcohol. *trans*-1,3-Dichloropropylene was reported to hydrolyze to 3-chloro-2-propen-1-ol and can be biologically oxidized to 3-chloropropenoic acid which is further oxidized to formylacetic acid. Decarboxylation of this compound yields CO_2. Chloroacetaldehyde, formyl chloride and chloroacetic acid were formed from the ozonation of dichloropropylene @ ≈ 23° and ≈ 730 mm. Chloroacetaldehyde and formyl chloride also formed from the reaction of dichloropropylene with hydroxyl radicals.

USES: The isomer mixture is used as a soil fumigant and a nematocide.

DICHLORVOS

SYNONYMS: Apavap; Astrobot; Atgard; Atgard C; Atgard V; Bay 19149; Benfos; Bibesol; Brevinyl; Brevinyl E50; Canogard; Cekusan; Chlorvinphos; Cyanophos;

Cypona; DDVF; DDVP; Dedevap; Deriban; Derribante; Devikol; Dichlorman; 2,2-Dichloroethenyl dimethyl phosphate; 2,2-Dichloroethenyl phosphoric acid dimethyl ester; Dichlorophos; 2,2-Dichlorovinyl dimethyl phosphate; 2,2-Dichlorovinyl dimethyl phosphoric acid ester; Dichlorovos; Dimethyl 2,2-dichloroethenyl phosphate; Dimethyl dichlorovinyl phosphate; Dimethyl 2,2-dichlorovinyl phosphate; *O,O*-Dimethyl *O*-(2,2-dichlorovinyl)phosphate; Divipan; Duo-kill; Duravos; ENT 20,738; Equigard; Equigel; Estrosel; Estrosol; Fecama; Fly-die; Fly fighter; Herkal; Herkol; Krecalvin; Lindan; Mafu; Mafu strip; Marvex; Mopari; NA 2783; NCI-C00113; Nerkol; Nogos; Nogos 50; Nogos G; No-pest; No-pest strip; NSC 6,738; Nuva; Nuvan; Nuvan 100EC; Oko; OMS 14; **Phosphoric acid 2,2-dichloroethenyl dimethyl ester**; Phosphoric acid 2,2-dichlorovinyl dimethyl ester; Phosvit; SD-1750; Szklarniak; Tap 9VP; Task; Task tabs; Tenac; Tetravos; UDVF; Unifos; Unifos 50 EC; Vapona; Vaponite; Vapora II; Verdican; Verdipor; Vinyl alcohol 2,2-dichlorodimethyl phosphate; Vinylofos; Vinylophos.

CHEMICAL DESIGNATIONS: CAS: 62-73-7; DOT: 2783; mf: $C_4H_7Cl_2O_4P$; fw: 220.98; RTECS: TC 0350000.

PROPERTIES: Colorless to yellow liq with a slight odor. bp: 140° @ 20 mm, *d*: 1.415 @ 25/4°, K_H: 5.0 x 10^{-3} atm·m^3/mol, log K_{oc}: 9.57 (calcd), log K_{ow}: 1.40, S_o: sol in glycerol (≈ 5 g/L); misc with alcohol and most non-polar solvents, S_w: ≈ 1 wt % @ 20°, vap d: 9.03 g/L @ 25°, 7.63 (air = 1), vp: 0.012 mm @ 20°.

TRANSFORMATION PRODUCTS: Dichlorvos incubated with sewage sludge for 1 wk @ 29° degraded to dichloroethanol, dichloroacetic acid, ethyl dichloroacetate and an inorganic phosphate. In addition, dimethyl phosphate formed in the presence or absence of microorganisms. Releases very toxic fumes of chlorides and phosphorous oxides when heated to decomposition.

EXPOSURE LIMITS: NIOSH REL: IDLH 200 mg/m^3; OSHA PEL: TWA 0.1 ppm; ACGIH TLV: TWA 0.1 ppm.

SYMPTOMS OF EXPOSURE: Miosos, ache eyes; rhin; head; chest, wheez, lar spas, salv; cyan; anor, nau, vomit, diarr, sweat; musc fasc, para, atax; cinvul; low BP; card.

USES: Insecticide and fumigant.

DIELDRIN

SYNONYMS: Alvit; Compound 497; Dieldrite; Dieldrix; ENT 16,225; HEOD; Hexachloroepoxyoctahydro-*endo-exo*-dimethanonaphthalene; 1,2,3,4,10,10-Hexachloro-6,7-epoxy-1,4,4a,5,6,7,8,8a-octahydro-1,4-*endo-exo*-5,8-dimethanonaphthalene; **3,4,5,6,9,9-Hexachloro-1a,2,2a,3,6,6a,7,7a-octahydro-2,7:3,6-dimethanonaphth[2,3-*b*]oxirene**; Illoxol; Insecticide number 497; NA 2,761; NCI-C00124; Octalox; Panoram D-31; Quintox; RCRA waste number P037.

CHEMICAL DESIGNATIONS: CAS: 60-57-1; DOT: 2761; mf: $C_{12}H_8Cl_6O$; fw: 380.91; RTECS: IO 1750000.

PROPERTIES: White cryst or pale tan flakes with an odorless to mild chemical odor. Mp: 175-176°, bp: decomp, *d*: 1.75 @ 20/4°, K_H: 5.8 x 10^{-5} atm·m^3/mol @ 25°, log K_{oc}: 4.08-4.55, log K_{ow}: 3.692-6.2, S_o: sol in ethanol and benzene, S_w: 186-

200 μg/L @ 20°, vap d: 54 ng/L @ 20°, vp: 1.78×10^{-7} mm @ 20°.

TRANSFORMATION PRODUCTS: Identified metabolites of dieldrin from soln cultures containing *Pseudomonas* sp. in soils include aldrin and dihydroxydihydroaldrin. Other unidentified byproducts included a ketone, an aldehyde and an acid. Photolysis of an aq soln by sunlight for 3 mo resulted in a 70% yield of photodieldrin. A solid film of dieldrin exposed to sunlight for 2 mo resulted in a 25% yield of photodieldrin. In addition to sunlight, UV light converts dieldrin to photodieldrin. Solid dieldrin exposed to UV light (λ < 300 nm) under a stream of oxygen yielded small amounts of photodieldrin.

EXPOSURE LIMITS: NIOSH REL: 0.15 mg/m^3; OSHA PEL: TWA 0.25 mg/m^3; ACGIH TLV: TWA 0.25 mg/m^3.

SYMPTOMS OF EXPOSURE: Head; dizz; nau, vomit, mal; sweat; myoclonic limbjerks; clonic, tonic, convul; coma.

USES: Insecticide; wool processing industry.

DIETHYLAMINE

SYNONYMS: Diethamine; *N,N*-Diethylamine; ***N*-Ethylethanamine**; UN 1154.

CHEMICAL DESIGNATIONS: CAS: 109-89-7; DOT: 1154; mf: $C_4H_{11}N$; fw: 73.14; RTECS: HZ 8750000.

PROPERTIES: Colorless liq with an ammonia-like odor. Mp: -48°, bp: 56.3°, *d*: 0.7056 @ 20/4°, fl p: -28°, lel: 1.8%, uel: 10.1%, pK_a: 11.090 @ 20°, 10.93 @ 25°, K_H: 2.56×10^{-5} atm·m^3/mol @ 25°, IP: 8.01 eV, log K_{ow}: 0.43-0.81, S_o: misc with alcohol, S_w: 815 g/L @ 14°, vap d: 2.99 g/L @ 25°, 2.52 (air = 1), vp: 200 mm @ 20°.

TRANSFORMATION PRODUCTS: Diethylamine reacted with NO_x in the dark forming diethylnitrosamine. In an outdoor chamber, photooxidation by natural sunlight yielded the following products: diethylnitramine, diethylformamide, diethylacetamide, ethylacetamide, O_3, acetaldehyde and peroxyacetylnitrate. Forms soluble salts with acids.

EXPOSURE LIMITS: NIOSH REL: IDLH 2,000 ppm; OSHA PEL: TWA 25 ppm; ACGIH TLV: TWA 10 ppm, STEL 25 ppm.

SYMPTOMS OF EXPOSURE: Eye, skin, resp irrit.

USES: In flotation agents, resins, dyes, pesticides, rubber chemicals and pharmaceuticals; selective solvent; polymerization and corrosion inhibitors; petroleum chemicals; electroplating; organic synthesis.

2-DIETHYLAMINOETHANOL

SYNONYMS: DEAE; Diethylaminoethanol; β-Diethylaminoethanol; *N*-Diethylaminoethanol; 2-*N*-Diethylaminoethanol; 2-Diethylaminoethyl alcohol; β-Diethylaminoethyl alcohol; Diethylethanolamine; *N,N*-Diethylethanolamine; Diethyl(2-hydroxyethyl)amine; *N,N*-Diethyl-*N*-(β-hydroxyethyl)amine; 2-Hydroxytriethylamine; UN 2686.

CHEMICAL DESIGNATIONS: CAS: 100-37-8; DOT: 2686; mf: $C_6H_{15}NO$; fw: 117.19; RTECS: KK 5075000.

PROPERTIES: Colorless liq with a faint ammonia-like odor. bp: 163°, *d*: 0.8800 @ 25/4°, fl p: 48°, lel: 6.7%, uel: 11.7%, S_o: sol in alcohol, benzene and ether, S_w: misc, vap d: 4.79 g/L @ 25°, 4.05 (air = 1), vp: 1.4 mm @ 20°.

EXPOSURE LIMITS: NIOSH REL: IDLH 500 ppm; OSHA PEL: TWA 10 ppm.

SYMPTOMS OF EXPOSURE: Nau, vomit; irrit resp, skin, eyes.

USES: Water-soluble salts; textile softeners; antirust formulations; fatty acid derivatives; pharmaceuticals; curing agent for resins; emulsifying agents in acid media; organic synthesis.

DIETHYL PHTHALATE

SYNONYMS: Anozol; **1,2-Benzenedicarboxylic acid diethyl ester**; DEP; Diethyl-*o*-phthalate; Estol 1550; Ethyl phthalate; NCI-C60048; Neantine; Palatinol A; Phthalol; Placidol E; RCRA waste number U088; Solvanol.

CHEMICAL DESIGNATIONS: CAS: 84-66-2; DOT: none assigned; mf: $C_{12}H_{14}O_4$; fw: 222.24; RTECS: TI 1050000.

PROPERTIES: Clear, colorless liq with a mild chemical odor. Mp: -40.5°, bp: 298°, *d*: 1.1175 @ 20/4°, fl p: 140°, lel: 0.7% @ 186°, K_H: 8.46 x 10^{-7} atm·m^3/mol, log K_{oc}: 1.84, log K_{ow}: 1.40-3.00, S_o: sol in acetone and benzene; misc with ethanol, ether, esters and ketones, S_w: 896-1,200 mg/L @ 25°, vap d: 9.08 g/L @ 25°, 7.67 (air = 1), vp: 0.22 ± 0.7 Pa @ 25°.

TRANSFORMATION PRODUCTS: Hydrolyzes in water to phthalic acid and ethanol. A proposed microbial degradation mechanism is as follows: 4-hydroxy-3-methylbenzyl alcohol to 4-hydroxy-3-methylbenzaldehyde to 3-methyl-4-hydroxybenzoic acid to 4-hydroxyisophthalic acid to protocatechuic acid to β-ketoadipic acid. In anaerobic sludge, diethyl phthalate degraded as follows: monoethyl phthalate to phthalic acid to protocatechuic acid followed by ring cleavage and mineralization. An aq soln containing titanium dioxide and subjected to UV radiation (λ > 290 nm) produced hydroxyphthalates and dihydroxyphthalates as intermediates. Pyrolysis of diethyl phthalate in a flow reactor @ 700° yielded the following products: ethanol, ethylene, benzene, naphthalene, phthalic anhydride and 2-phenylenaphthalene.

EXPOSURE LIMITS: ACGIH TLV: TWA 5 mg/m^3.

USES: Plasticizer; plastic manufacturing and processing; denaturant for ethyl alcohol; ingredient in insecticidal sprays and explosives (propellant); dye application agent; wetting agent; perfumery as fixative and solvent; solvent for nitrocellulose and cellulose acetate; camphor substitute.

1,1-DIFLUOROTETRACHLOROETHANE

SYNONYMS: 1,1-Difluoro-1,2,2,2-tetrachloroethane; 2,2-Difluoro-1,1,1,2-

tetrachloroethane; Freon 112a; Halocarbon 112a; Refrigerant 112a; **1,1,1,2-Tetrachloro-2,2-difluoroethane.**

CHEMICAL DESIGNATIONS: CAS: 76-11-9; DOT: 1078; mf: $C_2Cl_4F_2$; fw: 203.83; RTECS: KI 1425000.

PROPERTIES: Colorless solid. Mp: 40.6°, bp: 91.5°, *d*: 2.191 (calcd), S_o: sol in alcohol, ether and chloroform, vp: 40 mm @ 19.8°.

EXPOSURE LIMITS: NIOSH REL: IDLH 15,000 ppm; OSHA PEL: TWA 500 ppm.

SYMPTOMS OF EXPOSURE: CNS depres, pulm edema, skin irrit, eye irrit, drowsy, dysp.

USE: Organic synthesis.

1,2-DIFLUOROTETRACHLOROETHANE

SYNONYMS: 1,2-Difluoro-1,1,2,2-tetrachloroethane; F-112; Freon 112; Genetron 112; Halocarbon 112; Refrigerant 112; **1,1,2,2-Tetrachloro-1,2-difluoroethane.**

CHEMICAL DESIGNATIONS: CAS: 76-12-0; DOT: 1078; mf: $C_2Cl_4F_2$; fw: 203.83; RTECS: KI 1420000.

PROPERTIES: Colorless liq with an ether-like odor. Mp: 25°, bp: 93°, *d*: 1.6447 @ 25/4°, K_H: 1.07 x 10^{-1} atm·m^3/mol @ 20° (calcd), log K_{oc}: 2.78 (calcd), log K_{ow}: 3.39 (calcd), S_o: sol in alcohol, chloroform and ether, S_w: 0.01 wt % @ 20°, vap d: 8.33 g/L @ 25°, 7.04 (air = 1), vp: 40 mm @ 19.8°.

EXPOSURE LIMITS: NIOSH REL: IDLH 15,000 ppm; OSHA PEL: TWA 500 ppm.

SYMPTOMS OF EXPOSURE: Irrit skin, conj, pulm edema.

USES: Organic synthesis.

DIISOBUTYL KETONE

SYNONYMS: DIBK; *sym*-Diisopropylacetone; 2,6-Dimethylheptan-4-one; **2,6-Dimethyl-4-heptanone**; Isobutyl ketone; Isovalerone; UN 1157; Valerone.

CHEMICAL DESIGNATIONS: CAS: 108-83-8; DOT: 1157; mf: $C_9H_{18}O$; fw: 142.24; RTECS: MJ 5775000.

PROPERTIES: Colorless liq with a mild odor. Mp: -46 to -42°, bp: 168°, *d*: 0.8053 @ 20/4°, fl p: 48°, lel: 0.8% @ 93°, uel: 6.2% @ 93°, K_H: 6.36 x 10^{-4} atm·m^3/mol @ 20° (calcd), log K_{ow}: 2.58 (calcd), S_o: sol in alcohol, ether, and other organic solvents, S_w: 0.05 wt % @ 20°, vap d: 5.81 g/L @ 25°, 4.91 (air = 1), vp: 1.7 mm @ 20°.

EXPOSURE LIMITS: NIOSH REL: 10-h TWA 25 ppm, IDLH 2,000 ppm; OSHA PEL: TWA 50 ppm; ACGIH TLV: TWA 25 ppm.

SYMPTOMS OF EXPOSURE: Irrit eyes, nose, throat; head; dizz; derm; unconscious.

USES: Solvent for nitrocellulose, synthetic resins, rubber; lacquers; coating compositions; inks and stains; organic synthesis.

DIISOPROPYLAMINE

SYNONYMS: DIPA; ***N*-(1-Methylethyl)-2-propanamine**; UN 1158.

CHEMICAL DESIGNATIONS: CAS: 108-18-9; DOT: 1158; mf: $C_6H_{15}N$; fw: 101.19; RTECS: IM 4025000.

PROPERTIES: Colorless liq with an ammonia-like odor. Mp: -96.3°, -61°, bp: 84°, *d*: 0.7169 @ 20/4°, fl p: -6.7°, lel: 0.8%, uel: 7.1%, pK_a: 11.13 @ 21°, IP: 7.73 eV, S_o: sol in acetone, alcohol, benzene and ether, S_w: misc, vap d: 4.14 g/L @ 25°, 3.49 (air = 1), vp: 60 mm @ 20°.

TRANSFORMATION PRODUCTS: Reacts with acids forming soluble salts.

EXPOSURE LIMITS: NIOSH REL: IDLH 1,000 ppm; ACGIH TLV: TWA 5 ppm.

SYMPTOMS OF EXPOSURE: Nau, vomit; head; eye irrit, vis dist; pulm irrit.

USES: Intermediate; catalyst; organic synthesis.

N,N-DIMETHYLACETAMIDE

SYNONYMS: Acetdimethylamide; Acetic acid dimethylamide; Dimethylacetamide; Dimethylacetone amide; Dimethylamide acetate; DMA; DMAC; Hallucinogen; NSC 3,138; U-5954.

CHEMICAL DESIGNATIONS: CAS: 127-19-5; DOT: none assigned; mf: C_4H_9NO; fw: 115.18; RTECS: AB 7700000.

PROPERTIES: Colorless liq with a weak, ammonia-like odor. Mp: -20°, bp: 165° @ 758 mm, *d*: 0.9366 @ 25/4°, fl p: 65.6°, lel: 1.8% @ 100°, uel: 11.5% @ 160°, IP: 8.60 eV, 8.81 eV, log K_{ow}: -0.77, S_o: sol in acetone, alcohol and benzene; misc with aromatics, esters, ketones and ethers, S_w: misc, vp: 1.5 mm @ 20°.

TRANSFORMATION PRODUCTS: Releases toxic fumes of NO_x when heated to decomposition.

EXPOSURE LIMITS: NIOSH REL: IDLH 400 ppm; OSHA PEL: TWA 10 ppm.

SYMPTOMS OF EXPOSURE: Jaun, liv damage; depres, lethargy, halu, delusions; irrit skin.

USES: Solvent used in organic synthesis; paint removers; solvent for plastics, resins, gums and electrolytes; intermediate; catalyst.

DIMETHYLAMINE

SYNONYMS: DMA; ***N*-Methylmethanamine**; RCRA waste number U092; UN 1032; UN 1160.

CHEMICAL DESIGNATIONS: CAS: 124-40-3; DOT: 1032 (anhydrous), 1160 (aq soln); mf: C_2H_7N; fw: 45.08; RTECS: IP 8750000.

PROPERTIES: Colorless liq or gas with a strong ammonia-like odor. Mp: -93°, bp: 7.4°, *d*: 0.6804 @ 0/4°, fl p: -17.7° (25% soln), lel: 2.8%, uel: 14.4%, pK_a: 10.732 @ 25°, K_H: 1.77 x 10^{-5} atm·m^3/mol @ 25°, IP: 8.24 eV, log K_{ow}: -0.38, S_o: sol in alcohol and ether, S_w: misc, vap d: 1.84 g/L @ 25°, 1.56 (air = 1), vp: 1.7 atm @ 20°.

TRANSFORMATION PRODUCTS: Dimethylnitramine, nitrous acid, formaldehyde, dimethylformamide and CO were reported as photooxidation products of dimethylamine with NO_x. An additional compound was tentatively identified as tetramethylhydrazine. In an aq soln, chloramine reacted with dimethylamine forming *N*-chlorodimethylamine. After 2 days, degradation yields in an Arkport fine sandy loam (Varna, NY) and sandy soil (Lake George, NY) amended with sewage and nitrite-N were 50 and 20%, respectively. *N*-Nitrosodimethylamine was identified as the major metabolite. Reacts with hydroxyl radicals forming formaldehyde and/or amides.

EXPOSURE LIMITS: NIOSH REL: IDLH 2,000 ppm; OSHA PEL: TWA 10 ppm; ACGIH TLV: TWA 10 ppm.

SYMPTOMS OF EXPOSURE: Irrit eyes, throat; sneez, cough, dysp; pulm edema; conj; burns skin, muc memb; derm.

USES: Detergent soaps; accelerator for vulcanizing rubber; detection of magnesium; tanning; acid gas absorbent solvent; gasoline stabilizers; textile chemicals; pharmaceuticals; surfactants; manufacture of dimethylformamide and dimethylacetamide; rocket propellants; missile fuels; dehairing agent; electroplating.

p-DIMETHYLAMINOAZOBENZENE

SYNONYMS: Atul fast yellow R; Benzeneazodimethylaniline; Brilliant fast oil yellow; Brilliant fast spirit yellow; Brilliant fast yellow; Brilliant oil yellow; Butter yellow; Cerasine yellow CG; C.I. 11,020; C.I. solvent yellow 2; DAB; Dimethylaminobenzene; 4-Dimethylaminoazobenzene; 4-(*N,N*-Dimethylamino)-azobenzene; *N,N*-Dimethyl-4-aminozobenzene; *N,N*-Dimethyl-*p*-aminoazobenzene; Dimethylaminoazobenzol; 4-Dimethylaminoazobenzol; 4-Dimethylaminophenylazobenzene; *N,N*-Dimethyl-*p*-azoaniline; *N,N*-Dimethyl-4-(phenylazo)-benzamine; *N,N*-Dimethyl-*p*-(phenylazo)benzamine; ***N,N*-Dimethyl-4-(phenylazo)benzenamine**; *N,N*-Dimethyl-*p*-(phenylazo)benzenamine; Dimethyl yellow; Dimethyl yellow analar; Dimethyl yellow *N,N*-dimethylaniline; DMAB; Enial yellow 2G; Fast oil yellow B; Fast yellow; Fat yellow; Fat yellow A; Fat yellow AD OO; Fat yellow ES; Fat yellow ES extra; Fat yellow extra conc.; Fat yellow R; Fat yellow R (8186); Grasal brilliant yellow; Methyl yellow; Oil yellow; Oil yellow 20; Oil yellow 2,625; Oil yellow 7,463; Oil yellow BB; Oil yellow D; Oil yellow DN; Oil yellow FF; Oil yellow FN; Oil yellow G; Oil yellow G-2; Oil yellow 2G; Oil yellow GG; Oil yellow GR; Oil yellow II; Oil yellow N; Oil yellow PEL; Oleal yellow 2G; Organol yellow ADM; Orient oil yellow GG; PDAB; Petrol yellow WT; RCRA waste number U093; Resinol yellow GR; Resoform yellow GGA; Silotras yellow T2G; Somalia yellow A; Stear yellow JB; Sudan GG; Sudan yellow; Sudan yellow 2G; Sudan yellow 2GA; Toyo oil yellow G; USAF EK-338; Waxoline yellow AD; Waxoline yellow ADS; Yellow G soluble in grease.

CHEMICAL DESIGNATIONS: CAS: 60-11-7; DOT: none assigned; mf: $C_{14}H_{15}N_3$; fw: 225.30; RTECS: BX 7350000.

PROPERTIES: Yellow leaflets. Mp: 114-117°, bp: subl, *d*: 1.212 (calcd), log K_{oc}: 3.00 (calcd), log K_{ow}: 4.58, S_o: sol in acetic acid, acetone, ether, pyrimidine, alcohol and benzene, S_w: 13.6 mg/L @ 20-30°.

TRANSFORMATION PRODUCTS: Releases toxic NO_x when heated to decomposition.

SYMPTOMS OF EXPOSURE: Enlarged liv, hep and renal dysfunction, contact derm, cough, wheez, difficulty breath, bloody sputum, bronchial secretions, frequent urination, hema, dys.

USES: pH indicator; determining hydrochloric acid in gastric juice; coloring agent; organic research.

DIMETHYLANILINE

SYNONYMS: Dimethylaminobenzene; *N,N*-Dimethylaniline; ***N,N*-Dimethylbenzenamine**; Dimethylphenylamine; *N,N*-Dimethylphenylamine; NCI-C56428; UN 2253; Versneller NL 63/10.

CHEMICAL DESIGNATIONS: CAS: 121-69-7; DOT: 2253; mf: $C_8H_{11}N$; fw: 121.18; RTECS: BX 4725000.

PROPERTIES: Straw to brown colored liq with an amine-like odor. Mp: 2.45°, bp: 194°, *d*: 0.9557 @ 20/4°, fl p: 61°, lel: 1%, pK_a: 5.21 @ 25°, K_H: 4.98 x 10^{-6} atm·m^3/mol @ 20° (calcd), IP: 7.12 eV, log K_{ow}: 2.31, 2.62, S_o: sol in acetone, alcohol, benzene, chloroform and ether, S_w: 1,105.2 mg/L @ 25°, vap d: 4.95 g/L @ 25°, 4.18 (air = 1), vp: 0.5 mm @ 20°.

TRANSFORMATION PRODUCTS: Products identified from the gas-phase reaction of O_3 with *N,N*-dimethylaniline in synthetic air @ 23° were: *N*-methylformanilide, formaldehyde, formic acid, H_2O_2 and a nitrated salt having the formula: $[C_6H_6NH(CH_3)_2]^+NO_3^-$. Reacts with acids forming soluble salts.

EXPOSURE LIMITS: NIOSH REL: IDLH 100 ppm; OSHA PEL: TWA 5 ppm; ACGIH TLV: TWA 5 ppm, STEL 10 ppm.

SYMPTOMS OF EXPOSURE: Cyan, weak, dizz, atax.

USES: Manufacture of vanillin, Michler's ketone, methyl violet and other dyes; solvent; reagent for methyl alcohol, H_2O_2, methyl furfural, nitrate and formaldehyde; chemical intermediate; stabilizer; reagent.

2,2-DIMETHYLBUTANE

SYNONYMS: Neohexane; UN 1208.

CHEMICAL DESIGNATIONS: CAS: 75-83-2; DOT: 2457; mf: C_6H_{14}; fw: 86.18; RTECS: EJ 9300000.

PROPERTIES: Colorless liq. Mp: -99.9°, bp: 49.7°, *d*: 0.6485 @ 20/4°, fl p: -47.8°, lel: 1.2%, uel: 7.0%, K_H: 1.943 atm·m^3/mol @ 25°, IP: 10.06 eV, log K_{ow}: 3.82, S_o: sol in acetone, alcohol, benzene and ether, S_w: 18.4-23.8 ppm @ 25°, vap d: 3.52 g/L @ 25°, 2.98 (air = 1), vp: 319.1 mm @ 25°.

USES: Intermediate for agricultural chemicals; in high octane fuels.

2,3-DIMETHYLBUTANE

SYNONYMS: Biisopropyl; Diisopropyl; Isopropyldimethylmethane; UN 2457.

CHEMICAL DESIGNATIONS: CAS: 79-29-8; DOT: 2457; mf: C_6H_{14}; fw: 86.18; RTECS: EJ 9350000.

PROPERTIES: Colorless liq. Mp: -128.5°, *d*: 0.6616 @ 20/4°, S_o: sol in acetone, alcohol, benzene and ether, bp: 58°, fl p: -33°, lel: 1.2%, uel: 7.0%, K_H: 1.18 atm·m^3/mol @ 25° (calcd), IP: 10.02 eV, log K_{ow}: 3.85, S_w: 19.1 ppm @ 25°, vap d: 3.52 g/L @ 25°, 2.98 (air = 1), vp: 200 mm @ 20°.

TRANSFORMATION PRODUCTS: Major products reported from the photooxidation of 2,3-dimethylbutane with NO_x are CO and acetone. Minor products included formaldehyde, acetaldehyde and peroxyacyl nitrates. Synthetic air containing gaseous nitrous acid and exposed to artificial sunlight (λ = 300-450 nm) photooxidized 2,3-dimethylbutane into acetone, hexyl nitrate, peroxyacetal nitrate and a nitro aromatic compound tentatively identified as a propyl nitrate.

USES: Organic synthesis; gasoline component.

cis-1,2-DIMETHYLCYCLOHEXANE

SYNONYMS: *cis-o*-Dimethylcyclohexane; *cis*-1,2-Hexahydroxylene.

CHEMICAL DESIGNATIONS: CAS: 2207-01-4; DOT: 2263; mf: C_8H_{16}; fw: 112.22; RTECS: none assigned.

PROPERTIES: Colorless liq. Mp: -50.1°, bp: 129.7°, *d*: 0.7963 @ 20/4°, fl p: -12°, K_H: 3.54 x 10^{-1} atm·m^3/mol @ 25°, IP: 10.08 eV, log K_{ow}: 3.26 (calcd), S_o: sol in acetone, alcohol, benzene, ether and lignoin, S_w: 6.0 ppm @ 25°, vap d: 4.59 g/L @ 25°, 3.87 (air = 1), vp: 14.5 mm @ 25°.

USE: Organic synthesis.

trans-1,4-DIMETHYLCYCLOHEXANE

SYNONYM: *trans-p*-Dimethylcyclohexane.

CHEMICAL DESIGNATIONS: CAS: 6876-23-9; DOT: 2263; mf: C_8H_{16}; fw: 112.22; RTECS: none assigned.

PROPERTIES: Colorless liq. Mp: -37°, bp: 119.3°, *d*: 0.7626 @ 20/4°, fl p: ≈ 10° (isomeric mixture), K_H: 8.70 x 10^{-1} atm·m^3/mol @ 25° (calcd), IP: 10.08 eV, log K_{ow}: 3.41 (calcd), S_o: sol in acetone, alcohol, benzene, ether and lignoin, S_w: 3.84 ppm @ 25°, vap d: 4.59 g/L @ 25°, 3.87 (air = 1), vp: 22.65 mm @ 25°.

USE: Organic synthesis.

DIMETHYLFORMAMIDE

SYNONYMS: ***N,N*-Dimethylformamide**; DMF; DMFA; *N*-Formyldimethylamine; NCI-C60913; NSC 5,536; U-4224; UN 2265.

CHEMICAL DESIGNATIONS: CAS: 68-12-2; DOT: 2265; mf: C_3H_7NO; fw:

73.09; RTECS: LQ 2100000.

PROPERTIES: Colorless to light yellow mobile liq with a faint ammonia-like odor. Mp: -60.5°, bp: 149-156°, *d*: 0.9487 @ 20/4°, fl p: 57.8°, lel: 2.2% @ 100°, uel: 15.2%, IP: 9.12 eV, log K_{ow}: -1.01, S_o: misc with most organic solvents, S_w: a saturated soln in equilibrium with its own vapor had a concn of 5,294 g/L @ 25°, vap d: 2.99 g/L @ 25°, 2.52 (air = 1), vp: 2.7 mm @ 20°.

TRANSFORMATION PRODUCTS: Incubation of ^{14}C-*N,N*-Dimethylformamide (0.1-100 μg/L) in natural seawater resulted in the compound mineralizing to CO_2. The rate of CO_2 formation was inversely proportional to the initial concn.

EXPOSURE LIMITS: NIOSH REL: IDLH 3,500 ppm; OSHA PEL: TWA 10 ppm; ACGIH TLV: TWA 10 ppm.

SYMPTOMS OF EXPOSURE: Nau, vomit, colic; liv damage, heptaomegaly; high BP; face flush; derm.

USES: Solvent for liquids, gases and vinyl resins; polyacrylic fibers; gas carrier; catalyst in carboxylation reactions; organic synthesis.

1,1-DIMETHYLHYDRAZINE

SYNONYMS: UDMH; Dimazine; *asym*-Dimethylhydrazine; *unsym*-Dimethylhydrazine; *N,N*-Dimethylhydrazine.

CHEMICAL DESIGNATIONS: CAS: 57-14-7; DOT: none assigned; mf: $C_2H_8N_2$; fw: 60.10; RTECS: MV 2450000.

PROPERTIES: Colorless to yellow, fuming liq with an amine-like odor. Mp: -57.2°, bp: 63.9°, *d*: 0.7914 @ 22/4°, fl p: -15.0°, lel: 2%, uel: 95%, K_H: 2.45 x 10^{-9} atm·m^3/mol @ 25°, IP: 7.46 eV, 7.67 eV, log K_{oc}: -0.70 (calcd), log K_{ow}: -2.42, S_o: misc with alcohol, dimethylformamide, ether and hydrocarbons, S_w: misc, vap d: 2.46 g/L @ 25°, 2.07 (air = 1), vp: 103 mm @ 20°.

TRANSFORMATION PRODUCTS: Releases toxic NO_x when heated to decomposition. *N*-Nitrosodimethylamine was the major product of ozonation of 1,1-dimethylhydrazine in the dark. H_2O_2, methyl hydroperoxide and methyl diazene were also identified.

EXPOSURE LIMITS: NIOSH REL: 2-h C 0.15 mg/m^3; OSHA PEL: TWA 0.5 ppm; ACGIH TLV: TWA 0.5 ppm.

SYMPTOMS OF EXPOSURE: Irrit eyes; choking, chest pain, dysp; lethargy; nau; irrit skin; anoxia; convuls; liv inj.

USES: Rocket fuel formulations; stabilizer for organic peroxide fuel additives; absorbent for acid gases; plant control agent; photography; organic synthesis.

2,3-DIMETHYLPENTANE

SYNONYM: 3,4-Dimethylpentane.

CHEMICAL DESIGNATIONS: CAS: 565-59-3; DOT: none assigned; mf: C_7H_{16}; fw: 100.20; RTECS: none assigned.

PROPERTIES: Colorless liq. Bp: 89.8°, *d*: 0.6951 @ 20/4°, fl p: -6°, lel: 1.1%, uel: 6.7%, K_H: 1.73 atm·m^3/mol @ 25° (calcd), log K_{ow}: 3.26 (calcd), S_o: sol in acetone, alcohol, benzene, chloroform and ether, S_w: 5.25 ppm @ 25°, vap d: 4.10 g/L @ 25°, 3.46 (air = 1), vp: 68.9 mm @ 25°.

USES: Organic synthesis; gasoline component.

2,4-DIMETHYLPENTANE

SYNONYM: Diisopropylmethane.

CHEMICAL DESIGNATIONS: CAS: 108-08-7; DOT: none assigned; mf: C_7H_{16}; fw: 100.20; RTECS: none assigned.

PROPERTIES: Colorless liq. Mp: -119.2°, bp: 80.5°, *d*: 0.6727 @ 20/4°, fl p: -12.1°, K_H: 3.152 atm·m^3/mol @ 25°, log K_{ow}: 3.24 (calcd), S_o: sol in acetone, alcohol, benzene, chloroform and ether, S_w: 4.06-5.50 ppm @ 25°, vap d: 4.10 g/L @ 25°, 3.46 (air = 1), vp: 98.4 mm @ 25°.

USES: Organic synthesis; gasoline component.

3,3-DIMETHYLPENTANE

SYNONYMS: None.

CHEMICAL DESIGNATIONS: CAS: 562-49-2; DOT: none assigned; mf: C_7H_{16}; fw: 100.20; RTECS: none assigned.

PROPERTIES: Colorless liq. Mp: -134.4°, bp: 86.1°, *d*: 0.6936 @ 20/4°, fl p: -6°, K_H: 1.84 atm·m^3/mol @ 25° (calcd), log K_{ow}: 3.22 (calcd), S_o: sol in acetone, alcohol, benzene, chloroform and ether, S_w: 5.94 ppm @ 25°, vap d: 4.10 g/L @ 25°, 3.46 (air = 1), vp: 82.8 mm @ 25°.

USES: Organic synthesis; gasoline component.

2,4-DIMETHYLPHENOL

SYNONYMS: 4,6-Dimethylphenol; 2,4-DMP; 1-Hydroxy-2,4-dimethylbenzene; 4-Hydroxy-1,3-dimethylbenzene; RCRA waste number U101; 1,3,4-Xylenol; 2,4-Xylenol; *m*-Xylenol.

CHEMICAL DESIGNATIONS: CAS: 105-67-9; DOT: none assigned; mf: $C_8H_{10}O$; fw: 122.17; RTECS: ZE 5600000.

PROPERTIES: Colorless solid, slowly turning brown on exposure to air. Mp: 27-28°, bp: 210°, *d*: 0.9650 @ 20/4°, fl p: > 110°, pK_a: 10.63, K_H: 6.55 x 10^{-6} atm·m^3/mol @ 25° (est), log K_{oc}: 2.07 (est), log K_{ow}: 2.30-2.50, S_o: freely sol in ethanol, chloroform, ether and benzene, S_w: 4,200 mg/L @ 20°, vp: 9.8 x 10^{-2} mm @ 25.0°.

TRANSFORMATION PRODUCTS: Wet oxidation of 2,4-dimethylphenol @ 320° yielded formic and acetic acids.

USES: Wetting agent; dyestuffs; preparation of phenolic antioxidants; plastics, resins, solvent, disinfectant, pharmaceuticals, insecticides, fungicides and rubber chemicals manufacturing; lubricant and gasoline additive; possibly used a pesticide.

DIMETHYL PHTHALATE

SYNONYMS: Avolin; **1,2-Benzenedicarboxylic acid dimethyl ester**; Dimethyl-1,2-benzenedicarboxylate; Dimethylbenzeneorthodicarboxylate; DMP; ENT 262; Fermine; Methyl phthalate; Mipax; NTM; Palatinol M; Phthalic acid dimethyl ester; Phthalic acid methyl ester; RCRA waste number U102; Solvanom; Solvarone.

CHEMICAL DESIGNATIONS: CAS: 131-11-3; DOT: none assigned; mf: $C_{10}H_{10}O_4$; fw: 194.19; RTECS: TI 1575000.

PROPERTIES: Colorless, odorless, moderately viscous liq. Mp: 5.5°, bp: 283.8°, *d*: 1.1905 @ 20/4°, fl p: 146°, lel: 1.2% @ 146°, K_H: 4.2 x 10^{-7} atm·m^3/mol, IP: 9.75 eV, log K_{oc}: 0.88-2.28, log K_{ow}: 1.53-2.00, S_o: sol in ethanol, ether and benzene, S_w: 4,320 mg/L @ 25°, vap d: 7.94 g/L @ 25°, 6.70 (air = 1), vp: < 10^{-2} mm @ 20°.

TRANSFORMATION PRODUCTS: In anaerobic sludge, degradation occurred as follows: monomethyl phthalate to phthalic acid to protocatechuic acid followed by ring cleavage and mineralization. An aq soln containing titanium dioxide and subjected to UV light (λ > 290 nm) yielded mono and dihydroxyphthalates as intermediates. Hydrolyzes in water to phthalic acid and methyl alcohol.

EXPOSURE LIMITS: NIOSH REL: IDLH 9,300 mg/m^3; OSHA PEL: TWA 5 mg/m^3; ACGIH TLV: TWA 5 mg/m^3.

SYMPTOMS OF EXPOSURE: Irrit nasal passages, upper resp, stomach; eye pain.

USES: Plasticizer for cellulose acetate, nitrocellulose, resins, rubber, elastomers; ingredient in lacquers; coating agents; safety glass; insect repellant; molding powders; perfumes.

2,2-DIMETHYLPROPANE

SYNONYMS: Neopentane; *tert*-Pentane; Tetramethylmethane; UN 1265; UN 2044.

CHEMICAL DESIGNATIONS: CAS: 463-82-1; DOT: 2044; mf: C_5H_{12}; fw: 72.15; RTECS: none assigned.

PROPERTIES: Gas. Mp: -16.5°, bp: 9.5°, *d*: 0.6135 @ 20/4°, fl p: -65.0°, lel: 1.4%, uel: 7.5%, K_H: 2.18 atm·m^3/mol @ 25°, IP: 10.35 eV, log K_{ow}: 3.11, S_o: sol in alcohol and ether, S_w: 33.2 ppm @ 25°, vap d: 2.95 g/L @ 25°, 2.49 (air = 1), vp: 1,287 mm @ 25°.

USES: Butyl rubber; organic synthesis.

2,7-DIMETHYLQUINOLINE

SYNONYMS: None.

CHEMICAL DESIGNATIONS: CAS: 93-37-8; DOT: none assigned; mf: $C_{11}H_{11}N$; fw: 157.22; RTECS: none assigned.

PROPERTIES: Liq. Mp: 57-59°, bp: 262-265°, S_o: sol in alcohol, ether, chloroform and benzene, *d*: 1.054 (calcd), S_w: 1,795 ppm @ 25°.

USES: Organic synthesis; dye intermediate.

DIMETHYL SULFATE

SYNONYMS: Dimethyl monosulfate; DMS; Methyl sulfate; RCRA waste number U103; **Sulfuric acid dimethyl ester**; UN 1595.

CHEMICAL DESIGNATIONS: CAS: 77-78-1; DOT: 1595; mf: $C_2H_6O_4S$; fw: 126.13; RTECS: WS 8225000.

PROPERTIES: Colorless, oily liq with an onion-like odor. Mp: -31.7°, bp: 188.5° (decomp), *d*: 1.3283 @ 20/4°, fl p: 83.3°, K_H: 2.96 x 10^{-6} atm·m^3/mol @ 20° (calcd), log K_{oc}: 0.61 (calcd), log K_{ow}: -1.24, S_o: sol in alcohol, benzene, ether, dioxane and aromatic hydrocarbons, S_w: 28 g/L @ 18°, vap d: 5.16 g/L @ 25°, 4.35 (air = 1), vp: 0.5 mm @ 20°.

TRANSFORMATION PRODUCTS: Hydrolyzes in water to methyl alcohol and sulfuric acid ($t_{1/2}$ = 1.2 h).

EXPOSURE LIMITS: NIOSH REL: IDLH 10 ppm; OSHA PEL: TWA 1 ppm; ACGIH TLV: TWA 0.1 ppm.

SYMPTOMS OF EXPOSURE: Irrit eyes, nose, throat; head; gidd; photo; periob edema; dysphonia, aphonia, dysphagia cough; chest pain; cyan; vomit; diarr dys; ict, album hema.

USES: In organic synthesis as a methylating agent.

1,2-DINITROBENZENE

SYNONYMS: *o*-Dinitrobenzene; *o*-Dinitrobenzol; UN 1597.

CHEMICAL DESIGNATIONS: CAS: 528-29-0; DOT: 1597; mf: $C_6H_4N_2O_4$; fw: 168.11; RTECS: CZ 7450000.

PROPERTIES: Colorless needles. Mp: 118.5°, bp: 319° @ 775 mm, *d*: 1.565 @ 17/4°, fl p: 150°, K_H: < 1.47 x 10^{-3} atm·m^3/mol @ 20° (calcd), log K_{ow}: 1.58, S_o: sol in alcohol (≈ 16.7 g/L), benzene (50 g/L); freely sol in chloroform and ethyl acetate, S_w: 0.015 wt % @ 20°, vp: < 1 mm @ 20°.

TRANSFORMATION PRODUCTS: Under anaerobic and aerobic conditions using a sewage inoculum, 1,2- and 1,3-dinitrobenzene degraded to nitroaniline. Releases toxic NO_x when heated to decomposition.

EXPOSURE LIMITS: NIOSH REL: IDLH 200 mg/m^3; OSHA PEL: TWA 1 mg/m^3.

SYMPTOMS OF EXPOSURE: Anoxia, cyan; vis dist, central scotomas; bad taste, burning mouth, dry throat; yellowing hair, eyes, skin, anem.

USES: Organic synthesis; dyes.

1,3-DINITROBENZENE

SYNONYMS: Binitrobenzene; 2,4-Dinitrobenzene; *m*-Dinitrobenzene; 1,3-Dinitrobenzol; UN 1597.

CHEMICAL DESIGNATIONS: CAS: 99-65-0; DOT: 1597; mf: $C_6H_4N_2O_4$; fw: 168.11; RTECS: CZ 7350000.

PROPERTIES: Yellowish cryst. Mp: 90°, bp: 300-303°, *d*: 1.5751 @ 18/4°, fl p: explodes, K_H: 2.75 x 10^{-7} atm·m^3/mol @ 35° (calcd), log K_{oc}: 2.18 (calcd), log K_{ow}: 1.49, S_o: sol in acetone, ether, pyrimidine, alcohol (27 g/L); freely sol in benzene, chloroform and ethyl acetate, S_w: 469 mg/L @ 15°, vp: 8.15 x 10^{-4} mm @ 35°.

TRANSFORMATION PRODUCTS: Under anaerobic and aerobic conditions using a sewage inoculum, 1,2- and 1,3-dinitrobenzene degraded to nitroaniline. Releases toxic NO_x when heated to decomposition.

EXPOSURE LIMITS: NIOSH REL: IDLH 200 mg/m^3; OSHA PEL: TWA 1 mg/m^3.

SYMPTOMS OF EXPOSURE: Anoxia, cyan; vis dist, central scotomas; bad taste, buring mouth, dry throat; yellowing hair, eyes, skin, anem.

USES: Organic synthesis; dyes.

1,4-DINITROBENZENE

SYNONYMS: *p*-Dinitrobenzene; Dithane A-4; UN 1597.

CHEMICAL DESIGNATIONS: CAS: 100-25-4; DOT: 1597; mf: $C_6H_4N_2O_4$; fw: 168.11; RTECS: CZ 7525000.

PROPERTIES: White cryst. Mp: 174°, bp: 298° @ 777 mm, *d*: 1.625 @ 18/4°, fl p: explodes, K_H: 4.79 x 10^{-7} atm·m^3/mol @ 35° (calcd), log K_{ow}: 1.46, 1.49, S_o: sol in acetone, acetic acid, benzene, toluene and alcohol (3.3 g/L), S_w: 0.01 wt % @ 20°, vp: 2.25 x 10^{-4} mm @ 35°.

TRANSFORMATION PRODUCTS: Releases toxic NO_x when heated to decomposition.

EXPOSURE LIMITS: NIOSH REL: IDLH 200 mg/m^3; OSHA PEL: TWA 1 mg/m^3.

SYMPTOMS OF EXPOSURE: Anoxia, cyan; vis dist, central scotomas; bad taste, buring mouth, dry throat; yellowing hair, eyes, skin, anem.

USES: Organic synthesis; dyes.

4,6-DINITRO-*o*-CRESOL

SYNONYMS: Antinonin; Antinonnon; Arborol; Capsine; Chemsect DNOC; Degrassan; Dekrysil; Detal; Dinitrocresol; Dinitro-*o*-cresol; 2,4-Dinitro-*o*-cresol; 3,5-Dinitro-*o*-cresol; Dinitrodendtroxal; 3,5-Dinitro-2-hydroxytoluene; Dinitrol; Dinitromethyl cyclohexyltrienol; 2,4-Dinitro-2-methylphenol; 2,4-Dinitro-6-methylphenol; 4,6-Dinitro-2-methylphenol; Dinitrosol; Dinoc; Dinurania; DN; DNC; DN-dry mix no. 2; DNOC; Effusan; Effusan 3,436; Elgetol; Elgetol 30; Elipol; ENT 154; Extrar; Hedolit; Hedolite; K III; K IV; Kresamone; Krezotol 50; Lipan; **2-Methyl-4,6-dinitrophenol**; 6-Methyl-2,4-dinitrophenol; Nitrador; Nitrofan; Prokarbol; Rafex; Rafex 35; Raphatox; RCRA waste number P047; Sandolin; Sandolin A; Selinon; Sinox; Trifina; Trifocide; Winterwash.

CHEMICAL DESIGNATIONS: CAS: 534-52-1; DOT: 1598; mf: $C_7H_6N_2O_5$; fw: 198.14; RTECS: GO 9625000.

PROPERTIES: Yellow, odorless cryst. Mp: 86.5°, bp: 312°, pK_a: 4.35-4.46, K_H: 1.4 x 10^{-6} atm·m^3/mol @ 25°, log K_{oc}: 2.64 (calcd), log K_{ow}: 2.12-2.85, S_o @ 15° (mg/L): methanol (7.33), ethanol (9.12),chloroform (37.2), acetone (100.6); slightly sol in petroleum ether, S_w: 198 mg/L @ 20°, vp: 5 x 10^{-5} mm @ 20°.

EXPOSURE LIMITS: NIOSH REL: 10-h TWA 0.2 mg/m^3, IDLH 5 mg/m^3; OSHA PEL: TWA 0.2 mg/m^3.

SYMPTOMS OF EXPOSURE: Sense of well being, head, fvr, lass, profuse sweat, excess thirst, tacar, cough, short breath, coma.

USES: Dormant ovicidal spray for fruit trees (highly phototoxic and cannot be used successfully on actively growing plants); herbicide; insecticide.

2,4-DINITROPHENOL

SYNONYMS: Aldifen; Chemox PE; α-Dinitrophenol; DNP; 2,4-DNP; Fenoxyl carbon n; 1-Hydroxy-2,4-dinitrobenzene; Maroxol-50; Nitro kleenup; NSC 1,532; RCRA waste number P048; Solfo black B; Solfo black BB; Solfo black 2B supra; Solfo black G; Solfo black SB; Tetrasulphur black PB; Tetrosulphur PBR.

CHEMICAL DESIGNATIONS: CAS: 51-28-5; DOT: 0076; mf: $C_6H_4N_2O_5$; fw: 184.11; RTECS: SL 2800000.

PROPERTIES: Yellow cryst with a sweet, musty odor. Mp: 115-116°, bp: subl, *d*: 1.68 @ 20/4°, pK_a: 4.09 @ 25°, K_H: 1.57 x 10^{-8} atm·m^3/mol @ 18-20° (calcd), log K_{oc}: 1.25 (est), log K_{ow}: 1.51-1.67, S_o @ 15° (wt %): ethyl acetate (13.46), acetone (26.42), chloroform (5.11), pyridine (16.72), carbon tetrachloride (0.42), toluene (5.98), ether, benzene and alcohol (30.5 g/L), S_w: 6 g/L @ 25°, vp: 3.9 x 10^{-4} mm @ 20°.

TRANSFORMATION PRODUCTS: Ozonation of an aq soln containing 2,4-dinitrophenol (100 mg/L) yielded formic, acetic, glyoxylic and oxalic acids. When an aq soln containing 2,4-dinitrophenol and titanium dioxide was illuminated by UV light, ammonium and nitrate ions formed as the major products.

USES: Manufacture of pesticides, herbicides, explosives and wood preservatives; yellow dyes; preparation of picric acid and diaminophenol (photographic developer); indicator; reagent for potassium and ammonium ions; insecticide.

2,4-DINITROTOLUENE

SYNONYMS: 2,4-Dinitromethylbenzene; Dinitrotoluol; 2,4-Dinitrotoluol; DNT; 2,4-DNT; **1-Methyl-2,4-dinitrobenzene**; NCI-C01865; RCRA waste number U105.

CHEMICAL DESIGNATIONS: CAS: 121-14-2; DOT: 1600 (liq), 2038 (solid); mf: $C_7H_6N_2O_4$; fw: 182.14; RTECS: XT 1575000.

PROPERTIES: Yellow to red needles or yellow liq with a slight odor. Mp: 67-70°, bp: 250°, 300° (decomp), *d*: 1.379 @ 20/4°, fl p: 206.7°, K_H: 8.67 x 10^{-7}

atm·m^3/mol, log K_{oc}: 1.79 (calcd), log K_{ow}: 1.98, S_o: sol in acetone, ethanol, benzene, ether and pyrimidine, S_w: 270 mg/L @ 22°, vp: 5.1 x 10^{-3} mm @ 20°.

TRANSFORMATION PRODUCTS: Wet oxidation of 2,4-dinitrotoluene @ 320° yielded formic and acetic acids.

EXPOSURE LIMITS: ACGIH TLV: TWA 1.5 mg/m^3.

USES: Organic synthesis; intermediate for toluidine, dyes and explosives.

2,6-DINITROTOLUENE

SYNONYMS: 2,6-Dinitromethylbenzene; 2,6-Dinitrotoluol; 2,6-DNT; **2-Methyl-1,3-dinitrobenzene**; RCRA waste number U106.

CHEMICAL DESIGNATIONS: CAS: 606-20-2; DOT: 1600 (liq), 2038 (solid); mf: $C_7H_6N_2O_4$; fw: 182.14; RTECS: XT 1925000.

PROPERTIES: Yellow cryst. Mp: 60.5°, bp: 285°, *d*: 1.2833 @ 111/4°, fl p: 206.7° (calcd), K_H: 2.17 x 10^{-7} atm·m^3/mol, log K_{oc}: 1.79 (calcd), log K_{ow}: 2.00, S_o: sol in ethanol, S_w: ≈ 300 mg/L, vp: 3.5 x 10^{-4} mm @ 20°.

TRANSFORMATION PRODUCTS: Under anaerobic and aerobic conditions, a sewage inoculum degraded 2,6-dinitrotoluene to aminonitrotoluene.

USES: Organic synthesis; manufacture of explosives.

DI-*n*-OCTYL PHTHALATE

SYNONYMS: **1,2-Benzenedicarboxylic acid dioctyl ester**; 1,2-Benzenedicarboxylic acid di-*n*-octyl ester; *o*-Benzenedicarboxylic acid dioctyl ester; Celluflex DOP; Dinopol NOP; Dioctyl-*o*-benzenedicarboxylate; Dioctyl phthalate; *n*-Dioctyl phthalate; DNOP; DOP; Octyl phthalate; *n*-Octyl phthalate; Polycizer 162; PX-138; RCRA waste number U107; Vinicizer 85.

CHEMICAL DESIGNATIONS: CAS: 117-84-0; DOT: none assigned; mf: $C_{24}H_{38}O_4$; fw: 390.57; RTECS: TI 1925000.

PROPERTIES: Clear, viscous liq with a slight odor. Mp: -30°, bp: 386°, *d*: 0.978 @ 20/4°, fl p: 218.3°, K_H: 1.41 x 10^{-12} atm·m^3/mol @ 25° (calcd), log K_{oc}: 8.99 (calcd), log K_{ow}: 9.2 (calcd), S_w: 3.0 mg/L @ 25°, vap d: 16.00 g/L @ 25°, 13.52 (air = 1), vp: 1.4 x 10^{-4} mm @ 25° (est).

TRANSFORMATION PRODUCTS: *o*-Phthalic acid was tentatively identified as the major biodegradation product of di-*n*-octyl phthalate produced by the bacterium *Serratia marcescens*. Hydrolyzes in water to phthalic acid and *n*-octyl alcohol.

USE: Plasticizer for polyvinyl chloride (PVC) and other vinyl polymers.

DIOXANE

SYNONYMS: Diethylene dioxide; 1,4-Diethylene dioxide; Diethylene ether; Diethylene oxide; Diokan; 1,4-Dioxacyclohexane; Dioxan; Dioxane-1,4; **1,4-**

Dioxane; *p*-Dioxane; Dioxyethylene ether; Glycol ethylene ether; NCI-C03689; RCRA waste number U108; Tetrahydro-1,4-dioxin; Tetrahydro-*p*-dioxin; UN 1165.

CHEMICAL DESIGNATIONS: CAS: 123-91-1; DOT: 1165; mf: $C_4H_8O_2$; fw: 88.11; RTECS: JG 8225000.

PROPERTIES: Colorless liq with a pleasant odor. Mp: 11.8°, bp: 101.1°, *d*: 1.0337 @ 20/4°, fl p: 5-18°, lel: 2%, uel: 22%, K_H: 4.88 x 10^{-6} atm·m^3/mol @ 25°, IP: 9.13 eV, log K_{oc}: 0.54 (calcd), log K_{ow}: -0.42, -0.27, S_o: sol in acetone, alcohol, benzene and ether; misc with most organic solvents, S_w: misc, vap d: 3.60 g/L @ 25°, 3.04 (air = 1), vp: 29 mm @ 20°.

TRANSFORMATION PRODUCTS: Anticipated products from the reaction of dioxane with O_3 or hydroxyl radicals in the atmosphere are glyoxylic acid, oxygenated formates and $OHCOCH_2CH_2OCHO$. Irradiation of pure dioxane through quartz using a 450-W medium-pressure mercury lamp gave meso and racemic forms of 1-hydroxyethyldioxane, a pair of diastereomeric dioxane dimers, dioxanone, dioxanol, hydroxymethyldioxane and hydroxyethylidenedioxane. When dioxane is subjected to a megawatt ruby laser, 4% was decomposed yielding ethylene, CO, hydrogen and a trace of formaldehyde. Storage of dioxane in the presence of air resulted in the formation of 1,2-ethanediol monoformate and 1,2-ethane diformate.

EXPOSURE LIMITS: NIOSH REL: 30-min C 1 ppm; OSHA PEL: TWA 100 ppm; ACGIH TLV: TWA 25 ppm.

SYMPTOMS OF EXPOSURE: Drow, head; nau, vomit; irrit eyes, nose, throat; liv damage; kid failure; skin irrit.

USES: Solvent for cellulose acetate, benzyl cellulose, ethyl cellulose, waxes, resins, oils and many other organic compounds; cements, cosmetics, deodorants; fumigants; paint and varnish removers, cleaning and detergent preparations; wetting and dispersing agent in textile processing, dyes baths, stain and printing compositions; polishing compositions; stabilizer for chlorinated solvents; scintillation counter; organic synthesis.

1,2-DIPHENYLHYDRAZINE

SYNONYMS: *N,N'*-Bianiline; *N,N'*-Diphenylhydrazine; *sym*-Diphenylhydrazine; DPH; Hydrazobenzene; 1,1'-Hydrazobenzene; Hydrazodibenzene; NCI-C01854; RCRA waste number U109.

CHEMICAL DESIGNATIONS: CAS: 122-66-7; DOT: none assigned; mf: $C_{12}H_{12}N_2$; fw: 184.24; RTECS: MW 2625000.

PROPERTIES: Colorless to pale yellow cryst. Mp: 126-128° (decomp), *d*: 1.158 @ 16/4°, K_H: 4.11 x 10^{-11} atm·m^3/mol @ 25° (calcd), log K_{oc}: 2.82 (calcd), log K_{ow}: 2.94, S_o: sol in ethanol, S_w: 221 mg/L @ 25°, vp: 2.6 x 10^{-5} mm @ 25°.

TRANSFORMATION PRODUCTS: Wet oxidation of 1,2-diphenylhydrazine @ 320° yielded formic and acetic acids. Releases toxic fumes of NO_x when heated to decomposition.

USES: Manufacture of benzidine and starting material for pharmaceutical drugs.

DIURON

SYNONYMS: AF 101; Cekiuron; Crisuron; Dailon; DCMU; Diater; Dichlorfenidim; 3-(3,4-Dichlorophenol)-1,1-dimethylurea; 3-(3,4-Dichlorophenyl)-1,1-dimethylurea; ***N′*-(3,4-Dichlorophenyl)-*N,N*-dimethylurea**; 1,1-Dimethyl-3-(3,4-dichlorophenyl)urea; Di-on; Direx 4L; Diurex; Diurol; DMU; Dynex; Farmco diuron; Herbatox; HW 920; Karmex; Karmex diuron herbicide; Karmex DW; Marmer; NA 2,767; Sup'r flo; Telvar; Telvar diuron weed killer; Unidron; Urox D; USAF P-7; USAF XR-42; Vonduron.

CHEMICAL DESIGNATIONS: CAS: 330-54-1; DOT: 2767; mf: $C_9H_{10}Cl_2N_2O$; fw: 233.11; RTECS: YS 8925000.

PROPERTIES: White cryst solid. Mp: 150-155°, bp: 180° (decomp), *d*: 1.385 (calcd), pK_a: -1 to -2, K_H: 1.46 x 10^{-9} atm·m^3/mol @ 25-30° (calcd), log K_{oc}: 2.21-2.87, log K_{ow}: 1.97-2.68, S_o: acetone 5.3 wt % @ 27°, S_w: 40 ppm @ 20°, vp: 2 x 10^{-7} mm @ 30°.

TRANSFORMATION PRODUCTS: Radiolabeled diuron degraded in aerobic soils to 3-(3,4-dichlorophenyl)-1-methylurea and 3-(3,4-dichlorophenyl)urea. 3,4-Dichloroaniline was reported as a minor degradation product of diuron. Incubation of diuron in soils releases CO_2. The rate of CO_2 formation nearly tripled when the soil temperature was increased from 25 to 35°. Half-lives (soil temperature) of diuron in an Adkins loamy sand were: 705 (25), 414 (30) and 225 days (35). However, in a Semiahoo mucky peat, the half-lives were considerable higher: 3,991, 2,164 and 1,165 days @ 25, 30 and 35°, respectively.

USES: Pre-emergence herbicide; sugar cane flowering depressant.

n-DODECANE

SYNONYMS: Adakane 12; Bihexyl; Dihexyl; **Dodecane**; Duodecane.

CHEMICAL DESIGNATIONS: CAS: 112-40-3; DOT: none assigned; mf: $C_{12}H_{26}$; fw: 174.34; RTECS: JR 2125000.

PROPERTIES: Colorless liq. Mp: -12°, bp: 216.3°, *d*: 0.7487 @ 20/4°, fl p: 71.1°, lel: 0.6%, K_H: 24.2 atm·m^3/mol @ 25° (calcd), log K_{ow}: 5.64, 7.24, S_o: sol in acetone, alcohol, chloroform and ether, S_w: 3.4 μg/L @ 25°, vap d: 7.13 g/L @ 25°, 6.02 (air = 1), vp: 0.3 mm @ 20°.

TRANSFORMATION PRODUCTS: *n*-Dodecane may biodegrade in 2 ways. The first is the formation of dodecyl hydroperoxide which decomposes to *n*-dodecanol followed by oxidation to dodecanoic acid. The other pathway involves dehydrogenation to 1-dodecene which may react with water giving *n*-dodecanol.

USES: Solvent; jet fuel research; rubber industry; paper processing; standardized hydrocarbon; distillation chaser; gasoline component; organic synthesis.

α-ENDOSULFAN

SYNONYMS: Benzoepin; Beosit; Bio 5,462; Chlorthiepin; Crisulfan; Cyclodan;

Endocel; Endosol; Endosulfan; Endosulfan I; Endosulphan; ENT 23,979; FMC 5,462; 1,2,3,7,7-Hexachlorobicyclo[2.2.1]-2-heptene-5,6-bisoxymethylene sulfite; α,β-1,2,3,7,7-Hexachlorobicyclo[2.2.1]-2-heptene-5,6-bisoxymethylene sulfite; Hexachlorohexahydromethano-2,4,3-benzodioxathiepin-3-oxide; **(3α,5aβ,6α,9α,9aβ)-6,7,8,9,10,10-Hexachloro-1,5,5a,6,9,9a-hexahydro-6,9-methano-2,4,3-benzodioxathiepin-3-oxide**; 1,4,5,6,7,7-Hexachloro-5-norborene-2,3-dimethanol cyclic sulfite; Hildan; HOE 2,671; Insectophene; KOP-thiodan; Malix; NCI-C00566; NIA 5,462; Niagara 5,462; OMS-570; RCRA waste number P050; Thifor; Thimul; Thiodan; Thiofor; Thiomul; Thionex; Thiosulfan; Tionel; Tiovel.

CHEMICAL DESIGNATIONS: CAS: 959-98-8; DOT: 2761; mf: $C_9H_6Cl_6O_3S$; fw: 406.92; RTECS: RB 9275000.

PROPERTIES: Colorless to brown cryst with a sulfur dioxide odor. Mp: 106°, *d*: 1.745 @ 20/4°, K_H: 1.01 x 10^{-4} atm·m^3/mol @ 25°, log K_{oc}: 3.31 (calcd), log K_{ow}: 3.55, S_w: 530 ppb @ 25°, vp: 10^{-5} mm @ 25°.

TRANSFORMATION PRODUCTS: Metabolites of endosulfan identified in 7 soils were: enodiol, endohydroxy ether, endolactone and endosulfan sulfate. Endosulfan sulfate was the major biodegradation product in soils under aerobic, anaerobic and flooded conditions. In flooded soils, endolactone was detected only once whereas endodiol and endohydroxy ether were identified in all soils under these conditions. Under anaerobic conditions, endodiol formed in low amounts in 2 soils. Indigenous microorganisms obtained from a sandy loam degraded α-endosulfan to endosulfan diol. This diol was converted to endosulfan α-hydroxy ether and trace amounts of endosulfan ether and both were degraded to endosulfan lactone. Thin films of endosulfan on glass and irradiated by UV light (λ > 300 nm) produced endosulfan diol with minor amounts of endosulfan ether, lactone, α-hydroxyether and other unidentified compounds.

USE: Insecticide for vegetable crops.

β-ENDOSULFAN

SYNONYMS: Benzoepin; Beosit; Bio 5,462; Chlorthiepin; Crisulfan; Cyclodan; Endocel; Endosol; Endosulfan; Endosulfan II; Endosulphan; ENT 23,979; FMC 5,462; 1,2,3,7,7-Hexachlorobicyclo[2.2.1]-2-heptene-5,6-bisoxymethylene sulfite; α,β-1,2,3,7,7-Hexachlorobicyclo[2.2.1]-2-heptene-5,6-bisoxymethylene sulfite; Hexachlorohexahydromethano-2,4,3-benzodioxathiepin-3-oxide; **(3α,5aα,6β,9β,9aα)-6,7,8,9,10,10-Hexachloro-1,5,5a,6,9,9a-hexahydro-6,9-methano-2,4,3-benzodioxathiepin-3-oxide**; 1,4,5,6,7,7-Hexachloro-5-norborene-2,3-dimethanol cyclic sulfite; Hildan; HOE 2,671; Insectophene; KOP-thiodan; Malix; NCI-C00566; NIA 5,462; Niagara 5,462; OMS-570; RCRA waste number P050; Thifor; Thimul; Thiomul; Thiodan; Thiofor; Thionex; Thiosulfan; Tionel; Tiovel.

CHEMICAL DESIGNATIONS: CAS: 33213-65-9; DOT: 2761; mf: $C_9H_6Cl_6O_3S$; fw: 406.92; RTECS: RB 9275000.

PROPERTIES: Colorless to brown cryst with a sulfur dioxide odor. Mp: 207-

209°, *d*: 1.745 @ 20/20°, K_H: 1.91 x 10^{-5} atm·m^3/mol @ 25° (calcd), log K_{oc}: 3.37 (calcd), log K_{ow}: 3.62, S_w: 280 ppb @ 25°, vp: 10^{-5} mm @ 25°.

TRANSFORMATION PRODUCTS: Metabolites of endosulfan identified in 7 soils were: enodiol, endohydroxy ether, endolactone and endosulfan sulfate. Endosulfan sulfate was the major biodegradation product in soils under aerobic, anaerobic and flooded conditions. In flooded soils, endolactone was detected only once whereas endodiol and endohydroxy ether were identified in all soils under these conditions. Under anaerobic conditions, endodiol formed in low amounts in 2 soils. Indigenous microorganisms obtained from a sandy loam degraded β-endosulfan to endosulfan diol. This diol was converted to endosulfan α-hydroxy ether and trace amounts of endosulfan ether and both were degraded to endosulfan lactone. Thin films of endosulfan on glass and irradiated by UV light (λ > 300 nm) produced endosulfan diol with minor amounts of endosulfan ether, lactone, α-hydroxyether and other unidentified compounds. Gaseous β-endosulfan subjected to UV light (λ > 300 nm) produced endosulfan ether, endosulfan diol, endosulfan sulfate, endosulfan lactone, α-endosulfan and a dechlorinated ether. Irradiation of β-endosulfan in hexane by UV light produced the photoisomer α-endosulfan. Endosulfan detected in Little Miami River, OH was readily hydrolyzed and tentatively identified as endosulfan diol.

USE: Insecticide for vegetable crops.

ENDOSULFAN SULFATE

SYNONYMS: 6,7,8,9,10,10-Hexachloro-1,5,5a,6,9,9a-hexahydro-3,3-dioxide; 6,9-Methano-2,4,3-benzodioxathiepin.

CHEMICAL DESIGNATIONS: CAS: 1031-07-8; DOT: 2761; mf: $C_9H_6Cl_6O_4S$; fw: 422.92; RTECS: none assigned.

PROPERTIES: Solid. Mp: 198-201°, log K_{oc}: 3.37 (calcd), log K_{ow}: 3.66, S_w: 117 ppb.

TRANSFORMATION PRODUCTS: A mixed culture of soil microorganisms biodegraded endosulfan sulfate to endosulfan ether, endosulfan-α-hydroxy ether and endosulfan lactone. Indigenous microorganisms obtained from a sandy loam degraded endosulfan sulfate (a metabolite of α- and β-endosulfan) to endosulfan diol. This diol was converted to endosulfan α-hydroxy ether and trace amounts of endosulfan ether and both were degraded to endosulfan lactone.

USES: Not known.

ENDRIN

SYNONYMS: Compound 269; Endrex; ENT 17,251; Experimental insecticide no. 269; Hexachloroepoxyoctahydro-*endo-endo*-dimethanonaphthalene; 1,2,3,4,10,10-Hexachloro-6,7-epoxy-1,4,4a,5,6,7,8,8a-octahydro-*endo,endo*-1,4:5,8-dimethanonaphthalene; **3,4,5,6,9,9-Hexachloro-1a,2,2a,3,6,6a,7,7a-octahydro-2,7:3,6-dimethanonaphth[2,3-*b*]oxirene**; Hexadrin; Isodrin epoxide; Mendrin; NA 2,761;

NCI-C00157; Nendrin; RCRA waste number P051.

CHEMICAL DESIGNATIONS: CAS: 72-20-8; DOT: 2761; mf: $C_{12}H_8Cl_6O$; fw: 380.92; RTECS: IO 1575000.

PROPERTIES: White, odorless, cryst solid when pure; light tan color with faint chemical odor for tech grade. Mp: 200°, bp: 245° (decomp), *d*: 1.65 @ 25/4°, fl p: > 26.6° (xylene soln), lel: 1.1% in xylene, uel: 7.0% in xylene, K_H: 5.0 x 10^{-7} atm·m^3/mol, log K_{ow}: 3.209-5.339, S_o @ 25° (g/L): acetone (170), benzene (138), carbon tetrachloride (33), hexane (71), xylene (183), S_w: 220-260 ppb @ 25°, vp: 7 x 10^{-7} mm @ 25°.

TRANSFORMATION PRODUCTS: Microbial degradation in soil formed several ketones and aldehydes of which *keto*-endrin was the only metabolite identified. Photolysis of thin films of solid endrin using UV light (λ = 254 nm) produced δ-ketoendrin and endrin aldehyde and other compounds. Endrin exposed to sunlight for 17 days completely isomerized to δ-ketoendrin or 1,8-*exo*-9,10,11,11-hexachlorocyclo[6.2.1.1^{3,6}.0^{2,7}.0^{4,10}]dodecan-5-one. Irradiation of endrin by UV light (λ = 253.7 nm and 300 nm) or by natural sunlight in cyclohexane and hexane soln resulted in an 80% yield of 1,8-*exo*-9,11,11-pentachloropentacyclo[6.2.1.1^{3,6}.0^{2,7}.0^{4,10}]dodecan-5-one.

EXPOSURE LIMITS: NIOSH REL: IDLH 200 mg/m^3; OSHA PEL: TWA 0.1 mg/m^3; ACGIH TLV: TWA 0.1 mg/m^3.

SYMPTOMS OF EXPOSURE: Epileptiform convul, stupor, head, dizz, abdom discomfort, nau, vomit, insom, aggressive conf, lethargy, weak, anor.

USE: Insecticide.

ENDRIN ALDEHYDE

SYNONYMS: **2,2a,3,3,4,7-Hexachlorodecahydro-1,2,4-methenocyclopenta[*c,d*]pentalene-5-carboxaldehyde.**

CHEMICAL DESIGNATIONS: CAS: 7421-93-4; DOT: 2761; mf: $C_{12}H_8Cl_6O$; fw: 380.92; RTECS: none assigned.

PROPERTIES: Solid. Mp: 145-149°, bp: 235° (decomp), K_H: 3.86 x 10^{-7} atm·m^3/mol 25° (calcd), log K_{oc}: 4.43 (calcd), log K_{ow}: 5.6 (calcd), S_w: 260 ppb @ 25°, vp: 2 x 10^{-7} mm @ 25°.

USES: Not known.

EPICHLOROHYDRIN

SYNONYMS: 1-Chloro-2,3-epoxypropane; 3-Chloro-1,2-epoxypropane; (Chloromethyl)ethylene oxide; **(Chloromethyl)oxirane**; 2-(Chloromethyl)oxirane; 2-Chloropropylene oxide; γ-Chloropropylene oxide; 3-Chloro-1,2-propylene oxide; ECH; α-Epichlorohydrin; (*dl*)-α-Epichlorohydrin; Epichlorophydrin; 1,2-Epoxy-3-chloropropane; 2,3-Epoxypropyl chloride; Glycerol epichlorohydrin; RCRA waste number U041; UN 2023.

CHEMICAL DESIGNATIONS: CAS: 106-89-8; DOT: 2023; mf: C_3H_5ClO; fw:

92.53; RTECS: TX 4900000.

PROPERTIES: Colorless, mobile liq with a strong chloroform-like odor. Mp: -48°, bp: 115-117°, *d*: 1.1801 @ 20/4°, fl p: 31° (OC), lel: 3.8%, uel: 21.0%, K_H: 2.38-2.54 x 10^{-5} atm·m^3/mol @ 20° (calcd), IP: 10.64 eV, log K_{oc}: 1.00 (calcd), log K_{ow}: 0.45, S_o: sol in benzene; misc with alcohol, carbon tetrachloride, chloroform, ether and tetrachloroethylene, S_w: 60 g/L @ 20°, vap d: 3.78 g/L @ 25°, 3.19 (air = 1), vp: 12.5 mm @ 20°.

TRANSFORMATION PRODUCTS: Anticipated products from the reaction of epichlorohydrin with O_3 or hydroxyl radicals in the atmosphere are formaldehyde, glyoxylic acid and $ClCH_2C(O)OHCHO$. Hydrolyzes in water forming 1-chloro-2,3-hydroxypropane ($t_{1/2}$ = 8.2 days @ 20° and pH 7). Releases toxic fumes of NO_x when heated to decomposition.

EXPOSURE LIMITS: NIOSH REL: lowest feasible limit; OSHA PEL: TWA 5 ppm; ACGIH TLV: TWA 2 ppm.

SYMPTOMS OF EXPOSURE: Nau, vomit; abdom pain; resp distress; cough; cyan; irrit eyes, skin with deep pain.

USES: Solvent for natural and synthetic resins, cellulose esters and ethers, gums, paints, varnishes, lacquers and nail enamels; manufacture of epoxy resins, surface active agents, pharmaceuticals, insecticides, adhesives, coatings, plasticizers, glycidyl ethers, ion-exchange resins and fatty acid derivatives; organic synthesis.

EPN

SYNONYMS: ENT 17,798; EPN 300; Ethoxy-4-nitrophenoxy phenylphosphine sulfide; Ethyl *p*-nitrophenyl benzenethionophosphate; Ethyl *p*-nitrophenyl benzenethiophosphonate; Ethyl *p*-nitrophenyl ester; *O*-Ethyl *O*-4-nitrophenyl phenylphosphonothioate; Ethyl *p*-nitrophenyl phenylphosphonothioate; *O*-Ethyl *O*-*p*-nitrophenyl phenylphosphonothioate; Ethyl *p*-nitrophenyl thionobenzenephosphate; Ethyl *p*-nitrophenyl thionobenzenephosphonate; *O*-Ethyl phenyl *p*-nitrophenyl phenylphosphorothioate; Ethyl *p*-nitrophenyl thionobenzenephosphate; *O*-Ethyl phenyl *p*-nitrophenyl thiophosphonate; Phenylphosphonothioic acid *O*-ethyl *O*-*p*-nitrophenyl ester; **Phosphonothioic acid *0,O*-diethyl *O*-(3,5,6-trichloro-2-pyridinyl) ester**; Pin; Santox.

CHEMICAL DESIGNATIONS: CAS: 2104-64-5; DOT: none assigned; mf: $C_{14}H_{14}NO_4PS$; fw: 323.31; RTECS: TB 1925000.

PROPERTIES: Yellow solid or brown liq with an aromatic odor. Mp: 36°, bp: 215° @ 5 mm; *d*: 1.27 @ 25/4°, log K_{oc}: 3.12, log K_{ow}: 3.85, S_o: misc with acetone, benzene, methanol, isopropanol, toluene and xylene, vp: 0.126 mPa @ 25°.

TRANSFORMATION PRODUCTS: Releases toxic sulfur, phosphorous and NO_x when heated to decomposition.

EXPOSURE LIMITS: NIOSH REL: TWA 0.5 mg/m^3; OSHA PEL: TWA 0.5 mg/m^3.

SYMPTOMS OF EXPOSURE: Miosis, irrit eyes; rhin; head; tight chest, wheez, lar spas; salv; cyan; anor, nau, abdom cramp, diarr; para convul; low BP; card.

USES: Insecticide; acaricide.

ETHANOLAMINE

SYNONYMS: **2-Aminoethanol**; β-Aminoethyl alcohol; Colamine; β-Ethanolamine; Ethylolamine; Glycinol; 2-Hydroxyethylamine; β-Hydroxyethylamine; MEA; Monoethanolamine; Olamine; Thiofalco M-50; UN 2491; USAF EK-1597.

CHEMICAL DESIGNATIONS: CAS: 141-43-5; DOT: 2491; mf: C_2H_7NO; fw: 61.08; RTECS: KJ 5775000.

PROPERTIES: Colorless liq with a mild ammonia-like odor. Mp: 10.3°, bp: 170°, *d*: 1.0180 @ 20/4°, fl p: 85°, lel: 5.5%, uel: ≈ 17%, pK_a: 9.50 @ 25°, log K_{ow}: -1.31, S_o: misc with acetone and methanol; sol in benzene (1.4 wt %), carbon tetrachloride (0.2 wt %) and ether (2.1 wt %), S_w: misc, vap d: 2.50 g/L @ 25°, 2.11 (air = 1), vp: 0.4 mm @ 20°.

TRANSFORMATION PRODUCTS: Aq chlorination of ethanolamine at high pH produced *N*-chloroethanolamine which slowly degraded to unidentified products.

EXPOSURE LIMITS: NIOSH REL: IDLH 1,000 ppm; OSHA PEL: TWA 3 ppm.

SYMPTOMS OF EXPOSURE: Resp sys, skin, eye irrit, lethargy.

USES: Removing CO_2 and hydrogen sulfide from natural gas; in emulsifiers, hair waving solns, polishes; softening agent for hides; agricultural sprays; pharmaceuticals, chemical intermediates; corrosion inhibitor; rubber accelerator; non-ionic detergents used in dry cleaning, wool treatment.

2-ETHOXYETHANOL

SYNONYMS: Cellosolve; Cellosolve solvent; Dowanol EE; Ektasolve EE; Ethyl cellosolve; Ethylene glycol ethyl ether; Ethylene glycol monoethyl ether; Glycol ether EE; Glycol monoethyl ether; Hydroxy ether; Jeffersol EE; NCI-C54853; Oxitol; Poly-solv EE; UN 1171.

CHEMICAL DESIGNATIONS: CAS: 110-80-5; DOT: 1171; mf: $C_4H_{10}O_2$; fw: 90.12; RTECS: KK 8050000.

PROPERTIES: Colorless liq with a sweetish odor. Mp: -90°, -70°, bp: 135°, fl p: 44°, lel: 1.8%, uel: 14%, log K_{ow}: -0.54, S_o: misc with acetone, alcohol, ether and liq esters, S_w: misc, *d*: 0.9297 @ 20/4°, vap d: 3.68 g/L @ 25°, 3.11 (air = 1), vp: 3.8 mm @ 20°.

EXPOSURE LIMITS: NIOSH REL: lowest feasible limit; OSHA PEL: TWA 200 ppm; ACGIH TLV: TWA 5 ppm.

SYMPTOMS OF EXPOSURE: In animals: hematologic effects; liv, kid, lung damage, irrit eyes.

USES: Solvent for lacquers and dopes, nitrocellulose, natural and synthetic resins; in cleaning solns, varnish removers, dye baths; mutual solvent for formation of soluble oils; lacquers and lacquer thinners; anti-icing additive for aviation fuels.

2-ETHOXYETHYL ACETATE

SYNONYMS: Acetic acid 2-ethoxyethyl ester; Cellosolve acetate; CSAC; Ekasolve

EE acetate solvent; Ethoxyacetate; **2-Ethoxyethanol acetate**; Ethoxyethyl acetate; β-Ethoxyethyl acetate; Ethylene glycol ethyl ether acetate; Ethylene glycol monoethyl ether acetate; Glycol ether EE acetate; Glycol monoethyl ether acetate; Oxytol acetate; Poly-solv EE acetate; UN 1172.

CHEMICAL DESIGNATIONS: CAS: 111-15-9; DOT: 1172; mf: $C_6H_{12}O_3$; fw: 132.18; RTECS: KK 8225000.

PROPERTIES: Colorless liq with a faint, pleasant odor. Mp: -62.2°, bp: 156.1°, *d*: 0.975 @ 20/20°, fl p: 47.2°, lel: 1.7%, K_H: 9.07 x 10^{-7} atm·m^3/mol @ 20° (calcd), log K_{ow}: 0.50 (calcd), S_w: 230 g/L @ 20°, vap d: 5.40 g/L @ 25°, 4.56 (air = 1), vp: 2 mm @ 20°.

EXPOSURE LIMITS: NIOSH REL: IDLH 2,500 ppm; OSHA PEL: TWA 100 ppm; ACGIH TLV: TWA 5 ppm.

SYMPTOMS OF EXPOSURE: Irrit eyes, nose; vomit; kid damage; para.

USES: Automobile lacquers to reduce evaporation and to impart a high gloss; solvent for nitrocellulose, oils and resins; varnish removers; wood stains; textiles; leather.

ETHYL ACETATE

SYNONYMS: Acetic ether; **Acetic acid ethyl ester**; Acetidin; Acetoxyethane; Ethyl acetic ester; Ethyl ethanoate; RCRA waste number U112; UN 1173; Vinegar naphtha.

CHEMICAL DESIGNATIONS: CAS: 141-78-6; DOT: 1173; mf: $C_4H_8O_2$; fw: 88.11; RTECS: AH 5425000.

PROPERTIES: Colorless liq with a fruity odor. Mp: -83.6°, bp: 77.06°, *d*: 0.9003 @ 20/4°, fl p: -4.4°, lel: 2.0%, uel: 11.5%, K_H: 1.34 x 10^{-4} atm·m^3/mol @ 25°, IP: 10.11 eV, log K_{ow}: 0.66-0.81, S_o: sol in benzene; misc with acetone, alcohol, chloroform and ether, S_w: 79 g/L @ 20°, vap d: 3.60 g/L @ 25°, 3.04 (air = 1), vp: 72.8 mm @ 20°.

TRANSFORMATION PRODUCTS: Hydrolyzes in water forming ethyl alcohol and acetic acid ($t_{1/2}$ = 2.0 yr @ 25° and pH 7).

EXPOSURE LIMITS: NIOSH REL: IDLH 10,000 ppm; OSHA PEL: 400 ppm.

SYMPTOMS OF EXPOSURE: Irrit eyes, nose, throat; narco, derm.

USES: Manufacture of smokeless powder, photographic film and plates, artificial leather and silk, perfumes; pharmaceuticals; cleaning textiles; solvent for nitrocellulose, lacquers, varnishes and airplane dopes.

ETHYL ACRYLATE

SYNONYMS: Acrylic acid ethyl ester; Ethoxycarbonylethylene; Ethyl propenoate; Ethyl-2-propenoate; NCI-C50384; **2-Propenoic acid ethyl ester**; RCRA waste number U113; UN 1917.

CHEMICAL DESIGNATIONS: CAS: 140-88-5; DOT: 1917; mf: $C_5H_8O_2$; fw: 100.12; RTECS: AT 0700000.

PROPERTIES: Colorless liq with a sharp, penetrating odor. Mp: -71.2°, bp: 99.8°, *d*: 0.9234 @ 20/4°, fl p: 8.9°, lel: 1.8%, uel: 14%, K_H: 1.94-2.59 x 10^{-3} atm·m^3/mol @ 20° (calcd), log K_{ow}: 1.20 (calcd), S_o: sol in alcohol, chloroform and ether, S_w: 20 g/L @ 20°, vap d: 4.09 g/L @ 25°, 3.45 (air = 1), vp: 29.5 mm @ 20°.

TRANSFORMATION PRODUCTS: Polymerizes on standing and is catalyzed by heat, light and peroxides. Anticipated hydrolysis products include ethyl alcohol and acrylic acid.

EXPOSURE LIMITS: NIOSH REL: IDLH 2,000 ppm; OSHA PEL: TWA 25 ppm; ACGIH TLV: TWA 5 ppm.

SYMPTOMS OF EXPOSURE: Irrit eyes, resp sys, skin.

USES: Manufacture of water emulsion paints, textile and paper coatings, adhesives and leather finish resins.

ETHYLAMINE

SYNONYMS: Aminoethane; 1-Aminoethane; **Ethanamine**; Monoethylamine; UN 1036.

CHEMICAL DESIGNATIONS: CAS: 75-04-7; DOT: 1036; mf: C_2H_7N; fw: 45.08; RTECS: KH 2100000.

PROPERTIES: Colorless liq or gas with a strong ammonia-like odor. Mp: -81.1°, bp: 16.6°, *d*: 0.6829 @ 20/4°, fl p: < -17.8°, lel: 3.5%, uel: 14.0%, pK_a: 10.807 @ 20°, 10.63 @ 25°, K_H: 1.07 x 10^{-5} atm·m^3/mol @ 25°, IP: 8.86 eV, log K_{ow}: -2.74 (calcd), S_o: misc with alcohol and ether, S_w: a saturated soln in equilibrium with its own vapor had a concn of 5,176 g/L @ 25°, vap d: 1.84 g/L @ 25°, 1.56 (air = 1), vp: 1.18 atm @ 20°.

TRANSFORMATION PRODUCTS: Reacts with hydroxyl radicals possibly forming acetaldehyde or acetamide. Forms soluble salts with acids.

EXPOSURE LIMITS: NIOSH REL: IDLH 4,000 ppm; OSHA PEL: TWA 10 ppm.

SYMPTOMS OF EXPOSURE: Irrit eyes, burns skin, resp irrit, derm.

USES: Stabilizer for latex rubber; intermediate for dyestuffs and medicinals; oil refining; resin and detergent manufacturing; solvent in petroleum and vegetable oil refining; starting material for manufacturing amides; plasticizer; organic synthesis.

ETHYLBENZENE

SYNONYMS: EB; Ethylbenzol; NCI-C56393; Phenylethane; UN 1775.

CHEMICAL DESIGNATIONS: CAS: 100-41-4; DOT: 1175; mf: C_8H_{10}; fw: 106.17; RTECS: DA 0700000.

PROPERTIES: Clear, colorless liq with a sweet gasoline-like odor. Mp: -95.0°, bp: 136.2°, *d*: 0.8670 @ 20/4°, fl p: 15°, lel: 1.0%, uel: 6.7%, pK_a: > 15, K_H: 8.68 x 10^{-3} atm·m^3/mol @ 25°, IP: 8.76 eV, 9.12 eV, log K_{oc}: 2.41, log K_{ow}: 3.05-3.15, S_o: freely sol in most solvents, S_w: 152 mg/L @ 20°, vap d: 4.34 g/L @ 25°, 3.66 (air = 1), vp: 7.08 mm @ 20°.

TRANSFORMATION PRODUCTS: Irradiation of ethylbenzene (λ < 2537 Å) at low temperatures formed hydrogen, styrene and free radicals. Phenylacetic acid was reported to be the biooxidation product of ethylbenzene by *Nocardia* sp. in soil using *n*-hexadecane or *n*-octadecane as the substrate. In addition, *Methylosinus trichosporium* OB3b metabolized ethylbenzene to *o*- and *m*-hydroxybenzaldehyde with methane as the substrate. A culture of *Nocardia tartaricans* ATCC 31190, growing in a hexadecane medium, oxidized ethylbenzene to 1-phenethanol which further oxidized to acetophenone.

EXPOSURE LIMITS: NIOSH REL: IDLH 2,000 ppm; OSHA PEL: TWA 100 ppm; ACGIH TLV: TWA 100 ppm, STEL 125 ppm.

SYMPTOMS OF EXPOSURE: Irrit eyes, muc memb; head; derm; narco, coma.

USES: Manufacture of styrene, acetophenone, ethylcyclohexane, benzoic acid, 1-bromo-1-phenylethane, 1-chloro-1-phenylethane, 2-chloro-1-phenylethane, *p*-chloroethylbenzene, *p*-chlorostyrene and many other compounds; solvent.

ETHYL BROMIDE

SYNONYMS: Bromic ether; **Bromoethane**; Halon 2001; Hydrobromic ether; Monobromoethane; NCI-C55481; UN 1891.

CHEMICAL DESIGNATIONS: CAS: 74-96-4; DOT: 1891; mf: C_2H_5Br; fw: 108.97; RTECS: KH 6475000.

PROPERTIES: Colorless to yellow, volatile liq with an ether-like odor. Mp: -119°, bp: 38.4°, *d*: 1.4604 @ 20/4°, fl p: < -26°, lel: 6.7%, uel: 8.0%, K_H: 7.56 x 10^{-3} atm·m^3/mol @ 25°, IP: 10.29 eV, log K_{oc}: 2.67 (calcd), log K_{ow}: 1.57 (calcd), S_o: misc with alcohol, chloroform and ether, S_w: 0.9 wt % @ 20°, vap d: 4.05 g/L @ 25°, 3.76 (air = 1), vp: 375 mm @ 20°.

TRANSFORMATION PRODUCTS: Ethyl bromide hydrolyzes in water forming ethyl alcohol and bromide ions ($t_{1/2}$ = 30 days at 25° and pH 7). Groundwater under reducing conditions in the presence of hydrogen sulfide converted ethyl bromide to sulfur-containing products. A strain of *Acinetobacter* sp. isolated from activated sludge degraded ethyl bromide to ethanol and bromide ions. When *Methanococcus thermolithotrophicus*, *Methanococcus deltae* and *Methanobacterium thermoautotrophicum*, grown with H_2-CO_2 in the presence of ethyl bromide, methane and ethane were produced.

EXPOSURE LIMITS: NIOSH REL: IDLH 3,500 ppm; OSHA PEL: TWA 200 ppm; ACGIH TLV: TWA 200 ppm, STEL 250 ppm.

SYMPTOMS OF EXPOSURE: Irrit eyes, resp sys; CNS depres; irrit skin; pulm edema; liv, kid, disease, card arrhy, card arrest; dizz.

USES: In organic synthesis as an ethylating agent; refrigerant; solvent; grain and fruit fumigant; in medicine as an anesthetic.

ETHYLCYCLOPENTANE

SYNONYMS: None.

CHEMICAL DESIGNATIONS: CAS: 1640-89-7; DOT: none assigned; mf:

C_7H_{14}; fw: 98.19; RTECS: GY 4450000.

PROPERTIES: Colorless liq. Mp: -138.4°, bp: 103.5°, *d*: 0.7665 @ 20/4°, fl p: 15°, lel: 1.1%, uel: 6.7%, K_H: 2.10 x 10^{-2} atm·m^3/mol @ 25° (calcd), log K_{ow}: 1.90 (calcd), S_o: sol in acetone, alcohol, benzene, ether and petroleum, S_w: 245 ppm @ 25°, vap d: 4.01 g/L @ 25°, 3.39 (air = 1), vp: 39.9 mm @ 25°.

USES: Organic research.

ETHYLENE CHLOROHYDRIN

SYNONYMS: **2-Chloroethanol**; δ-Chloroethanol; 2-Chloroethyl alcohol; β-Chloroethyl alcohol; Ethylene chlorhydrin; Glycol chlorohydrin; Glycol monochlorohydrin; 2-Monochloroethanol; NCI-C50135; UN 1135.

CHEMICAL DESIGNATIONS: CAS: 107-07-3; DOT: none assigned; mf: C_2H_5ClO; fw: 80.51; RTECS: KK 0875000.

PROPERTIES: Colorless liq with an ether-like odor. Mp: -67.5°, bp: 128°, *d*: 1.2003 @ 20/4°, fl p: 40° (OC), lel: 4.9%, uel: 15.9%, IP: 10.90 eV, S_o: sol in alcohol, S_w: misc, vap d: 3.29 g/L @ 25°, 2.78 (air = 1), vp: 4.9 mm @ 20°.

EXPOSURE LIMITS: NIOSH REL: IDLH 10 ppm; OSHA PEL: TWA 5 ppm; ACGIH TLV: C 1 ppm.

SYMPTOMS OF EXPOSURE: Irrit muc memb; nau, vomit; verti, inco; numb, vision dist; head; thirst; delirium; low BP, coll, shock, coma.

USES: Solvent for cellulose acetate, ethylcellulose; manufacturing insecticides, ethylene oxide and ethylene glycol; treating sweet potatoes before planting; organic synthesis (introduction of the hydroxyethyl group).

ETHYLENEDIAMINE

SYNONYMS: 1,2-Diaminoethane; Dimethylenediamine; **1,2-Ethanediamine**; 1,2-Ethylenediamine; NCI-C60402; UN 1604.

CHEMICAL DESIGNATIONS: CAS: 107-15-3; DOT: 1604; mf: $C_2H_8N_2$; fw: 60.10; RTECS: KH 8575000.

PROPERTIES: Colorless, volatile, hygroscopic liq with an ammonia-like odor. Mp: 10°, bp: 116.5°, *d*: 0.963 @ 21/4°, fl p: 33.9°, lel: 4.2%, uel: 11.1%, pK_a: @ 20°: pK_1 = 10.075, pK_2 = 6.985, K_H: 1.73 x 10^{-9} atm·m^3/mol @ 25°, S_o: sol in alcohol; slightly sol in benzene and ether, S_w: misc, vap d: 2.46 g/L @ 25°, 2.07 (air = 1), vp: 10 mm @ 20°.

TRANSFORMATION PRODUCTS: May absorb CO_2 forming carbonates.

EXPOSURE LIMITS: NIOSH REL: IDLH 2,000 ppm; OSHA PEL: TWA 10 ppm; ACGIH TLV: TWA 10 ppm.

SYMPTOMS OF EXPOSURE: Nasal irrit, primary irrit, sens derm; irrit resp sys, asth; liv, kid damage.

USES: Stabilizing rubber latex; solvent for albumin, casein, shellac and sulfur; neutralizing oils; in antifreeze as a corrosion inhibitor; emulsifier; adhesives; textile lubricants; fungicides; manufacturing chelating agents such as EDTA (ethylenediaminetetraacetic acid); dimethylolethylene-urea resins; organic synthesis.

ETHYLENE DIBROMIDE

SYNONYMS: Acetylene dibromide; Bromofume; Celmide; DBE; Dibromoethane; **1,2-Dibromoethane**; *sym*-Dibromoethane; α,β-Dibromoethane; Dowfume 40; Dowfume EDB; Dowfume W-8; Dowfume W-85; Dowfume W-90; Dowfume W-100; EDB; EDB-85; E-D-BEE; ENT 15,349; Ethylene bromide; Ethylene bromide glycol dibromide; 1,2-Ethylene dibromide; Fumo-gas; Glycol bromide; Glycol dibromide; Iscobrome D; Kopfume; NCI-C00522; Nephis; Pestmaster; Pestmaster EDB-85; RCRA waste number U067; Soilbrom-40; Soilbrom-85; Soilbrom-90; Soilbrom-90EC; Soilbrom-100; Soilbrome-85; Soilfume; UN 1605; Unifume.

CHEMICAL DESIGNATIONS: CAS: 106-93-4; DOT: 1605; mf: $C_2H_4Br_2$; fw: 187.86; RTECS: KH 9275000.

PROPERTIES: Liq with a chloroform-like odor. Mp: 9.8°, bp: 131.3°, *d*: 2.1792 @ 20/4°, K_H: 7.06 x 10^{-4} atm·m^3/mol @ 25°, IP: 9.45 eV, log K_{oc}: 1.56-2.21, log K_{ow}: 1.76, S_o: sol in acetone, alcohol, benzene and ether, S_w: 4,321 mg/L @ 20°, vap d: 7.68 g/L @ 25°, 6.49 (air = 1), vp: 11 mm @ 20°.

TRANSFORMATION PRODUCTS: Hydrolyzes in water to ethylene glycol and bromoethanol. Dehydrobromination of ethylene dibromide to vinyl bromide was observed in various aq buffer solns (pH 7 to 11) over the temperature range of 45 to 90° (est $t_{1/2}$ = 2.5 yr @ 25° and pH 7). Complete biodegradation by soil cultures resulted in the formation of ethylene and bromide ions. A mutant of strain *Acinetobacter* sp. GJ70 isolated from activated sludge degraded ethylene dibromide to ethylene glycol and bromide ions. When *Methanococcus thermolithotrophicus*, *Methanococcus deltae* and *Methanobacterium thermoautotrophicum* grown with H_2-CO_2 in the presence of ethylene dibromide, methane and ethylene were produced. Anticipated products from the reaction of ethylene dibromide with O_3 or hydroxyl radicals in the atmosphere are bromoacetaldehyde, formaldehyde, bromoformaldehyde and Br•. In a shallow aquifer material, ethylene dibromide aerobically degraded to CO_2, microbial biomass and nonvolatile water-soluble compound(s).

EXPOSURE LIMITS: NIOSH REL: TWA 0.045 ppm, 15-min C 0.13 ppm; OSHA PEL: TWA 20 ppm, C 30 ppm, 5-min peak 50 ppm.

SYMPTOMS OF EXPOSURE: Irrit resp sys, eyes; derm with vesic.

USES: In anti-knock gasolines; grain and fruit fumigant; waterproofing preparations; insecticide; medicines; general solvent; organic synthesis.

ETHYLENIMINE

SYNONYMS: Aminoethylene; Azacyclopropane; Azirane; **Aziridine**; Dihydro-1*H*-azirine; Dihydroazirine; Dimethyleneimine; Dimethylenimine; EI; ENT 50,324; Ethyleneimine; Ethylimine; RCRA waste number P054; TL 337; UN 1185.

CHEMICAL DESIGNATIONS: CAS: 151-56-4; DOT: 1185; mf: C_2H_5N; fw: 43.07; RTECS: KX 5075000.

PROPERTIES: Colorless liq with a very strong ammonia odor. Mp: -71.5°, bp: [illegible]°, *d*: 0.8321 @ 20/4°, fl p: -11.1°, lel: 3.3%, uel: 46%, pK_a: 8.04, K_H: 1.33 x 10^{-7} [illegible]·m^3/mol @ 25°, IP: 9.9 eV, log K_{oc}: 0.11 (calcd), log K_{ow}: -1.01, S_o: sol in acetone, alcohol, benzene and ether, S_w: misc, vap d: 1.76 g/L @ 25°, 1.49 (air = 1), vp: 160 mm @ 20°.

TRANSFORMATION PRODUCTS: Polymerizes easily. Hydrolyzes in water forming ethanolamine (est $t_{1/2}$ = 154 days @ 25° and pH 7). The vacuum UV photolysis (λ = 147 nm) and γ radiolysis of ethylenimine resulted in the formation of acetylene, methane, ethane, ethylene, hydrogen cyanide, methyl radicals and hydrogen. Photolysis of ethylenimine vapor @ krypton and xenon lines yielded ethylene, ethane, methane, acetylene, propane, *n*-butane, hydrogen, ammonia, ethylenimino radicals.

EXPOSURE LIMITS: ACGIH TLV: TWA 0.5 ppm.

SYMPTOMS OF EXPOSURE: Nau, vomit; head; dizz; pulm edema,; liv and kid damage; eye burns; skin sens; nose and throat irrit.

USES: Manufacture of triethylenemelamine; fuel oil and lubricant refining; ion exchange; protective coatings; adhesives; pharmaceuticals; polymer stabilizers; surfactants; organic synthesis.

ETHYL ETHER

SYNONYMS: Aether; Anaesthetic ether; Anesthesia ether; Anesthetic ether; Diethyl ether; Diethyl oxide; Ether; Ethoxyethane; Ethyl oxide; **1,1′-Oxybisethane**; RCRA waste number U117; Solvent ether; Sulfuric ether; UN 1155.

CHEMICAL DESIGNATIONS: CAS: 60-29-7; DOT: 1155; mf: $C_4H_{10}O$; fw: 74.12; RTECS: KI 5775000.

PROPERTIES: Colorless, volatile liq with a sweet, pungent odor. Mp: -116.2°, bp: 34.5°, *d*: 0.7138 @ 20/4°, fl p: -45°, lel: 1.9%, uel: 36.0%, K_H: 1.28 x 10^{-3} atm·m^3/mol @ 25°, IP: 9.56 eV, log K_{ow}: 0.77-0.89, S_o: sol in acetone; misc with lower aliphatic alcohols, benzene, chloroform, petroleum ether and many oils, S_w: 69 g/L @ 20°, vap d: 3.03 g/L @ 25°, 2.55 (air = 1), vp: 442 mm @ 20°.

EXPOSURE LIMITS: NIOSH REL: IDLH 19,000 ppm; OSHA PEL: TWA 400 ppm; ACGIH TLV: TWA 400 ppm, STEL 500 ppm.

SYMPTOMS OF EXPOSURE: Dizz; drow; head, excited, narco; nau, vomit; irrit eyes, upper resp, skin.

USES: Solvent for oils, waxes, perfumes, alkaloids, fats and gums; organic synthesis (Grignard and Wurtz reactions); extractant; manufacture of gun powder, ethylene and other organic compounds; analytical chemistry; perfumery; alcohol denaturant; primer for gasoline engines; anesthetic.

ETHYL FORMATE

SYNONYMS: Areginal; Ethyl formic ester; Ethyl methanoate; **Formic acid ethyl ester**; Formic ether; UN 1190.

CHEMICAL DESIGNATIONS: CAS: 109-94-4; DOT: 1190; mf: $C_3H_6O_2$; fw: 74.08; RTECS: LQ 8400000.

PROPERTIES: Colorless, clear liq with a pleasant, fruity odor. Mp: -80.5°,

54.5°, *d*: 0.9168 @ 20/4°, fl p: -20°, lel: 2.7%, uel: 13.5%, K_H: 2.23 x 10^{-4} atm·m^3/mol @ 25°, IP: 10.61 eV, log K_{ow}: 0.36 (calcd), S_o: sol in acetone; misc with alcohol, benzene and ether, S_w: 105 g/L @ 20°, vap d: 3.03 g/L @ 25°, 2.56 (air = 1), vp: 192 mm @ 20°.

TRANSFORMATION PRODUCTS: Slowly hydrolyzes in water forming formic acid and ethanol.

EXPOSURE LIMITS: NIOSH REL: IDLH 8,000 ppm; OSHA PEL: TWA 100 ppm; ACGIH TLV: TWA 100 ppm.

SYMPTOMS OF EXPOSURE: Irrit eyes, resp sys; narco.

USES: Solvent for nitrocellulose and cellulose acetate; artificial flavor for lemonades and essences; fungicide and larvacide for cereals, tobacco, dried fruits; acetone substitute; organic synthesis.

ETHYL MERCAPTAN

SYNONYMS: 2-Aminoethanethiol; **Ethanethiol**; Ethyl hydrosulfide; Ethyl sulfhydrate; Ethyl thioalcohol; LPG ethyl mercaptan 1010; UN 2363; Thioethanol; Thioethyl alcohol.

CHEMICAL DESIGNATIONS: CAS: 75-08-1; DOT: 2363; mf: C_2H_6S; fw: 62.13; RTECS: KI 9625000.

PROPERTIES: Colorless liq with a strong skunk-like odor. Mp: -144.4°, bp: 35°, *d*: 0.8391 @ 20/4°, fl p: -48.3°, lel: 2.8%, uel: 18.0%, pK_a: 10.50 @ 20°, K_H: 2.74 x 10^{-3} atm·m^3/mol @ 25°, IP: 9.285 eV, log K_{ow}: 1.49 (calcd), S_o: sol in acetone, alcohol and ether, S_w: 6.76 g/L @ 20°, vap d: 2.54 g/L @ 25°, 2.14 (air = 1), vp: 442 mm @ 20°.

TRANSFORMATION PRODUCTS: In the presence of NO, ethyl mercaptan reacted with hydroxyl radicals forming ethyl thionitrite.

EXPOSURE LIMITS: NIOSH REL: 15-min C 0.5 ppm, IDLH 2,500 ppm; OSHA PEL: C 10 ppm; ACGIH TLV: TWA 0.5 ppm.

SYMPTOMS OF EXPOSURE: Head; nau; irrit muc memb, throat; in animals: inco; para; pulm irrit; liv and kid damage.

USES: Odorant for natural gas; manufacture of plastics, antioxidants, pesticides; adhesive stabilizer; chemical intermediate.

4-ETHYLMORPHOLINE

SYNONYM: *N*-Ethylmorpholine.

CHEMICAL DESIGNATIONS: CAS: 100-74-3; DOT: none assigned; mf: $C_6H_{13}NO$; fw: 115.18; RTECS: QE 4025000.

PROPERTIES: Colorless liq with an ammonia-like odor. Mp: -65 to -63°, bp: 138-139° @ 763 mm, *d*: 0.9886 @ 20/4°, fl p: 27°, lel: 1.0%, uel: 9.8%, S_o: sol in acetone, alcohol, benzene and ether, S_w: misc, vap d: 4.71 g/L @ 25°, 3.98 (air = 1), vp: 5 mm @ 20°.

TRANSFORMATION PRODUCTS: Releases toxic NO_x when heated to decomposition.

EXPOSURE LIMITS: NIOSH REL: IDLH 2,000 ppm; OSHA PEL: TWA 20 ppm; ACGIH TLV: 5 ppm.

SYMPTOMS OF EXPOSURE: Eye, nose, throat irrit; vis dist; severe eye irrit from splashes.

USES: Intermediate for pharmaceuticals, dyestuffs, emulsifying agents and rubber accelerators; solvent for dyes, resins and oils; catalyst for making polyurethane foams.

2-ETHYLTHIOPHENE

SYNONYM: None.

CHEMICAL DESIGNATIONS: CAS: 872-55-9; DOT: none assigned; mf: C_6H_8S; fw: 112.19; RTECS: none assigned.

PROPERTIES: Liq with a pungent odor. Bp: 104°, 134°, *d*: 0.9930 @ 20/4°, fl p: 21°, IP: 8.8 eV, log K_{ow}: 2.83 (calcd), S_o: sol in alcohol and ether, S_w: 292 ppm @ 25°, vap d: 4.59 g/L @ 25°, 3.87 (air = 1), vp: 60.9 mm @ 60.3°.

USE: Ingredient in crude petroleum.

FORMALDEHYDE

SYNONYMS: BFV; FA; Fannoform; Formalin; Formalin 40; Formalith; Formic aldehyde; Formol; Fyde; HOCH; Ivalon; Karsan; Lysoform; Methanal; Methyl aldehyde; Methylene glycol; Methylene oxide; Morbicid; NCI-C02799; Oxomethane; Oxymethylene; Paraform; Polyoxymethylene glycols; RCRA waste number U122; Superlysoform; UN 1198; UN 2209.

CHEMICAL DESIGNATIONS: CAS: 50-00-0; DOT: 1198; mf: CH_2O; fw: 30.03; RTECS: LP 8925000.

PROPERTIES: Clear, colorless liq with a pungent odor. Mp: -92°, bp: -21°, 98-99° (40% aq soln), *d*: 0.815 @ -20/4°, fl p: 50° (15% methanol-free), lel: 7.0%, uel: 73.0%, K_H: 1.67 x 10^{-7} atm·m^3/mol, IP: 10.88 eV, log K_{oc}: 0.56 (calcd), log K_{ow}: 0.00, S_o: sol in acetone, alcohol, benzene and ether, S_w: misc @ 25°, vap d: 1.23 g/L @ 25°, 1.067 (air = 1), vp: 760 mm @ -19.5°.

TRANSFORMATION PRODUCTS: Biodegradation products reported include formic acid and ethanol each of which can further degrade to CO_2. Oxidizes in air to formic acid. Trioxymethylene may precipitate under cold temperatures. Polymerizes easily. Anticipated products from the reaction of formaldehyde with O_3 or hydroxyl radicals in the atmosphere are CO and CO_2. Major products reported from the photooxidation of formaldehyde with NO_x are CO, CO_2 and H_2O_2. In synthetic air, photolysis of formaldehyde gave hydrogen chloride and CO. Photooxidation of formaldehyde in NO_x-free air (λ = 2900-3500 Å) formed H_2O_2, alkylhydroperoxides, CO and lower molecular weight aldehydes. In the presence of NO_x, photooxidation products reported include O_3, H_2O_2 and peroxyacyl nitrates. Formaldehyde reacted with hydrogen chloride in moist air forming bis(chloromethyl)ether. This compound may also form from an acidic soln containing chloride ion and formaldehyde. Irradiation of gaseous formaldehyde containing an excess of NO_2 over chlorine yielded O_3, CO, N_2O_5, nitryl chloride, nitric and hydrochloric acids. Peroxy nitric acid was the major photolysis product when chlorine exceeded NO_2 concentration.

EXPOSURE LIMITS: NIOSH REL: 15-min C 0.1 ppm; OSHA PEL: TWA 3 ppm, C 5 ppm, 30-min C 10 ppm; ACGIH TLV: TWA 1 ppm.

SYMPTOMS OF EXPOSURE: Irrit eyes, nose, throat; lac; burns nose; cough, bronspas; pulm irrit; derm, nau, vomit; loss of consciousness.

USES: Manufacture of phenolic, melamine, urea and acetal resins; fertilizers; dyes; ethylene glycol; embalming fluids; textiles; fungicides; air fresheners; cosmetics; in medicine as a disinfectant and germicide; preservative; hardening agent; in oil wells as a corrosion inhibitor; industrial sterilant; reducing agent.

FLUORANTHENE

SYNONYMS: 1,2-Benzacenaphthene; Benzo[*jk*]fluorene; Idryl; 1,2-(1,8-Naphthylene)benzene; 1,2-(1,8-Naphthalenediyl)benzene; RCRA waste number U120.

CHEMICAL DESIGNATIONS: CAS: 206-44-0; DOT: none assigned; mf: $C_{16}H_{10}$; fw: 202.26; RTECS: LL 4025000.

PROPERTIES: Colorless cryst. Mp: 107°, bp: 375°, *d*: 1.252 @ 0/4°, pK_a: > 15, K_H: 1.69 x 10^{-2} atm·m^3/mol @ 25°, IP: 8.54 eV, log K_{oc}: 4.62, log K_{ow}: 5.15-5.20, S_o: sol in acetic acid, benzene, chloroform, carbon disulfide, ethanol and ether, S_w: 166 μg/L @ 20°, vp: 5.0 x 10^{-6} mm @ 25°.

TRANSFORMATION PRODUCTS: 2-Nitrofluoranthene was the principal product formed from the gas-phase reaction of fluoranthene with hydroxyl radicals in a NO_x atmosphere. Minor products found include 7- and 8-nitrofluoranthene. The reaction of fluoranthene with NO_x to form 3-nitrofluoranthene was reported to occur in urban air from St. Louis, MO. Chlorination of fluoranthene in polluted humus poor lake water gave a large number of mono-, di- and trichlorofluoranthene derivatives. At pH < 4, chlorination of fluoranthene produced 3-chlorofluoranthene as the major product. It was suggested that the chlorination of fluoranthene in tap water accounted for the presence of chloro- and dichlorofluoranthenes.

EXPOSURE LIMITS: No individual standards have been set, however, as a constituent in coal tar pitch volatiles, the following exposure limits have been established: NIOSH REL: 10-h TWA 0.1 mg/m^3 (cyclohexane-extractable fraction); OSHA PEL: TWA 0.2 mg/m^3 (benzene-soluble fraction); ACGIH TLV: TWA 0.2 mg/m^3 (benzene solubles).

USE: Research chemical.

FLUORENE

SYNONYMS: 2,3-Benzindene; *o*-Biphenylenemethane; *o*-Biphenylmethane;

EXPOSURE LIMITS: NIOSH REL: 10-h TWA 50 ppm, IDLH 250 ppm; OSHA PEL: TWA 50 ppm; ACGIH TLV: TWA 10 ppm, STEL 15 ppm.

SYMPTOMS OF EXPOSURE: Dizz; nau, diarr, vomit; resp depres; body temp depres.

USES: Solvent for dyes and resins; preparation of furfuryl esters; furan polymers; solvent for textile printing; manufacturing wetting agents and resins; penetrant; flavoring; corrosion-resistant sealants and cements; viscosity reducer for viscous epoxy resins.

GLYCIDOL

SYNONYMS: Epihydric alcohol; Epihydrin alcohol; 2,3-Epoxypropanol; 2,3-Epoxy-1-propanol; Epoxypropyl alcohol; Glycide; Glycidyl alcohol; 3-Hydroxy-1,2-epoxypropane; Hydroxymethyl ethylene oxide; 2-Hydroxymethyloxiran; 3-Hydroxypropylene oxide; NCI-C55549; **Oxiranemethanol**; Oxiranylmethanol.

CHEMICAL DESIGNATIONS: CAS: 556-52-5; DOT: none assigned; mf: $C_3H_6O_2$; fw: 74.08; RTECS: UB 4375000.

PROPERTIES: Colorless liq. Mp: -45°, bp: 166-167° (decomp), *d*: 1.1143 @ 25/4°, fl p: 72.2°, log K_{ow}: -0.95, S_o: sol in acetone, alcohol, benzene, chloroform and ether, S_w: misc, vap d: 3.03 g/L @ 25°, 2.56 (air = 1), vp: 0.9 mm @ 25°.

TRANSFORMATION PRODUCTS: May hydrolyze in water forming glycerin.

EXPOSURE LIMITS: NIOSH REL: IDLH 500 ppm; OSHA PEL: TWA 50 ppm; ACGIH TLV: TWA 25 ppm.

SYMPTOMS OF EXPOSURE: Irrit eyes, nose, throat, skin, narco.

USES: Demulsifier; dye-leveling agent; stabilizer for natural oils and vinyl polymers; organic synthesis.

HEPTACHLOR

SYNONYMS: Aahepta; Agroceres; Basaklor; 3-Chlorochlordene; Drinox; Drinox H-34; E 3,314; ENT 15,152; GPKh; H-34; Heptachlorane; 3,4,5,6,7,8,8-Heptachlorodicyclopentadiene; 3,4,5,6,7,8,8a-Heptachlorodicyclopentadiene; 1(3a),4,5,6,7,8,8-Heptachloro-3a(1),4,7,7a-tetrahydro-4,7-methanoindene; 1,4,5,6,7,8,8-Heptachloro-3a,4,7,7a-tetrahydro-4,7-methanoindene; **1,4,5,6,7,8,8-Heptachloro-3a,4,7,7a-tetrahydro-4,7-methanol-1*H*-indene**; 1,4,5,6,7,8,8-Heptachloro-3a,4,7,7a-tetrahydro-4,7-*endo*-methanoindene; 1,4,5,6,7,8,8a-Heptachloro-3a,4,7,7a-tetrahydro-4,7-methanoindene; 1,4,5,6,7,8,8-Heptachloro-3a,4,7,7a-tetrahydro-4,7-methyleneindene; 1,4,5,6,7,10,10-Heptachloro-4,7,8,9-tetrahydro-4,7-methyleneindene; 1,4,5,6,7,10,10-Heptachloro-4,7,8,9-tetrahydro-4,7-*endo*-methyleneindene; 3,4,5,6,7,8,8a-Heptachloro-α-dicyclopentadiene; Heptadichlorocyclopentadiene; Heptagran; Heptagranox; Heptamak; Heptamul; Heptasol; Heptox; NA 2,761; NCI-C00180; Soleptax; RCRA waste number P059; Rhodiachlor; Velsicol 104; Velsicol heptachlor.

CHEMICAL DESIGNATIONS: CAS: 76-44-8; DOT: 2761; mf: $C_{10}H_5Cl_7$; fw:

373.32; RTECS: PC 0700000.

PROPERTIES: Cryst white to light tan, waxy solid with a camphor-like odor. Mp: 46-74°, 95-96°, bp: 135-145° @ 1-1.5 mm, *d*: 1.66 @ 20/4°, K_H: 2.3 x 10^{-3} atm·m^3/mol, log K_{oc}: 4.38, log K_{ow}: 4.40-5.5, S_o: @ 27°: acetone (750 g/L), benzene (1,060 g/L), carbon tetrachloride (1,120 g/L), cyclohexanone (1,190 g/L), alcohol (45 g/L), xylene (1,020 g/L); sol in ether, kerosene and ligroin, S_w: 180 ppb @ 25°, vp: 3 x 10^{-4} mm @ 20°.

TRANSFORMATION PRODUCTS: Many soil microorganisms were found to oxidize heptachlor to heptachlor epoxide. In addition, hydrolysis produced hydroxychlordene with subsequent epoxidation yielding 1-hydroxy-2,3-epoxychlordene. Heptachlor reacted with reduced hematin forming chlordene which decomposed to hexachlorocyclopentadiene and cyclopentadiene. In a model ecosystem containing plankton, *Daphnia magna*, mosquito larva (*Culex pipiens quinquefasciatus*), fish (*Cambusia affinis*), alga (*Oedogonium cardiacum*) and snail (*Physa* sp.), heptachlor degraded to 1-hydroxychlordene, 1-hydroxy-2,3-epoxychlordene, hydroxychlordene epoxide, heptachlor epoxide and 5 unidentified compounds. Sunlight and UV light converts heptachlor to photoheptachlor.

EXPOSURE LIMITS: NIOSH REL: IDLH 100 mg/m^3; OSHA PEL: TWA 0.5 mg/m^3; ACGIH TLV: TWA 0.5 mg/m^3.

SYMPTOMS OF EXPOSURE: In animals: tremors, convul, liv damage.

USE: Insecticide for termite control.

HEPTACHLOR EPOXIDE

SYNONYMS: ENT 25,584; Epoxy heptachlor; HCE; 1,4,5,6,7,8,8-Heptachloro-2,3-epoxy-2,3,3a,4,7,7a-hexahydro-4,7-methanoindene; 1,2,3,4,5,6,7,8,8-Heptachloro-2,3-epoxy-3a,4,7,7a-tetrahydro-4,7-methanoindene; **5a,6,6a-Hexahydro-2,5-methano-2*H*-indeno[1,2-*b*]oxirene**; 2,3,4,5,6,7,7-Heptachloro-1a,1b,5,5a,6,6a-hexahydro-2,5-methano-2*H*-oxireno[*a*]indene; Velsicol 53-CS-17.

CHEMICAL DESIGNATIONS: CAS: 1024-57-3; DOT: 2761; mf: $C_{10}H_5Cl_7O$; fw: 389.32; RTECS: PB 9450000.

PROPERTIES: Liq. Mp: 157-160°, K_H: 3.2 x 10^{-5} atm·m^3/mol @ 25°, log K_{oc}: 4.32 (calcd), log K_{ow}: 3.65, 5.40, S_w: 275 μg/L @ 25°, vp: 2.6 x 10^{-6} mm @ 20°.

TRANSFORMATION PRODUCTS: In a model ecosystem containing plankton, *Daphnia magna*, mosquito larva (*Culex pipiens quinquefasciatus*), fish (*Cambusia affinis*), alga (*Oedogonium cardiacum*) and snail (*Physa* sp.), heptachlor epoxide degraded to hydroxychlordene epoxide. Irradiation of heptachlor epoxide by a 450-W high-pressure mercury lamp gave 2 half-cage isomers, each containing a ketone functional group.

USES: Not known.

n-HEPTANE

SYNONYMS: Dipropylmethane; Gettysolve-C; **Heptane**; Heptyl hydride; UN

1206.

CHEMICAL DESIGNATIONS: CAS: 142-82-5; DOT: 1206; mf: C_7H_{16}; fw: 100.20; RTECS: MI 7700000.

PROPERTIES: Colorless liq with a faint, pleasant odor. Mp: -90.6°, bp: 98.4°, *d*: 0.6837 @ 20/4°, fl p: -3.9°, lel: 1.1%, uel: 6.7%, K_H: 2.035 atm·m^3/mol @ 25°, IP: 9.90 eV, log K_{ow}: 4.66, S_o: sol in acetone, alcohol, chloroform, ether and petroleum ether, S_w: 2.24-3.37 ppm @ 25°, vap d: 4.10 g/L @ 25°, 3.46 (air = 1), vp: 40 mm @ 20°.

TRANSFORMATION PRODUCTS: *n*-Heptane may biodegrade in 2 ways. The first is the formation of heptyl hydroperoxide which decomposes to *n*-heptanol followed by oxidation to heptanoic acid. The other pathway involves dehydrogenation to 1-heptene which may react with water giving *n*-heptanol.

EXPOSURE LIMITS: NIOSH REL: TWA 350 mg/m^3; OSHA PEL: TWA 400 ppm, STEL 500 ppm; ACGIH TLV: TWA 400 ppm, STEL 500 ppm.

SYMPTOMS OF EXPOSURE: Li-head, giddy, stupor; no appetite, nau; derm; chem pneu; unconsciousness.

USES: Standard in testing knock of gasoline engines and for octane rating determinations; anesthetic; solvent; organic synthesis.

2-HEPTANONE

SYNONYMS: Amyl methyl ketone; *n*-Amyl methyl ketone; Methyl amyl ketone; Methyl *n*-amyl ketone; UN 1110.

CHEMICAL DESIGNATIONS: CAS: 110-43-0; DOT: 1110; mf: $C_7H_{14}O$; fw: 114.19; RTECS: MJ 5075000.

PROPERTIES: Colorless liq with a banana-like odor. Mp: -35.5°, -27°, bp: 151.4°, *d*: 0.8111 @ 20/4°, fl p: 47°, lel: 1.1%, uel: 7.9%, K_H: 1.44 x 10^{-4} atm·m^3/mol @ 25°, IP: 9.33 eV, log K_{ow}: 1.98, S_o: sol in alcohol and ether; misc with organic solvents, S_w: 0.43 wt % @ 20°, vap d: 4.67 g/L @ 25°, 3.94 (air = 1), vp: 2.6 mm @ 20°.

EXPOSURE LIMITS: NIOSH REL: 10-h TWA 100 ppm, IDLH 4,000 ppm; OSHA PEL: TWA 100 ppm; ACGIH TLV: TWA 50 ppm.

SYMPTOMS OF EXPOSURE: Irrit eyes, muc memb; head; narco, coma; derm.

USES: Ingredient in artificial carnation oils; industrial solvent; synthetic flavoring; solvent for nitrocellulose lacquers; organic synthesis.

3-HEPTANONE

SYNONYMS: Butyl ethyl ketone; *n*-Butyl ethyl ketone; Ethyl butyl ketone; Heptan-3-one.

CHEMICAL DESIGNATIONS: CAS: 106-35-4; DOT: none assigned; mf: $C_7H_{14}O$; fw: 114.19; RTECS: MJ 5250000.

PROPERTIES: Colorless liq with a fruity odor. Mp: -39°, bp: 147° @ 765 mm, *d*: 0.8183 @ 20/4°, fl p: 41°, lel: 1.4%, uel: 8.8%, K_H: 4.20 x 10^{-5} atm·m^3/mol @

20° (calcd), IP: 9.15 eV, log K_{ow}: 1.32 (calcd), S_o: sol in alcohol and ether, S_w: 14,300 mg/L @ 20°, vap d: 4.67 g/L @ 25°, 3.94 (air = 1), vp: 4 mm @ 20°.

EXPOSURE LIMITS: NIOSH REL: IDLH 3,000 ppm; OSHA PEL: TWA 50 ppm; ACGIH TLV: TWA 50 ppm.

SYMPTOMS OF EXPOSURE: Irrit eyes, muc memb; head, narco, coma; derm.

USES: Solvent mixtures for coatings and finishes.

cis-2-HEPTENE

SYNONYMS: **(*Z*)-2-Heptene**; *cis*-2-Heptylene.

CHEMICAL DESIGNATIONS: CAS: 6443-92-1; DOT: 2278; mf: C_7H_{14}; fw: 98.19; RTECS: none assigned.

PROPERTIES: Colorless liq. Bp: 98.5°, *d*: 0.708 @ 20/4°, fl p: -6°, K_H: 4.13 x 10^{-1} atm·m^3/mol @ 20° (calcd), log K_{ow}: 2.88 (calcd), S_o: sol in acetone, alcohol, benzene, chloroform and ether, S_w: 15 ppm @ 25°, vap d: 4.01 g/L @ 25°, 3.39 (air = 1), vp: 48 mm @ 25°.

USE: Organic synthesis.

trans-2-HEPTENE

SYNONYMS: **(*E*)-2-Heptene**; *trans*-2-Heptylene.

CHEMICAL DESIGNATIONS: CAS: 14686-13-6; DOT: 2278; mf: C_7H_{14}; fw: 98.19; RTECS: none assigned.

PROPERTIES: Colorless liq. Mp: -109.5°, bp: 98°, *d*: 0.7012 @ 20/4°, fl p: -1°, K_H: 4.22 x 10^{-1} atm·m^3/mol @ 25° (calcd), log K_{ow}: 2.88 (calcd), S_o: sol in acetone, alcohol, benzene, chloroform and ether, S_w: 15 ppm @ 25°, vap d: 4.01 g/L @ 25°, 3.39 (air = 1), vp: 49 mm @ 25°.

USES: Organic synthesis.

HEXACHLOROBENZENE

SYNONYMS: Amatin; Anticarie; Bunt-cure; Bunt-no-more; Co-op hexa; Granox NM; HCB; Hexa C.B.; Julin's carbon chloride; No bunt; No bunt 40; No bunt 80; No bunt liquid; Pentachlorophenyl chloride; Perchlorobenzene; Phenyl perchloryl; RCRA waste number U127; Sanocide; Smut-go; Snieciotox; UN 2729.

CHEMICAL DESIGNATIONS: CAS: 118-74-1; DOT: 2729; mf: C_6Cl_6; fw: 284.78; RTECS: DA 2975000.

PROPERTIES: Monoclinic white cryst. Mp: 230°, bp: 323-326°, *d*: 2.049 @ 20/4°, fl p: 242°, K_H: 7.1 x 10^{-3} atm·m^3/mol @ 20°, log K_{oc}: 2.56-4.54, log K_{ow}: 3.93-6.42, S_o: sol in acetone, benzene, ether and chloroform, S_w: 40 µg/L @ 20°, vp: 1.089 x 10^{-5} mm @ 20°.

TRANSFORMATION PRODUCTS: Reductive monodechlorination occurred in an anaerobic sewage sludge yielding principally 1,3,5-trichlorobenzene. Other compounds identified included pentachlorobenzene, 1,2,3,5-tetrachlorobenzene and dichlorobenzenes. Solid hexachlorobenzene exposed to artificial sunlight for 5 mo photolyzed at a very slow rate with no decomposition products identified. The sunlight irradiation of hexachlorobenzene (20 g) in a 100 ml borosilicate glass-stoppered Erlenmeyer flask for 56 days yielded 64 ppm pentachlorobiphenyl. A CO_2 yield < 0.1% was observed when hexachlorobenzene adsorbed on silica gel was irradiated with light (λ > 290 nm) for 17 h. In activated sludge, only 1.5% of the applied hexachlorobenzene mineralized to CO_2 after 5 days. In a 5-day experiment, ^{14}C-labelled hexachlorobenzene applied to soil-water suspensions under aerobic and anaerobic conditions gave $^{14}CO_2$ yields of 0.4 and 0.2%, respectively. Irradiation ($\lambda \geq$ 285 nm) of hexachlorobenzene (1.1-1.2 mM/L) in an acetonitrile-water mixture containing acetone (concn = 0.553 mM/L) as a sensitizer gave the following products (% yield): pentachlorobenzene (71.0), 1,2,3,4-tetrachlorobenzene (0.6), 1,2,3,5-tetrachlorobenzene (2.2) and 1,2,4,5-tetrachlorobenzene (3.7). Without acetone, the identified photolysis products (% yield) included 1,2,3,4,5-pentachlorobenzene (76.8), 1,2,3,5-tetrachlorobenzene (1.2), 1,2,4,5-tetrachlorobenzene (1.7) and 1,2,4-trichlorobenzene (0.2).

USES: Intermediate in the manufacture of pentachlorophenol; seed fungicide; wood preservative.

HEXACHLOROBUTADIENE

SYNONYMS: Dolen-pur; GP-40-66:120; HCBD; Hexachlorbutadiene; 1,1,2,3,4,4-Hexachlorobutadiene; 1,3-Hexachlorobutadiene; Hexachloro-1,3-butadiene; **1,1,2,3,4,4-Hexachloro-1,3-butadiene**; Perchlorobutadiene; RCRA waste number U128; UN 2279.

CHEMICAL DESIGNATIONS: CAS: 87-68-3; DOT: 2279; mf: C_4Cl_6; fw: 260.76; RTECS: EJ 0700000.

PROPERTIES: Clear, yellowish-green liq with a mild to pungent odor. Mp: -21°, bp: 215°, *d*: 1.5542 @ 20/4°, K_H: 4.3 x 10^{-3} atm·m^3/mol @ 20°, log K_{oc}: 6.10 (calcd), log K_{ow}: 4.78, S_o: sol in ethanol and ether, S_w: 2 mg/L @ 20°, vap d: 10.66 g/L @ 25°, 9.00 (air = 1), vp: 0.15 mm @ 20°.

TRANSFORMATION PRODUCTS: In a model ecosystem containing plankton, *Daphnia magna*, mosquito larva (*Culex pipiens quinquefasciatus*), fish (*Cambusia affinis*), alga (*Oedogonium cardiacum*) and snail (*Physa* sp.), hexachlorocyclopentadiene degraded slightly but no products were identified.

EXPOSURE LIMITS: OSHA PEL: TWA 20 ppb.

USES: Solvent for elastomers, natural and synthetic rubber; heat-transfer liquid; transformer and hydraulic fluid; wash liquor for removing C_4 and higher hydrocarbons; sniff gas recovery agent in chlorine plants; chemical intermediate for fluorinated lubricants; fluid for gyroscopes; fumigant for grapes.

HEXACHLOROCYCLOPENTADIENE

SYNONYMS: C-56; Graphlox; HCCP; HCCPD; HCPD; Hex; **1,2,3,4,5,5-Hexa-**

chloro-1,3-cyclopentadiene; HRS 1,655; NCI-C55607; PCL; Perchlorocyclopentadiene; RCRA waste number U130; UN 2646.

CHEMICAL DESIGNATIONS: CAS: 77-47-4; DOT: 2646; mf: C_5Cl_6; fw: 272.77; RTECS: GY 1225000.

PROPERTIES: Greenish-yellow liq with a harsh, unpleasant odor. Mp: -9°, bp: 236-238°, *d*: 1.7119 @ 20/4°, K_H: 1.6 x 10^{-2} atm·m^3/mol, log K_{oc}: 3.63, log K_{ow}: 4.00-5.04, S_o: based on structurally similar compounds, hexachlorocyclopentadiene should be sol in benzene, ethanol, chloroform and other liq halogenated solvents, S_w: 1.8 mg/L @ 25°, vap d: 11.15 g/L @ 25°, 9.42 (air = 1), vp: 8.1 x 10^{-2} mm @ 25°.

TRANSFORMATION PRODUCTS: The major photolysis and hydrolysis products identified in distilled water are pentachlorocyclopentenone and hexachlorocyclopentenone. In mineralized water, the products identified include *cis*- and *trans*-pentachlorobutadiene, tetrachlorobutenyne and pentachloropentadienoic acid. In a similar experiment, irradiation of hexachlorocyclopentadiene in water by mercury-vapor lamps resulted in the formation of 2,3,4,4,5-pentachloro-2-cyclopentenone. This was found to hydrolyze partially to hexachloroindenone. Other photodegradation products identified include hexachloro-2-cyclopentenone and hexachloro-3-cyclopentenone as major products. Secondary photodegradation products reported include pentachloro-*cis*-2,4-pentadienoic acid, *Z*- and *E*-pentachlorobutadiene and tetrachlorobutyne. Anticipated products from the reaction of hexachlorocyclopentadiene with O_3 or hydroxyl radicals in the atmosphere are phosgene, diacylchlorides, ketones and Cl•. Phosgene is hydrolyzed readily to hydrogen chloride and CO_2.

EXPOSURE LIMITS: ACGIH TLV: TWA 10 ppb.

USES: Intermediate in the synthesis of dyes, cyclodiene pesticides, (aldrin, dieldrin, endosulfan), fungicides and pharmaceuticals; manufacture of chlorendic anhydride and chlorendic acid.

HEXACHLOROETHANE

SYNONYMS: Avlothane; Carbon hexachloride; Carbon trichloride; Distokal; Distopan; Distopin; Egitol; Ethane hexachloride; Ethylene hexachloride; Falkitol; Fasciolin; HCE; 1,1,1,2,2,2-Hexachloroethane; Hexachloroethylene; Mottenhexe; NA 9,037; NCI-C04604; Perchloroethane; Phenohep; RCRA waste number U131.

CHEMICAL DESIGNATIONS: CAS: 67-72-1; DOT: 9037; mf: C_2Cl_6; fw: 236.74; RTECS: KI 4025000.

PROPERTIES: Rhombic, triclinic or cubic colorless cryst with a camphor-like odor. Mp: subl, bp: 184.4°, *d*: 2.091 @ 20/4°, K_H: 2.18 x 10^{-3} atm·m^3/mol @ 25°, log K_{oc}: 3.34, log K_{ow}: 3.58-4.62, S_o: sol in ethanol, benzene, chloroform and ether, S_w: 50 ppm @ 20°, vap d: 6.3 kg/m^3 @ the sublimation point and 1 bar (0.987 atm), vp: 0.18 mm @ 20°.

TRANSFORMATION PRODUCTS: Under aerobic conditions or in experimental systems containing mixed cultures, hexachloroethane was reported to degrade to tetrachloroethane. In an uninhibited anoxic-sediment water suspension,

suspension, hexachloroethane degraded to tetrachloroethylene ($t_{1/2}$ = 19.7 min). Hydrolyzes very slowly in water to pentachloroethanol.

EXPOSURE LIMITS: NIOSH REL: lowest feasible level; OSHA PEL: TWA 1 ppm; ACGIH TLV: TWA 1 ppm.

SYMPTOMS OF EXPOSURE: Irrit eyes.

USES: Plasticizer for cellulose resins; moth repellant; camphor substitute in cellulose solvent; manufacture of smoke candles and explosives; solvent; rubber vulcanization accelerator; refining aluminum alloys.

n-HEXANE

SYNONYMS: Gettysolve-B; **Hexane**; Hexyl hydride; NCI-C60571; RCRA waste number U114; UN 1208.

CHEMICAL DESIGNATIONS: CAS: 110-54-3; DOT: 1208; mf: C_6H_{14}; fw: 86.18; RTECS: MN 9275000.

PROPERTIES: Colorless liq with a faint odor. Mp: -95°, bp: 69°, *d*: 0.6603 @ 20/4°, fl p: -21.7°, lel: 1.1%, uel: 7.5%, K_H: 1.184 atm·m^3/mol @ 25°, IP: 10.18 eV, log K_{ow}: 3.90, 4.11, S_o: misc with alcohol, chloroform and ether, S_w: 9.47 ppm @ 25°, vap d: 3.52 g/L @ 25°, 2.98 (air = 1), vp: 124 mm @ 20°.

TRANSFORMATION PRODUCTS: An aq soln irradiated by UV light @ 50° for 1 day resulted in a 50.51% yield of CO_2. Synthetic air containing gaseous nitrous acid and exposed to artificial sunlight (λ = 300-450 nm) photooxidized *n*-hexane into 2 isomers of hexyl nitrate and peroxyacetal nitrate. *n*-Hexane may biodegrade in 2 ways. The first is the formation of hexyl hydroperoxide which decomposes to *n*-hexanol followed by oxidation to hexanoic acid. The other pathway involves dehydrogenation to 1-hexene which may react with water giving *n*-hexanol.

EXPOSURE LIMITS: NIOSH REL: 10-h TWA 100 ppm, 15-min C 510 ppm, IDLH 5,000 ppm; OSHA PEL: TWA 500 ppm; ACGIH TLV: TWA 50 ppm.

SYMPTOMS OF EXPOSURE: Li-head; nau; head; numb, musc weak; irrit eyes, nose; derm; chem pneu; gidd.

USES: Determining refractive index of minerals; paint diluent; dyed hexane is used in thermometers instead of mercury; polymerization reaction medium; calibrations; solvent for vegetable oils; alcohol denaturant; organic synthesis.

2-HEXANONE

SYNONYMS: Butyl ketone; Butyl methyl ketone; *n*-Butyl methyl ketone; Hexanone-2; MBK; Methyl *n*-butyl ketone; MNBK; Propylacetone; RCRA waste number U159.

CHEMICAL DESIGNATIONS: CAS: 591-78-6; DOT: none assigned; mf: $C_6H_{12}O$; fw: 100.16; RTECS: MP 1400000.

PROPERTIES: Colorless liq. Mp: -56.9°, bp: 128°, *d*: 0.8113 @ 20/4°, fl p: -9°, lel: 1.2%, uel: 8.0%, K_H: 1.75 x 10^{-3} atm·m^3/mol @ 25° (calcd), IP: 9.35 eV, log

K_{oc}: 2.13 (calcd), log K_{ow}: 1.38, S_o: sol in acetone, ethanol and ether, S_w: 35 g/L @ 25°, vap d: 4.09 g/L @ 25°, 3.46 (air = 1), vp: 2 mm @ 20°.

EXPOSURE LIMITS: NIOSH REL: 10-h TWA 1 ppm, IDLH 5,000 ppm; OSHA PEL: TWA 100 ppm; ACGIH TLV: TWA 5 ppm.

USES: Solvent of paints, varnishes, nitrocellulose lacquers; denaturant for ethyl alcohol; organic synthesis.

1-HEXENE

SYNONYMS: Butylethylene; Hex-1-ene; Hexylene; UN 2370.

CHEMICAL DESIGNATIONS: CAS: 592-41-6; DOT: 2370; mf: C_6H_{12}; fw: 84.16; RTECS: MP 6600100.

PROPERTIES: Colorless liq. Mp: -139.8°, -98.5°, bp: 63.3°, *d*: 0.6731 @ 20/4°, fl p: -26.1°, K_H: 4.35 x 10^{-1} atm·m^3/mol @ 25°, IP: 9.45 eV, log K_{ow}: 2.25, S_o: sol in alcohol, benzene, chloroform, ether and petroleum, S_w: 50 ppm @ 25°, vap d: 3.4403 g/L @ 25°, 2.9156 (air = 1), vp: 186.0 mm @ 25°.

TRANSFORMATION PRODUCTS: Biooxidation of 1-hexene may occur yielding 5-hexen-1-ol which may further oxidize to give 5-hexenoic acid. Washed cell suspensions of bacteria belonging to the genera *Mycobacterium*, *Nocardia*, *Xanthobacter* and *Pseudomonas* and growing on selected alkenes metabolized 1-hexene to 1,2-epoxyhexane.

USES: Synthesis of perfumes, flavors, dyes and resins; polymer modifier.

sec-HEXYL ACETATE

SYNONYMS: Acetic acid 1,3-dimethylbutyl ester; 1,3-Dimethylbutyl acetate; *sec*-Hexyl acetate; MAAC; Methylamyl acetate; Methylisoamyl acetate; Methylisobutylcarbinol acetate; **4-Methyl-2-pentanol acetate**; 4-Methyl-2-pentyl acetate; UN 1233.

CHEMICAL DESIGNATIONS: CAS: 108-84-9; DOT: 1233; mf: $C_8H_{16}O_2$; fw: 144.21; RTECS: SA 7525000.

PROPERTIES: Colorless liq with a fruity odor. Mp: -63.9°, bp: 146.3°, 158°, *d*: 0.8658 @ 20/4°, fl p: 43.3° (OC), lel: 0.9% @ 43.3°, K_H: 4.38-5.84 x 10^{-3} atm·m^3/mol @ 20° (calcd), log K_{ow}: 3.37 (calcd), S_o: sol in alcohol and ether, S_w: 0.013 wt % @ 20°, vap d: 5.893.03 g/L @ 25°, 4.982.56 (air = 1), vp: 4 mm @ 20°.

TRANSFORMATION PRODUCTS: Anticipated hydrolysis products include 4-methyl-2-pentanol and acetic acid.

EXPOSURE LIMITS: NIOSH REL: IDLH 4,000 ppm; OSHA PEL: TWA 50 ppm.

SYMPTOMS OF EXPOSURE: Eye irrit, head, narcosis.

USES: Solvent for nitrocellulose and other lacquers.

HYDROQUINONE

SYNONYMS: Arctuvin; *p*-Benzenediol; **1,4-Benzenediol**; Benzohydroquinone;

Benzoquinol; Black and white bleaching cream; Dihydroxybenzene; *p*-Dihydroxybenzene; 1,4-Dihydroxybenzene; *p*-Dioxobenzene; Eldopaque; Eldoquin; Hydroquinol; Hydroquinole; α-Hydroquinone; *p*-Hydroquinone; 4-Hydroxyphenol; *p*-Hydroxyphenol; NCI-C55834; Quinol; β-Quinol; Quinone; Tecquinol; Tenox HQ; Tequinol; UN 2662; USAF EK-356.

CHEMICAL DESIGNATIONS: CAS: 123-31-9; DOT: 2662; mf: $C_6H_6O_2$; fw: 110.11; RTECS: MX 3500000.

PROPERTIES: Colorless to pale brown hexagonal cryst. Mp: 173-174°, bp: 285-287°, *d*: 1.358 @ 20/4°, fl p: 165°, pK_a: @ 25°: pK_1 = 10.0, pK_2 = 12.0, K_H: < 2.07 x 10^{-9} atm·m^3/mol @ 20-25° (calcd), log K_{oc}: 0.98 (calcd), log K_{ow}: 0.50-0.59, S_o: sol in acetone, alcohol and ether; slightly sol in benzene, S_w: 70 g/L @ 25°, vp: < 10^{-3} mm @ 20°.

TRANSFORMATION PRODUCTS: Ozonolysis products reported are *p*-quinone and dibasic acids. A CO_2 yield of 53.7% was achieved when hydroquinone adsorbed on silica gel was irradiated with light (λ > 290 nm) for 17 h. In activated sludge, 7.5% mineralized to CO_2 after 5 days. Under methanogenic conditions, inocula from a municipal sewage treatment plant digester degraded hydroquinone to phenol prior to being mineralized to CO_2 and methane. In various pure cultures, hydroquinone degraded to the following intermediates: benzoquinone, 2-hydroxy-1,4-benzoquinone and β-ketoadipic acid. Hydroquinone also degraded in activated sludge but no products were identified. Chlorine dioxide reacted with hydroquinone in an aq soln forming *p*-benzoquinone.

EXPOSURE LIMITS: NIOSH REL: 15-min C 2 mg/m^3, IDLH 200 mg/m^3; OSHA PEL: TWA 2 mg/m^3; ACGIH TLV: TWA 2 mg/m^3.

SYMPTOMS OF EXPOSURE: Irrit eyes, conj kera; excitement; colored urine; nau; dizz; suffocation, rapid breath; musc twitch, delirium; coll.

USES: Antioxidant; photographic reducer and developer for black and white film; determination of phosphate; dye intermediate; medicine; in monomeric liquids to prevent polymerization; stabilizer in paints and varnishes; motor fuels and oils.

INDAN

SYNONYMS: 2,3-Dihydroindene; **2,3-Dihydro-1*H*-indene**; Hydrindene.

CHEMICAL DESIGNATIONS: CAS: 496-11-7; DOT: none assigned; mf: C_9H_{10}; fw: 118.18; RTECS: NK 3750000.

PROPERTIES: Liq. Mp: -51.4°, bp: 178°, *d*: 0.9639 @ 20/4°, fl p: 50°, log K_{oc}: 2.48 (calcd), log K_{ow}: 2.38 (calcd), S_o: sol in alcohol and ether, S_w: 88.9 ppm @ 25°, vap d: 4.83 g/L @ 25°, 4.08 (air = 1).

USE: Organic synthesis.

INDENO[1,2,3-*cd*]PYRENE

SYNONYMS: Indenopyren; IP; 1,10-(*o*-Phenylene)pyrene; 2,3-Phenylenepyrene;

2,3-*o*-Phenylenepyrene; 3,4-(*o*-Phenylene)pyrene; 2,3-Phenylene-*o*-pyrene; 1,10-(1,2-Phenylene)pyrene; *o*-Phenylpyrene; RCRA waste number U137.

CHEMICAL DESIGNATIONS: CAS: 193-39-5; DOT: none assigned; mf: $C_{22}H_{12}$; fw: 276.34; RTECS: NK 9300000.

PROPERTIES: Solid. Mp: 160-163°, bp: 536°, pK_a: > 15, K_H: 2.96 x 10^{-20} atm·m^3/mol @ 25° (calcd), log K_{oc}: 7.49 (calcd), log K_{ow}: 5.97, 7.70, S_o: sol in most solvents, S_w: 62 μg/L, vp: 1.01 x 10^{-10} mm @ 25°.

EXPOSURE LIMITS: No individual standards have been set, however, as a constituent in coal tar pitch volatiles, the following exposure limits have been established: NIOSH REL: 10-h TWA 0.1 mg/m^3 (cyclohexane-extractable fraction); OSHA PEL: TWA 0.2 mg/m^3 (benzene-soluble fraction); ACGIH TLV: TWA 0.2 mg/m^3 (benzene solubles).

USES: Derived from industrial and experimental coal gasification operations where the maximum concn detected in coal tar streams was 1.7 mg/m^3.

INDOLE

SYNONYMS: 1-Azaindene; 1-Benzazole; Benzopyrrole; Benzo[*b*]pyrrole; 1-Benzo[*b*]pyrrole; 2,3-Benzopyrrole; **1*H*-Indole**; Ketole.

CHEMICAL DESIGNATIONS: CAS: 120-72-9; DOT: none assigned; mf: C_8H_7N; fw: 117.15; RTECS: NL 2450000.

PROPERTIES: Colorless to yellow scales with an unpleasant odor. Turns red on exposure to light and air. Mp: 52.5°, bp: 254°, *d*: 1.22, fl p: > 110°, log K_{oc}: 1.69 (calcd), log K_{ow}: 1.81-2.25, S_o: sol in alcohol, benzene, ether and lignoin, S_w: 3,558 ppm @ 25°.

TRANSFORMATION PRODUCTS: The aq chlorination of indole by hypochlorite/hypochlorous acid, chlorine dioxide and chloramines produced oxindole, isatin and possibly 3-chloroindole. In 9% anaerobic municipal sludge, indole degraded to 1,3-dihydro-2*H*-indol-2-one (oxindole) which further degraded to methane and CO_2.

USES: Chemical reagent; medicine; flavoring agent; perfumery.

INDOLINE

SYNONYMS: 2,3-Dihydroindole; **2,3-Dihydro-1*H*-indole**.

CHEMICAL DESIGNATIONS: CAS: 496-15-1; DOT: none assigned; mf: C_8H_9N; fw: 119.17; RTECS: none assigned.

PROPERTIES: Dark brown liq. Bp: 228-230°, *d*: 1.069 @ 20/4°, fl p: 92°, log K_{oc}: 1.42 (calcd), log K_{ow}: 0.16 (calcd), S_o: sol in acetone, benzene and ether, S_w: 10,800 ppm @ 25°.

USES: Organic synthesis.

1-IODOPROPANE

SYNONYMS: Propyl iodide; *n*-Propyl iodide.

CHEMICAL DESIGNATIONS: CAS: 107-08-4; DOT: 2392; mf: C_3H_7I; fw:

169.99; RTECS: TZ 4100000.

PROPERTIES: Colorless liq. Mp: -101°, bp: 102.4°, *d*: 1.7489 @ 20/4°, K_H: 9.09 x 10^{-3} atm·m^3/mol @ 25°, IP: 9.26 eV, log K_{oc}: 2.16 (calcd), log K_{ow}: 2.49 (calcd), S_o: sol in benzene and chloroform; misc with alcohol and ether, S_w: 1,040 mg/kg @ 30°, vap d: 6.95 g/L @ 25°, 5.87 (air = 1), vp: 43.1 mm @ 25°.

TRANSFORMATION PRODUCTS: A strain of *Acinetobacter* sp. isolated from activated sludge degraded 1-iodopropane to 1-propanol and iodide ions. Anticipated hydrolysis products include *n*-propyl alcohol and hydroiodic acid.

USES: Organic synthesis.

ISOAMYL ACETATE

SYNONYMS: Acetic acid isopentyl ester; Acetic acid 3-methylbutyl ester; Banana oil; Isoamyl ethanoate; Isopentyl acetate; Isopentyl alcohol acetate; **3-Methyl-1-butanol acetate**; 3-Methylbutyl acetate; 3-Methyl-1-butyl acetate; 3-Methylbutyl ethanoate; Pear oil.

CHEMICAL DESIGNATIONS: CAS: 123-92-2; DOT: none assigned; mf: $C_7H_{14}O_2$; fw: 130.19; RTECS: NS 9800000.

PROPERTIES: Colorless liq with a banana or pear-like odor. Mp: -78.5°, bp: 142°, *d*: 0.8670 @ 20/4°, fl p: 25°, lel: 1.0% @ 100°, uel: 7.5%, K_H: 5.87 x 10^{-2} atm·m^3/mol @ 25°, log K_{oc}: 1.95 (calcd), log K_{ow}: 2.30 (calcd), S_o: sol in acetone; misc with alcohol, amyl alcohol, ether and ethyl acetate, S_w: 0.2 wt % @ 20°, vap d: 5.32 g/L @ 25°, 4.49 (air = 1), vp: 4 mm @ 20°.

TRANSFORMATION PRODUCTS: Anticipated hydrolysis products include 3-methyl-1-butanol and acetic acid.

EXPOSURE LIMITS: NIOSH REL: IDLH 3,000 ppm; OSHA PEL: TWA 100 ppm; ACGIH TLV: TWA 100 ppm.

SYMPTOMS OF EXPOSURE: Irrit eyes, nose, throat; narco; derm.

USES: Artificial pear flavor in mineral waters and syrups; dyeing and finishing textiles; solvent for tannins, lacquers, nitrocellulose, camphor, oil colors and celluloid; manufacture of artificial leather, pearls or silk, photographic films, swelling bath sponges; celluloid cements, waterproof varnishes; bronzing liquids; metallic paints; perfumery; masking undesirable odors.

ISOAMYL ALCOHOL

SYNONYMS: Fermentation amyl alcohol; Fusel oil; Isoamylol; Isobutyl carbinol; Isopentanol; Isopentyl alcohol; 2-Methyl-4-butanol; 3-Methylbutanol; 3-Methyl-butan-1-ol; **3-Methyl-1-butanol**; Primary isoamyl alcohol; Primary isobutyl alcohol; UN 1105.

CHEMICAL DESIGNATIONS: CAS: 123-51-3; DOT: 1105; mf: $C_5H_{12}O$; fw: 88.15; RTECS: EL 5425000.

PROPERTIES: Clear, colorless liq with a pungent odor. Mp: -117.2°, bp: 132°, *d*: 0.8092 @ 20/4°, fl p: 42.8°, lel: 1.2% @ 100°, uel: 9.0% @ 100°, K_H: 8.89 x 10^{-6}

atm·m^3/mol @ 20° (calcd), log K_{ow}: 1.16, S_o: sol in acetone; misc with alcohol, benzene, chloroform, ether, glacial acetic acid, oils and petroleum ether, S_w: 30 g/L @ 20°, vap d: 3.60 g/L @ 25°, 3.04 (air = 1), vp: 2.3 mm @ 20°.

EXPOSURE LIMITS: NIOSH REL: IDLH 8,000 ppm; OSHA PEL: TWA 100 ppm; ACGIH TLV: TWA 100 ppm, STEL 125 ppm.

SYMPTOMS OF EXPOSURE: Irrit eyes, nose, throat; narcosis; head, dizz; dysp, nau, vomit, diarr; skin cracking.

USES: Determining fat content in milk; solvent for alkaloids, fats, oils; manufacturing isovaleric acid, isoamyl or amyl compounds, mercury fulminate, artificial silk, smokeless powders, lacquers, pyroxylin; dehydrating celloid in solns; photographic chemicals; pharmaceutical products; microscopy.

ISOBUTYL ACETATE

SYNONYMS: Acetic acid isobutyl ester; **Acetic acid 2-methylpropyl ester**; 2-Methylpropyl acetate; 2-Methyl-1-propyl acetate; *β*-Methylpropyl ethanoate; UN 1213.

CHEMICAL DESIGNATIONS: CAS: 110-19-0; DOT: 1213; mf: $C_6H_{12}O_2$; fw: 116.16; RTECS: AI 4025000.

PROPERTIES: Colorless liq with a fruity odor. Mp: -98.58°, bp: 117.2°, *d*: 0.8712 @ 20/4°, fl p: 17.8°, lel: 1.3%, uel: 10.5%, K_H: 4.85 x 10^{-4} atm·m^3/mol @ 25° (calcd), IP: 9.97 eV, log K_{ow}: 1.76 (calcd), S_o: sol in acetone, alcohol and ether, S_w: 6,300 mg/L @ 25°, vap d: 4.75 g/L @ 25°, 4.01 (air = 1), vp: 13 mm @ 20°.

TRANSFORMATION PRODUCTS: Anticipated hydrolysis products include 2-methylpropanol and acetic acid.

EXPOSURE LIMITS: NIOSH REL: IDLH 7,500 ppm; OSHA PEL: TWA 150 ppm; ACGIH TLV: TWA 150 ppm, STEL 187 ppm.

SYMPTOMS OF EXPOSURE: Head; drow; irrit eyes, upper resp, skin; anes.

USES: Solvent for nitrocellulose; in thinners, sealers and topcoat lacquers; flavoring; perfumery.

ISOBUTYL ALCOHOL

SYNONYMS: Fermentation butyl alcohol; 1-Hydroxymethylpropane; IBA; Isobutanol; Isopropylcarbinol; 2-Methylpropanol; 2-Methylpropanol-1; **2-Methyl-1-propanol**; 2-Methyl-1-propan-1-ol; 2-Methylpropyl alcohol; RCRA waste number U140; UN 1212.

CHEMICAL DESIGNATIONS: CAS: 78-83-1; DOT: 1212; mf: $C_4H_{10}O$; fw: 74.12; RTECS: NP 9625000.

PROPERTIES: Colorless liq with a sweet odor. Mp: -107.8°, bp: 108.1°, *d*: 0.8018 @ 20/4°, fl p: 27.8°, lel: 1.2% @ 100°, uel: 10.9% @ 100°, K_H: 9.25 x 10^{-6} atm·m^3/mol @ 20° (calcd), IP: 10.09 eV, log K_{ow}: 0.65, 0.83, S_o: sol in acetone; misc with alcohol and ether, S_w: 94.87 g/L @ 20°, vap d: 3.03 g/L @ 25°, 2.56 (air = 1), vp: 9 mm @ 20°.

EXPOSURE LIMITS: NIOSH REL: IDLH 8,000 ppm; OSHA PEL: TWA 100 ppm; ACGIH TLV: TWA 50 ppm.

SYMPTOMS OF EXPOSURE: Irrit eyes, throat; head; drow; skin irrit, cracking.

USES: Preparation of esters for the flavoring industry; solvent in varnish removers and paint; intermediate for amino coating resins; paint removers; liquid chromatography; fluorometric determinations; organic synthesis.

ISOBUTYLBENZENE

SYNONYMS: **(2-Methylpropyl)benzene**; 2-Methyl-1-phenylpropane; 1-Phenyl-2-methylpropane.

CHEMICAL DESIGNATIONS: CAS: 538-93-2; DOT: none assigned; mf: $C_{10}H_{14}$; fw: 134.22; RTECS: DA 3550000.

PROPERTIES: Colorless liq. Mp: -51.5°, bp: 172.8°, *d*: 0.8532 @ 20/4°, fl p: 60°, lel: 0.8%, uel: 6.0%, K_H: 1.09 x 10^{-2} atm·m^3/mol @ 25°, log K_{oc}: 3.90 (calcd), log K_{ow}: 4.11, S_o: sol in acetone, alcohol, benzene, ether and petroleum hydrocarbons, S_w: 33.71 mg/L @ 25°, vap d: 5.49 g/L @ 25°, 4.63 (air = 1), vp: 2.06 mm @ 25°.

TRANSFORMATION PRODUCTS: Oxidation of isobutylbenzene by *Pseudomonas desmolytica* S44B1 and *Pseudomonas convexa* S107B1 yielded 3-isobutylcatechol and (+)-2-hydroxy-8-methyl-6-oxononanoic acid.

USES: Perfume synthesis; flavoring; pharmaceutical intermediate.

ISOPHORONE

SYNONYMS: Isoacetophorone; Isoforon; Isoforone; Isooctaphenone; Isophoron; NCI-C55618; 1,1,3-Trimethyl-3-cyclohexene-5-one; Trimethylcyclohexenone; **3,5,5-Trimethyl-2-cyclohexen-1-one**.

CHEMICAL DESIGNATIONS: CAS: 78-59-1; DOT: 1224; mf: $C_9H_{14}O$; fw: 138.21; RTECS: GW 7700000.

PROPERTIES: Colorless liq with a camphor-like odor. Mp: -8.1°, bp: 215.2°, *d*: 0.9229 @ 20/4°, fl p: 84.4°, lel: 0.8%, uel: 3.8%, K_H: 5.8 x 10^{-6} atm·m^3/mol (calcd), log K_{oc}: 1.49 (calcd), log K_{ow}: 1.67, 1.70, S_o: sol in acetone, ethanol and ether, S_w: 12 g/L @ 25°, vap d: 5.65 g/L @ 25°, 4.77 (air = 1), vp: 0.38 mm @ 20°.

TRANSFORMATION PRODUCTS: The pure culture *Aspergillus niger* biodegraded isophorone to 3,5,5-trimethyl-2-cyclohexene-1,4-dione, 3,5,5-trimethylcyclohexane-1,4-dione, (*S*)-4-hydroxy-3,5,5-trimethyl-2-cyclohex-1-one and 3-hydroxymethyl-5,5-dimethyl-2-cyclohexen-1-one.

EXPOSURE LIMITS: NIOSH REL: 10-h TWA 4 ppm, IDLH 800 ppm; OSHA PEL: TWA 25 ppm; ACGIH TLV: C 5 ppm.

SYMPTOMS OF EXPOSURE: Irrit eyes, nose, throat; narco; derm; head; dizz.

USES: Solvent for paints, tin coatings, agricultural chemicals and synthetic resins; excellent solvent for vinyl resins, cellulose esters and ethers; pesticides; storing lacquers; pesticide manufacturing; intermediate in the manufacture of 3,5-xylenol, 2,3,5-trimethylcyclohexanol and 3,5-dimethylaniline.

ISOPROPYL ACETATE

SYNONYMS: Acetic acid isopropyl ester; **Acetic acid 1-methylethyl ester**; 2-Acetoxypropane; 2-Propyl acetate; UN 1220.

CHEMICAL DESIGNATIONS: CAS: 108-21-4; DOT: 1220; mf: $C_5H_{10}O_2$; fw: 102.13; RTECS: AI 4930000.

PROPERTIES: Colorless liq with an aromatic odor. Mp: -73.4°, bp: 90°, *d*: 0.8718 @ 20/4°, fl p: 2°, lel: 1.8%, uel: 8.0%, K_H: 2.81 x 10^{-4} atm·m^3/mol @ 25°, IP: 9.98 eV, log K_{ow}: 1.03 (calcd), S_o: sol in acetone; misc with alcohol and ether, S_w: 30.9 g/L @ 20°, vap d: 4.17 g/L @ 25°, 3.53 (air = 1), vp: 47.5 mm @ 20°.

TRANSFORMATION PRODUCTS: Hydrolyzes in water forming isopropyl alcohol and acetic acid ($t_{1/2}$ = 8.4 yr @ 25° and pH 7).

EXPOSURE LIMITS: NIOSH REL: IDLH 16,000 ppm; OSHA PEL: TWA 250 ppm, STEL 310 ppm; ACGIH TLV: TWA 250 ppm.

SYMPTOMS OF EXPOSURE: Irrit eyes, nose; narcosis; derm; skin irrit.

USES: Solvent for plastics, oils, fats and cellulose derivatives; perfumes; paints, lacquers and printing inks; perfumery; organic synthesis.

ISOPROPYLAMINE

SYNONYMS: 2-Aminopropane; 1-Methylethylamine; Monoisopropylamine; **2-Propanamine**; *sec*-Propylamine; 2-Propylamine; UN 1221.

CHEMICAL DESIGNATIONS: CAS: 75-31-0; DOT: 1221; mf: C_3H_9N; fw: 59.11; RTECS: NT 8400000.

PROPERTIES: Colorless liq with a penetrating ammonia-like odor. Mp: -95.2°, bp: 32.4°, *d*: 0.6891 @ 20/4°, fl p: -37.2°, lel: 2.0%, uel: 10.4%, pK_a: 10.53 @ 25°, IP: 8.72 eV, log K_{ow}: -0.03, S_o: sol in acetone, benzene and chloroform; misc with alcohol and ether, S_w: misc, vap d: 2.42 g/L @ 25°, 2.04 (air = 1), vp: 478 mm @ 20°.

TRANSFORMATION PRODUCTS: Releases toxic NO_x when heated to decomposition. Forms soluble salts with acids.

EXPOSURE LIMITS: NIOSH REL: IDLH 4,000 ppm; OSHA PEL: TWA 5 ppm.

SYMPTOMS OF EXPOSURE: Irrit eyes, nose, throat, skin; pulm edema; vis dist; skin, eye burns; derm.

USES: Intermediate in synthesis of rubber accelerators, dyes, pharmaceuticals, insecticides, bactericides, textiles and surface-active agents; solvent; dehairing agent; solubilizer for 2,4-D.

ISOPROPYLBENZENE

SYNONYMS: Cumene; Cumol; Isopropylbenzol; **(1-Methylethyl)benzene**; 2-Phenylpropane; RCRA waste number U055; UN 1221.

CHEMICAL DESIGNATIONS: CAS: 98-82-8; DOT: 1918; mf: C_9H_{12}; fw:

120.19; RTECS: GR 8575000.

PROPERTIES: Colorless liq. Mp: -96°, bp: 152.4°, 164.6°, *d*: 0.8618 @ 20/4°, fl p: 39°, lel: 0.9%, uel: 6.5%, K_H: 1.47 x 10^{-2} atm·m^3/mol @ 25°, IP: 8.69 eV, 9.13 eV, log K_{oc}: 3.45 (calcd), log K_{ow}: 3.51, 3.66, S_o: sol in acetone, alcohol, benzene, ether and carbon tetrachloride, S_w: 48.3 ppm @ 25°, vap d: 4.91 g/L @ 25°, 4.15 (air = 1), vp: 3.2 mm @ 20°.

TRANSFORMATION PRODUCTS: Photooxidation of isopropylbenzene with NO_x yielded nitric acid and benzaldehyde. A *n*-hexane soln containing isopropylbenzene and spread as a thin film (4 mm) on cold water (10°) was irradiated by a mercury medium pressure lamp. In 3 h, 22% photooxidized into α,α-dimethylbenzyl alcohol, 2-phenylpropionaldehyde and allylbenzene. When isopropylbenzene was incubated with *Pseudomonas putida*, the substrate was converted to *ortho*-dihydroxy compounds in which the isopropyl part of the compound remained intact. Oxidation of isopropylbenzene by *Pseudomonas desmolytica* S44B1 and *Pseudomonas convexa* S107B1 yielded 3-isopropylcatechol and a ring fission product, (+)-2-hydroxy-7-methyl-6-oxooctanoic acid.

SYMPTOMS OF EXPOSURE: Irrit eyes, muc memb; head; derm; narco; coma.

USES: Manufacture of acetone, acetophenone, diisopropylbenzene, α-methylstyrene and phenol, polymerization catalysts; constituent of motor fuel, asphalt and naphtha; catalyst for acrylic and polyester-type resins; octane booster for gasoline; solvent.

ISOPROPYL ETHER

SYNONYMS: Diisopropyl ether; Diisopropyl oxide; DIPE; IPE; 2-Isopropoxypropane; **2,2′-Oxybis[propane]**; UN 1159.

CHEMICAL DESIGNATIONS: CAS: 108-20-3; DOT: 1159; mf: $C_6H_{14}O$; fw: 102.18; RTECS: TZ 5425000.

PROPERTIES: Colorless liq with a penetrating, sweet, ether-like odor. Mp: -85.9°, -60°, bp: 68°, *d*: 0.7241 @ 20/4°, fl p: -27.8°, lel: 1.4%, uel: 7.9%, K_H: 9.97 x 10^{-3} atm·m^3/mol @ 25°, IP: 9.20 eV, log K_{ow}: 1.68 (calcd), S_o: sol in acetone; misc with alcohol and ether, S_w: 9 g/L @ 20°, vap d: 4.18 g/L @ 25°, 3.53 (air = 1), vp: 119 mm @ 20°.

EXPOSURE LIMITS: NIOSH REL: TWA 500 ppm; OSHA PEL: TWA 250 ppm.

SYMPTOMS OF EXPOSURE: Irrit eyes, nose; resp discomfort; derm in animals: drowsy, dizz, unconsciousness.

USES: Solvent for waxes, resins, animal, vegetable and mineral oils; paint and varnish removers; rubber cements; extracting acetic acid from soln; spotting compositions; rubber cements; organic synthesis.

KEPONE

SYNONYMS: Chlordecone; CIBA 8,514; Compound 1189; 1,2,3,5,6,7,8,9,10,10-

Decachloro[5.2.1.$0^{2,6}$.$0^{3,9}$.$0^{5,8}$]decano-4-one; Decachloroketone; Decachloro-1,3,4-metheno-2*H*-cyclobuta[*cd*]pentalen-2-one; Decachlorooctahydrokepone-2-one; Decachlorooctahydro-1,3,4-metheno-2*H*-cyclobuta[*cd*]pentalen-2-one; **1,1a,3,3a,4,5,5a,5b,6-Decachlorooctahydro-1,3,4-metheno-2*H*-cyclobuta[*cd*]pentalen-2-one**; Decachloropentacyclo[5.2.1.$0^{2,6}$.$0^{3,9}$.$0^{5,8}$]decan-3-one; Decachloropentacyclo[5.2.1.$0^{2,6}$.$0^{4,10}$.$0^{5,9}$]decan-3-one; Decachlorotetracyclodecanone; Decachlorotetrahydro-4,7-methanoindeneone; ENT 16,391; GC-1189; General chemicals 1189; Merex; NA 2,761; NCI-C00191; RCRA waste number U142.

CHEMICAL DESIGNATIONS: CAS: 143-50-0; DOT: none assigned; mf: $C_{10}Cl_{10}O$; fw: 490.68; RTECS: PC 8575000.

PROPERTIES: Cryst. Mp: subl, bp: 350° (decomp), K_H: 3.11 x 10^{-2} atm·m^3/mol @ 25° (calcd), log K_{oc}: 4.74 (calcd), log K_{ow}: 4.07 (calcd), S_o: sol in acetic acid, alcohols and ketones, S_w: 7,600 μg/L @ 24°, vp: 2.25 x 10^{-7} mm @ 25°.

TRANSFORMATION PRODUCTS: Readily reacts with moisture forming hydrates. Kepone-contaminated soils from a site in Hopewell, VA were analyzed by GC/MS. 8-Chloro and 9-chloro homologs identified suggested these were photodegradation products of kepone.

USES: Insecticide; fungicide.

LINDANE

SYNONYMS: Aalindan; Aficide; Agrisol G-20; Agrocide; Agrocide 2; Agrocide 6G; Agrocide 7; Agrocide III; Agrocide WP; Agronexit; Ameisenatod; Ameisenmittel merck; Aparasin; Aphtiria; Aplidal; Arbitex; BBX; Ben-hex; Bentox 10; Benzene hexachloride; Benzene-γ-hexachloride; γ-Benzene hexachloride; Bexol; BHC; γ-BHC; Celanex; Chloran; Chloresene; Codechine; DBH; Detmol-extrakt; Detox 25; Devoran; Dol granule; ENT 7,796; Entomoxan; Exagama; Forlin; Gallogama; Gamacid; Gamaphex; Gamene; Gamiso; Gammahexa; Gammalin; Gammexene; Gammopaz; Gexane; HCCH; HCH; γ-HCH; Heclotox; Hexa; γ-Hexachlor; Hexachloran; γ-Hexachloran; Hexachlorane; γ-Hexachlorane; γ-Hexachlorobenzene; 1,2,3,4,5,6-Hexachlorocyclohexane; **1α,2α,3β,4α,5α,6-β-Hexachlorocyclohexane**; 1,2,3,4,5,6-Hexachloro-γ-cyclohexane; γ-Hexachlorocyclohexane; γ-1,2,3,4,5,6-Hexachlorocyclohexane; Hexatox; Hexaverm; Hexicide; Hexyclan; HGI; Hortex; Inexit; γ-Isomer; Isotox; Jacutin; Kokotine; Kwell; Lendine; Lentox; Lidenal; Lindafor; Lindagam; Lindagrain; Lindagranox; γ-Lindane; Lindapoudre; Lindatox; Lindosep; Lintox; Lorexane; Milbol 49; Mszycol; NA 2,761; NCI-C00204; Neo-scabicidol; Nexen FB; Nexit; Nexit-stark; Nexol-E; Nicochloran; Novigam; Omnitox; Ovadziak; Owadziak; Pedraczak; Pflanzol; Quellada; RCRA waste number U129; Silvanol; Spritz-rapidin; Spruehpflanzol; Streunex; Tap 85; TBH; Tri-6; Viton.

CHEMICAL DESIGNATIONS: CAS: 58-89-9; DOT: 2761; mf: $C_6H_6Cl_6$; fw: 290.83; RTECS: GV 4900000.

PROPERTIES: Colorless and odorless solid. Mp: 112.5°, bp: 323.4°, *d*: 1.5691 @ 23.6/4°, K_H: 2.43 x 10^{-7} atm·m^3/mol @ 23°, log K_{oc}: 2.38-3.52, log K_{ow}: 3.20-

3.89, S_o: @ 20°: acetone (30.31 wt %), benzene (22.42 wt %), chloroform (19.35 wt %), ether (17.22 wt %), ethanol (6.02 wt %), S_w: 7.52 ppm @ 25°, vap d: 518 ng/L @ 20°, vp: 9.4 x 10^{-6} mm @ 20°.

TRANSFORMATION PRODUCTS: In moist soils, lindane biodegraded to γ-pentachlorocyclohexene. Under anaerobic conditions, degradation by soil bacteria yielded γ-3,4,5,6-tetrachloro-1-cyclohexane and α-BHC. Other reported biodegradation products include penta- and tetrachloro-1-cyclohexanes and penta- and tetrachlorobenzenes. Incubation of lindane for 6 wk in a sandy loam soil under flooded conditions yielded γ-3,4,5,6-tetrachlorocyclohexane, γ-2,3,4,5,6-pentachlorocyclohex-1-ene and small amounts of 1,2,4-trichlorobenzene, 1,2,3,4-tetrachlorobenzene, 1,2,3,5- and/or 1,2,4,5-tetrachlorobenzene. Incubation of lindane in moist soil for 8 wk yielded the following metabolites: γ-3,4,5,6-tetrachlorocyclohexene, γ-1,2,3,4,5-pentachlorocyclohex-1-ene, pentachlorobenzene, 1,2,3,4-tetrachlorobenzene, 1,2,3,5- and/or 1,2,4,5-tetrachlorobenzene, 1,2,4-trichlorobenzene, 1,3,5-trichlorobenzene, *m*- and/or *p*-dichlorobenzene. Photolysis produces β-BHC. Microorganisms isolated from a loamy sand soil degraded lindane and some of the metabolites identified were pentachlorobenzene, 1,2,4,5-tetrachlorobenzene, 1,2,3,5-tetrachlorobenzene, γ-2,3,4,5,6-pentachloro-1-cyclohexane (γ-PCCH), γ-3,4,5,6-tetrachloro-1-cyclohexane (γ-TCCH) and β-3,4,5,6-tetrachloro-1-cyclohexane (β-TCCH). In a laboratory experiment, a *Pseudomonas* culture transformed lindane to γ-TCCH, γ-PCCH and α-BHC. γ-TCCH was also reported as a product of lindane degradation by *Clostridium sphenoides*. Lindane appeared to be metabolized by several grasses to β-BHC. Indigenous microbes in soil partially degraded lindane to CO_2. In a 42-day experiment, ^{14}C-labelled lindane applied to soil-water suspensions under aerobic and anaerobic conditions gave $^{14}CO_2$ yields of 1.9 and 3.0%, respectively.

EXPOSURE LIMITS: NIOSH REL: IDLH 1,000 mg/m^3; OSHA PEL: TWA 0.5 mg/m^3; ACGIH TLV: TWA 0.5 mg/m^3.

SYMPTOMS OF EXPOSURE: Irrit eyes, nose, throat; head; nau; clonic convul; resp probs; cyan; aplastic anem; skin irrit; musc spas.

USES: Pesticide and insecticide.

MALATHION

SYNONYMS: American Cyanamide 4,049; *S*-1,2-Bis(carbethoxy)ethyl-*O,O*-dimethyl dithiophosphate; *S*-1,2-Bis(ethoxycarbonyl)ethyl-*O,O*-dimethyl phosphorodithioate; *S*-1,2-Bis(ethoxycarbonyl)ethyl-*O,O*-dimethyl thiophosphate; Calmathion; Carbethoxy malathion; Carbetovur; Carbetox; Carbofos; Carbophos; Celthion; Chemathion; Cimexan; Compound 4049; Cythion; Detmol MA; Detmol MA 96%; *S*-1,2-Dicarbethoxyethyl-*O,O*-dimethyl dithiophosphate; Dicarboethoxyethyl-*O,O*-dimethyl phosphorodithioate; 1,2-Di(ethoxycarbonyl)ethyl-*O,O*-dimethyl phosphorodithioate; *S*-1,2-Di(ethoxycarbonyl)ethyl dimethyl phosphorothiolothionate; Diethyl (dimethoxyphosphinothioythio) butanedioate; Diethyl (dimethoxyphosphinothioylthio) succinate; Diethyl mercaptosuccinate, *O,O*-dimethyl phosphorodithioate; Diethyl mercaptosuccinate, *O,O*-dimethyl thiophosphate; Diethylmercaptosuccinic acid *O,O*-dimethyl phosphorodithioate; **[(Dimethoxyphosphinothioyl)thio]butanedioic acid diethyl ester**; *O,O*-Dimethyl-*S*-1,2-bis(ethoxycarbonyl)ethyldithiophosphate; *O,O*-Dimethyl-*S*-(1,2-dicarbethoxyethyl)dithiophosphate; *O,O*-Dimethyl-*S*-(1,2-dicarbethoxyethyl)phosphorodithioate; *O,O*-Dimethyl-*S*-(1,2-dicarbethoxyethyl)thiothionophosphate; *O,O*-Dimethyl-*S*-1,2-di(ethoxycarbamyl)ethyl phosphorodithioate; *O,O*-Dimethyldithiophosphate dimethylmercaptosuccinate; EL 4,049; Emmatos; Emmatos extra; ENT 17,034; Ethiolacar; Etiol; Experimental insecticide 4049; Extermathion; Formal; Forthion; Fosfothion; Fosfotion; Four thousand forty-nine; Fyfanon; Hilthion; Hilthion 25WDP; Insecticide no. 4,049; Karbofos; Kop-thion; Kypfos; Malacide; Malafor; Malakill; Malagran; Malamar; Malamar 50; Malaphele; Malaphos; Malasol; Malaspray; Malathion E50; Malathion LV concentrate; Malathion ULV concentrate; Malathiozoo; Malathon; Malathyl LV concentrate & ULV concentrate; Malatol; Malatox; Maldison; Malmed; Malphos; Maltox; Maltox MLT; Mercaptosuccinic acid diethyl ester; Mercaptothion; MLT; Moscardia; NA 2,783; NCI-C00215; Oleophosphothion; Orthomalathion; Phosphothion; Prioderm; Sadofos; Sadophos; SF 60; Siptox I; Sumitox; Tak; TM-4049; Vegfru malatox; Vetiol; Zithiol.

CHEMICAL DESIGNATIONS: CAS: 121-75-5; DOT: 2783; mf: $C_{10}H_{19}O_6PS_2$; fw: 330.36; RTECS: WM 8400000.

PROPERTIES: Yellow to brown liq with a garlic odor. Mp: 2.9°, bp: 156-157° @ 0.7 mm, *d*: 1.23 @ 25/4°, fl p: > 162.8° (OC), K_H: 4.89 x 10^{-9} atm·m^3/mol @ 25°, log K_{oc}: 2.61, log K_{ow}: 2.84, 2.89, S_o: misc with most organic solvents, S_w: 145 ppm @ 20°, vap d: 13.50 g/L @ 25°, 11.40 (air = 1), vp: 1.25 x 10^{-6} mm @ 20°.

TRANSFORMATION PRODUCTS: Hydrolyzes in water forming *cis*-diethyl fumarate, *trans*-diethyl fumarate, thiomalic acid and dimethyl thiophosphate. The reported hydrolysis half-lives @ pH 7.4 @ 20 and 37.5° were 10.5 and 1.3 days, respectively. Malathion was degraded by soil microcosms isolated from an agricultural area on Kauai, HI. Degradation half-lives in the laboratory and field experiments were 8.2 and 2 h, respectively. Dimethyl phosphorodithioic acid and diethyl fumarate were identified as degradation products. Malathion also degraded in groundwater and seawater but at a slower rate (avg $t_{1/2}$= 4.7 d). Microorganisms isolated from papermill effluents were responsible for the formation of malathion monocarboxylic acid. Malaoxon and phosphoric were reported as ozonation products of malathion in drinking water. At 87° and pH 2.5, malathion degraded in water to malathion α-monoacid and malathion β-monoacid. From the extrapolated acid degradation constant @ 27°, the $t_{1/2}$ was calculated to be > 4 yr. Under alkaline conditions (pH 8 and 27°), malathion degraded in water to malathion monoacid, diethyl fumarate, ethyl hydrogen fumarate and *O,O*-dimethyl phosphorodithioic acid. At pH 8, the reported $t_{1/2}$ @ 0, 27 and 40° are 40 days, 36 and 1 h, respectively.

EXPOSURE LIMITS: NIOSH REL: 10-h TWA 15 mg/m^3, IDLH 5,000 mg/m^3; OSHA PEL: TWA 15 mg/m^3; ACGIH TLV: TWA 10 mg/m^3.

SYMPTOMS OF EXPOSURE: Miosis; eye, skin irrit; rhin; head; tight chest, wheez, lar spas; salv; anor, nau, vomit, abdom cramp; diarr, atax.

USE: Insecticide.

MALEIC ANHYDRIDE

SYNONYMS: *cis*-Butenedioic anhydride; **2,5-Furanedione**; Maleic acid anhydride; RCRA waste number U147; Toxilic anhydride; UN 2215.

CHEMICAL DESIGNATIONS: CAS: 108-31-6; DOT: 2215; mf: $C_4H_2O_3$; fw: 98.06; RTECS: ON 3675000.

PROPERTIES: White cryst or colorless needles. Mp: 60°, bp: 197-199°, *d*: 1.48 @ 20/4°, fl p: 101.7°, lel: 1.4%, uel: 7.1%, S_o @ 25° (wt %): acetone (227), benzene (50), carbon tetrachloride (0.60), chloroform (52.5), ethyl acetate (112), ligroin (0.25), toluene (23.4), *o*-xylene (19.4), alcohol and dioxane, vp: 5 x 10^{-5} mm @ 20°.

TRANSFORMATION PRODUCTS: Reacts with water forming maleic acid. Anticipated products from the reaction of maleic anhydride with O_3 or hydroxyl radicals in the atmosphere are CO, CO_2, aldehydes and esters.

EXPOSURE LIMITS: NIOSH REL: IDLH 500 ppm; OSHA PEL: TWA 0.25 ppm.

SYMPTOMS OF EXPOSURE: Conj; photophobia, double vision; nas, upper resp irrit; bron, asth; derm.

USES: In organic synthesis (Diels-Alder reactions); manufacture of agricultural chemicals, dye intermediates, pharmaceuticals and alkyd-type resins; manufacture of fumaric and tartaric acids; pesticides; preservative for oils and fats.

MESITYL OXIDE

SYNONYMS: Isobutenyl methyl ketone; Isopropylidene acetone; Methyl isobutenyl ketone; 2-Methyl-2-penten-4-one; **4-Methyl-2-penten-2-one**; MIBK; UN 1229.

CHEMICAL DESIGNATIONS: CAS: 141-79-7; DOT: 1229; mf: $C_6H_{10}O$; fw: 98.14; RTECS: SB 4200000.

PROPERTIES: Clear, pale yellow liq with a strong peppermint or honey-like odor. Mp: -59 to -41.5°, bp: 129.7°, *d*: 0.8653 @ 20/4°, fl p: 30.6°, lel: 1.4%, uel: 7.2%, K_H: 4.01 x 10^{-6} atm·m^3/mol @ 20°, IP: 9.08 eV, log K_{ow}: 1.25 (calcd), S_o: sol in acetone, alcohol and ether, S_w: 28 g/L @ 20°, vap d: 4.01 g/L @ 25°, 3.39 (air = 1), vp: 8.7 mm @ 20°.

EXPOSURE LIMITS: NIOSH REL: 10-h TWA 10 ppm, IDLH 5,000 ppm; OSHA PEL: TWA 25 ppm; ACGIH TLV: TWA 15 ppm.

SYMPTOMS OF EXPOSURE: Irrit eyes, skin, muc memb; narco, coma.

USES: Solvent for nitrocellulose, gums and resins; roll-coating inks, varnishes, lacquers, stains and enamels; organic synthesis; insect repellent; ore flotation.

METHOXYCHLOR

SYNONYMS: 2,2-Bis(*p*-anisyl)-1,1,1-trichloroethane; 1,1-Bis(*p*-ethoxyphenyl)-2,2,2-trichloroethane; 2,2-Bis(*p*-methoxyphenyl)-1,1,1-trichloroethane; Chemform; 2,2-Di-*p*-anisyl-1,1,1-trichloroethane; Dimethoxy-DDT; *p,p'*-Dimethoxydiphenyltrichloroethane; Dimethoxy-DT; 2,2-Di-(*p*-methoxyphenyl)-1,1,1-trichloroethane; Di(*p*-methoxyphenyl)trichloromethyl methane; DMDT; 4,4'-DMDT; *p,p'*-

DMDT; DMTD; ENT 1,716; Maralate; Marlate; Marlate 50; Methoxcide; Methoxo; 4,4'-Methoxychlor; *p,p'*-Methoxychlor; Methoxy-DDT; Metox; Moxie; NCI-C00497; RCRA waste number U247; 1,1,1-Trichloro-2,2-bis(*p*-anisyl)ethane; 1,1,1-Trichloro-2,2-bis(*p*-methoxyphenol)ethanol; 1,1,1-Trichloro-2,2-bis(*p*-methoxyphenyl)ethane; 1,1,1-Trichloro-2,2-di(4-methoxyphenyl)ethane; **1,1'-(2,2,2-Trichloroethylidene)bis(4-methoxybenzene).**

CHEMICAL DESIGNATIONS: CAS: 72-43-5; DOT: 2761; mf: $C_{16}H_{15}Cl_3O_2$; fw: 345.66; RTECS: KJ 3675000.

PROPERTIES: White, gray or pale yellow cryst or powder. May be dissolved in an organic solvent or petroleum distillate for application. Pungent to mild fruity odor. Mp: 89-98°, bp: decomp, *d*: 1.41 @ 25/4°, log K_{oc}: 4.90, 4.95, log K_{ow}: 3.31-5.08, S_o: sol in ethanol and chloroform, S_w: 40 μg/L @ 24°.

TRANSFORMATION PRODUCTS: Degradation by *Aerobacter aerogenes* under aerobic or anaerobic conditions yielded 1,1-dichloro-2,2-bis(*p*-methoxyphenyl)ethylene and 1,1-dichloro-2,2-bis(*p*-methoxyphenyl)ethane. Hydrolysis at common aquatic pH's produced anisoin, anisil and 2,2-bis(*p*-methoxyphenyl)-1,1-dichloroethylene (est $t_{1/2}$ = 270 days @ 25° and pH 7.1). In air-saturated distilled water, direct photolysis of methoxychlor (λ > 280 nm) produced 1,1-bis(*p*-methoxyphenyl)-2,2-dichloroethylene (DMDE) which photolyzed to *p*-methoxybenzaldehyde (est $t_{1/2}$ = 4.5 mo).

EXPOSURE LIMITS: NIOSH REL: IDLH 7,500 mg/m^3; OSHA PEL: TWA 15 mg/m^3; ACGIH TLV: TWA 10 mg/m^3.

SYMPTOMS OF EXPOSURE: None known in humans; in animals: trembling, convul, kid and liv damage.

USES: Insecticide to control mosquito larvae and house flies; to control ectoparasites on cattle, sheep and goats; recommended for use in dairy barns.

METHYL ACETATE

SYNONYMS: Devoton; **Acetic acid methyl ester**; Methyl ethanoate; Tereton; UN 1231.

CHEMICAL DESIGNATIONS: CAS: 79-20-9; DOT: 1231; mf: $C_3H_6O_2$; fw: 74.08; RTECS: AI 9100000.

PROPERTIES: Colorless liq with a pleasant odor. Mp: -98.1°, bp: 57°, *d*: 0.9330 @ 20/4°, fl p: -10.0°, lel: 3.1%, uel: 16%, K_H: 9.09 x 10^{-5} atm·m^3/mol @ 25°, IP: 10.27 eV, log K_{ow}: 0.18, S_o: sol in acetone, benzene and chloroform; misc with alcohol and ether, S_w: 319 g/L @ 20°, vap d: 3.03 g/L @ 25°, 2.56 (air = 1), vp: 170 mm @ 20°.

TRANSFORMATION PRODUCTS: Anticipated hydrolysis products include methyl alcohol and acetic acid.

EXPOSURE LIMITS: NIOSH REL: IDLH 10,000 ppm; OSHA PEL: TWA 200 ppm.

SYMPTOMS OF EXPOSURE: Irrit nose, throat; head; drow; optic atrophy.

USES: Solvent for resins, lacquers, oils, acetylcellulose, nitrocellulose; paint removers; synthetic flavoring.

METHYL ACRYLATE

SYNONYMS: Acrylic acid methyl ester; Curithane 103; Methoxycarbonylethylene; Methyl propenate; Methyl propenoate; Methyl-2-propenoate; Propenoic acid methyl ester; **2-Propenoic acid methyl ester**; UN 1919.

CHEMICAL DESIGNATIONS: CAS: 96-33-3; DOT: 1919; mf: $C_4H_6O_2$; fw: 86.09; RTECS: AT 2800000.

PROPERTIES: Clear, colorless liq with a heavy, sweet odor. Mp: -90°, bp: 80.5°, *d*: 0.9561 @ 20/4°, fl p: -10.6°, lel: 2.8%, uel: 25.0%, K_H: 1.23-1.44 x 10^{-4} atm·m^3/mol @ 20° (calcd), IP: 9.19 eV, log K_{ow}: 0.67 (calcd), S_o: sol in acetone, alcohol, benzene and ether, S_w: 60 g/L @ 20°, vap d: 3.52 g/L @ 25°, 2.97 (air = 1), vp: 68.2 mm @ 20°.

TRANSFORMATION PRODUCTS: Polymerizes on standing and is accelerated by heat, light and peroxides.

EXPOSURE LIMITS: NIOSH REL: IDLH 1,000 ppm; OSHA PEL: TWA 10 ppm.

SYMPTOMS OF EXPOSURE: Irrit eyes, upper resp, skin.

USES: Manufacturing plastic films, leather finish resins, textile and paper coatings; amphoteric surfactants; chemical intermediate.

METHYLAL

SYNONYMS: Anesthenyl; **Dimethoxymethane**; Formal; Formaldehyde dimethylacetal; Methyl formal; Methylene dimethyl ether; UN 1234.

CHEMICAL DESIGNATIONS: CAS: 109-87-5; DOT: 1234; mf: $C_3H_8O_2$; fw: 76.10; RTECS: PA 8750000.

PROPERTIES: Colorless liq with a pungent, chloroform-like odor. Mp: -104.8°, bp: 45.5°, *d*: 0.8593 @ 20/4°, fl p: -32° (OC), lel: 1.6%, uel: 13.8%, K_H: 1.73 x 10^{-4} atm·m^3/mol @ 25°, IP: 10.00 eV, log K_{ow}: 0.00, S_o: sol in acetone and benzene; misc with alcohol, ether and oils, S_w: 33 wt % @ 20°, vap d: 3.11 g/L @ 25°, 2.63 (air = 1), vp: 330 mm @ 20°.

EXPOSURE LIMITS: NIOSH REL: IDLH 10,000 ppm; OSHA PEL: TWA 1,000 ppm.

SYMPTOMS OF EXPOSURE: Mild irrit eyes, upper resp; anes; irrit skin.

USES: Artificial resins; perfumery; solvent; adhesives; protective coatings; special fuel; organic synthesis (Grignard and Reppe reactions).

METHYL ALCOHOL

SYNONYMS: Carbinol; Colonial spirit; Columbian spirits; Columbian spirits (wood alcohol); **Methanol**; Methyl hydroxide; Methylol; Monohydroxymethane; NA 1,230; Pyroxylic spirit; RCRA waste number U154; Wood alcohol; Wood naphtha; Wood spirit.

CHEMICAL DESIGNATIONS: CAS: 67-56-1; DOT: 1230; mf: CH_4O; fw:

32.04; RTECS: PC 1400000.

PROPERTIES: Colorless liq with a characteristic odor. Mp: -97.8°, bp: 65°, *d*: 0.7914 @ 20/4°, fl p: 11.1°, lel: 6%, uel: 36%, pK_a: ≈ 16, K_H: 4.66 x 10^{-6} atm·m^3/mol @ 25°, IP: 10.84 eV, log K_{ow}: -0.82 to -0.66, S_o: sol in acetone and chloroform; misc with benzene, ethanol, ether, ketone and many other organic solvents, S_w: a saturated soln in equilibrium with its own vapor had a concn of 1,163 g/L @ 25°, vap d: 1.31 g/L @ 25° 1.11 (air = 1), vp: 92 mm @ 20°.

TRANSFORMATION PRODUCTS: Photooxidation of methyl alcohol in an oxygen rich atmosphere (20%) yielded formaldehyde and hydroxyperoxyl radicals. With chlorine, formaldehyde, CO, H_2O_2 and formic acid were detected. In a 5-day experiment, ^{14}C-labelled methyl alcohol applied to soil-water suspensions under aerobic and anaerobic conditions gave $^{14}CO_2$ yields of 53.4 and 46.3%, respectively. In a smog chamber, methyl alcohol reacted with NO_2 to give methyl nitrite and nitric acid. The formation of these products was facilitated when this experiment was accompanied by UV light.

EXPOSURE LIMITS: NIOSH REL: 10-h TWA 200 ppm, 15-min C 800 ppm, IDLH 25,000 ppm; OSHA PEL: TWA 200 ppm.

SYMPTOMS OF EXPOSURE: Eye irrit, head, drowsy, li-head, nau, vomit, vis dist, eye burns, digestive dist, failure of vision.

USES: Solvent for nitrocellulose, ethyl cellulose, polyvinyl butyral, rosin, shellac, manila resin, dyes; fuel for utility plants; home heating oil extender; preparation of methyl esters, formaldehyde, methacrylates, methylamines, dimethyl terephthalate, polyformaldehydes; methyl halides, ethylene glycol; in gasoline and diesel oil antifreezes; octane booster in gasoline; source of hydrocarbon for fuel cells; extractant for animal and vegetable oils; denaturant for ethanol; softening agent for certain plastics; dehydrator for natural gas.

METHYLAMINE

SYNONYMS: Aminomethane; Carbinamine; Mercurialin; **Methanamine**; Monomethylamine; UN 1061; UN 1235.

CHEMICAL DESIGNATIONS: CAS: 74-89-5; DOT: 1061; mf: CH_5N; fw: 31.06; RTECS: PF 6300000.

PROPERTIES: Colorless gas with a strong ammonia-like odor. Mp: -93.5°, bp: -6.3°, *d*: 0.6628 @ 20/4°, fl p: -10° (gas), 1.1° (30% soln), lel: 4.9%, uel: 20.7%, pK_a: 10.657 @ 25°, K_H: 1.81 x 10^{-2} atm·m^3/mol @ 25° (calcd), IP: 8.97 eV, log K_{ow}: -0.57, S_o: sol in acetone, alcohol, chloroform, lignoin, benzene (105 g/L @ 25°); misc with ether, S_w: 9,590 ml/L @ 25°, vap d: 1.27 g/L @ 25°, 1.07 (air = 1), vp: 3.1 atm @ 20°.

TRANSFORMATION PRODUCTS: In an aq soln, chloramine reacted with methylamine forming *N*-chloromethylamine. Reacts with acids forming soluble salts.

EXPOSURE LIMITS: NIOSH REL: IDLH 100 ppm; OSHA PEL: TWA 10 ppm.

SYMPTOMS OF EXPOSURE: Irrit eyes, resp sys; cough; burns, derm, conj.

USES: Tanning; intermediate for accelerators, dyes, pharmaceuticals, insecticides, fungicides, tanning, surface active agents, fuel additive, dyeing of acetate textiles; polymerization inhibitor; ingredient in paint removers; photographic developer; solvent; rocket propellant; fuel additive; solvent; organic synthesis.

METHYLANILINE

SYNONYMS: Anilinomethane; MA; (Methylamino)benzene; *N*-Methylaminobenzene; *N*-Methylaniline; ***N*-Methylbenzenamine**; Methylphenylamine; *N*-Methylphenylamine; Monomethylaniline; *N*-Monomethylaniline; *N*-Phenylmethylamine; UN 2294.

CHEMICAL DESIGNATIONS: CAS: 100-61-8; DOT: 2294; mf: C_7H_9N; fw: 107.16; RTECS: BY 4550000.

PROPERTIES: Yellow to pale brown liq with a faint ammonia-like odor. Mp: -57°, bp: 196.25°, *d*: 0.9891 @ 20/4°, fl p: 79.4°, pK_a: 4.848 @ 25°, K_H: 1.19 x 10^{-5} atm·m^3/mol @ 25° (calcd), IP: 7.32 eV, log K_{ow}: 1.40-1.82, S_o: sol in alcohol and ether, S_w: 5.624 g/L @ 25°, vap d: 4.38 g/L @ 25°, 3.70 (air = 1), vp: 0.3 mm @ 20°.

TRANSFORMATION PRODUCTS: Reacts slowly with humic acids or humates forming quinoidal structures.

EXPOSURE LIMITS: NIOSH REL: IDLH 100 ppm; OSHA PEL: TWA 2 ppm; ACGIH TLV: TWA 0.5 ppm.

SYMPTOMS OF EXPOSURE: Weak; dizz; head; dysp cyan; methemoglobinemia; pulm edema; liv and kid damage.

USES: Solvent; acid acceptor; organic synthesis.

2-METHYLANTHRACENE

SYNONYM: β-Methylanthracene.

CHEMICAL DESIGNATIONS: CAS: 613-12-7; DOT: none assigned; mf: $C_{15}H_{12}$; fw: 192.96; RTECS: CB 0680000.

PROPERTIES: Solid. Mp: 209°, bp: 358.5°, *d*: 1.165 (calcd), log K_{oc}: 5.12 (calcd), log K_{ow}: 5.52 (calcd), S_o: sol in benzene and chloroform, S_w: 14.5 ppb @ 18.3°.

USES: Organic synthesis.

METHYL BROMIDE

SYNONYMS: Brom-o-gas; Brom-o-gaz; **Bromomethane**; Celfume; Dawson 100; Dowfume; Dowfume MC-2; Dowfume MC-2 soil fumigant; Dowfume MC-33; Edco; Embafume; Fumigant-1; Halon 1001; Iscobrome; Kayafume; MB; M-B-C Fumigant; MBX; MEBR; Metafume; Methogas; Monobromomethane; Pestmaster; Profume; R 40B1; RCRA waste number U029; Rotox; Terabol; Terr-o-gas 100;

UN 1062; Zytox.

CHEMICAL DESIGNATIONS: CAS: 74-83-9; DOT: 1062; mf: CH_3Br; fw: 94.94; RTECS: PA 4900000.

PROPERTIES: Colorless liq with an odor similar to chloroform @ high concn. Mp: -93.6°, bp: 3.55°, *d*: 1.6755 @ 20/4°, fl p: lel: 10%, uel: 16%, K_H: 3.18 x 10^{-2} atm·m^3/mol, IP: 10.54 eV, log K_{oc}: 1.92 (calcd), log K_{ow}: 1.00-1.19, S_o: sol in ethanol, ether, chloroform, carbon disulfide, carbon tetrachloride and benzene, S_w: 13.200 g/L @ 25°, vap d: 3.88 g/L @ 25°, 3.28 (air = 1), vp: 1,420 @ 20°.

TRANSFORMATION PRODUCTS: Hydrolyzes in water forming methanol and hydrobromic acid (est $t_{1/2}$ = 20 days @ 25° and pH 7).

EXPOSURE LIMITS: NIOSH REL: lowest feasible limit; OSHA PEL: C 20 ppm; ACGIH TLV: TWA 5 ppm.

SYMPTOMS OF EXPOSURE: Head; vis dist; verti; nau, vomit, malaise; hand tremor; convuls; dysp; irrit eyes, skin irrit, vesic.

USES: Soil, space and food fumigant; organic synthesis; fire extinguishing agent; refrigerant; disinfestation of potatoes, tomatoes and other crops; solvent for extracting vegetable oils.

2-METHYL-1,3-BUTADIENE

SYNONYMS: Hemiterpene; Isoprene; β-Methylbivinyl; 2-Methylbutadiene; UN 1218.

CHEMICAL DESIGNATIONS: CAS: 78-79-5; DOT: none assigned; mf: C_5H_8; fw: 68.12; RTECS: NT 4037000.

PROPERTIES: Colorless, volatile liq. Mp: -146°, bp: 34°, *d*: 0.6810 @ 20/4°, fl p: -53.9°, lel: 1.5%, uel: 8.9%, K_H: 7.7 x 10^{-2} atm·m^3/mol @ 25°, IP: 8.845 eV, log K_{ow}: 1.76 (calcd), S_o: sol in acetone and benzene; misc with alcohol and ether, S_w: 642 ppm @ 25°, vap d: 2.78 g/L @ 25°, 2.35 (air = 1), vp: 493 mm @ 20°.

TRANSFORMATION PRODUCTS: Methyl vinyl ketone and methacrolein were reported as major photooxidation products for the reaction of 2-methyl-1,3-butadiene with hydroxyl radicals. Formaldehyde, NO_2, NO and HO_2 were reported as minor products. Synthetic air containing gaseous nitrous acid and exposed to artificial sunlight (λ = 300-450 nm) photooxidized 2-methyl-1,3-butadiene into formaldehyde, methyl nitrate, peroxyacetal nitrate and a compound tentatively identified as methyl vinyl ketone. Oxidizes and polymerizes on standing in air.

USES: Manufacture of butyl and synthetic rubber; gasoline component; organic synthesis.

2-METHYLBUTANE

SYNONYMS: Ethyldimethylmethane; Isoamylhydride; Isopentane; UN 1265.

CHEMICAL DESIGNATIONS: CAS: 78-78-4; DOT: none assigned; mf: C_5H_{12}; fw: 72.15; RTECS: EK 4430000.

PROPERTIES: Colorless liq with a pleasant odor. Mp: -159.9°, bp: 27.8°, *d*: 0.6201 @ 20/4°, fl p: -57°, lel: 1.4%, uel: 7.6%, K_H: 1.35 atm·m^3/mol @ 25° (calcd), IP: 10.32 eV, log K_{ow}: 2.23, S_o: sol in acetone, alcohol, ether, hydrocarbons and oils, S_w: 48 mg/L @ 20°, vap d: 2.95 g/L @ 25°, 2.49 (air = 1), vp: 687.4 mm @ 25°.

TRANSFORMATION PRODUCTS: Synthetic air containing gaseous nitrous acid and exposed to artificial sunlight (λ = 300-450 nm) photooxidized 2-methylbutane into acetone, acetaldehyde, methyl nitrate, peroxyacetal nitrate, propyl nitrate and pentyl nitrate.

USES: Solvent; blowing agent for polystyrene; manufacturing chlorinated derivatives.

3-METHYL-1-BUTENE

SYNONYMS: Isopentene; Isopropylethylene; α-Isoamylene; UN 2561.

CHEMICAL DESIGNATIONS: CAS: 563-45-1; DOT: 2561; mf: C_5H_{10}; fw: 70.13; RTECS: EM 7600000.

PROPERTIES: Colorless liq with a disagreeable odor. Mp: -168.5°, bp: 20°, *d*: 0.6272 @ 20/4°, fl p: -57°, lel: 1.5%, uel: 9.1%, K_H: 5.35 x 10^{-1} atm·m^3/mol @ 25°, IP: 9.51 eV, log K_{ow}: 2.30 (calcd), S_o: sol in alcohol, benzene and ether, S_w: 130 mg/L @ 20°, vap d: 2.87 g/L @ 25°, 2.42 (air = 1), vp: 902.1 mm @ 25°.

USES: In high octane fuels; organic synthesis.

METHYL CELLOSOLVE

SYNONYMS: Dowanol EM; EGM; EGME; Ektasolve; Ethylene glycol methyl ether; Ethylene glycol monomethyl ether; Glycol ether EM; Glycol methyl ether; Glycol monomethyl ether; Jeffersol EM; MECS; **2-Methoxyethanol**; Methoxyhydroxyethane; Methyl ethoxol; Methyl glycol; Methyl oxitol; Poly-solv EM; Prist; UN 1188.

CHEMICAL DESIGNATIONS: CAS: 109-86-4; DOT: 1188; mf: $C_3H_8O_2$; fw: 76.10; RTECS: KL 5775000.

PROPERTIES: Colorless liq with a mild odor. Mp: -85.1°, bp: 125° @ 768 mm, *d*: 0.9647 @ 20/4°, fl p: 39°, lel: 2.5%, uel: 14%, S_o: sol in acetone and benzene; misc with alcohol, dimethylformamide, ether and glycerol, S_w: misc, vap d: 3.11 g/L @ 25°, 2.63 (air = 1), vp: 6.2 mm @ 20°.

EXPOSURE LIMITS: NIOSH REL: lowest feasible limit; OSHA PEL: TWA 25 ppm; ACGIH TLV: 5 ppm.

SYMPTOMS OF EXPOSURE: Head, drowsy, weak, atax, tremor, som, anemic pal, irrit eyes.

USES: Solvent for natural and synthetic resins, cellulose acetate, nitrocellulose and some dyes; nail polishes; dyeing leather, sealing moistureproof cellophane; lacquers, varnishes, enamels, wood stains; in solvent mixtures; perfume fixative; jet fuel deicing additive.

METHYL CELLOSOLVE ACETATE

SYNONYMS: Acetic acid 2-methoxyethyl ester; Ethylene glycol methyl ether acetate; Ethylene glycol monomethyl ether acetate; Glycol ether EM acetate; Glycol monomethyl ether acetate; **2-Methoxyethanol acetate**; 2-Methoxyethyl acetate; Methyl glycol acetate; Methyl glycol monoacetate; UN 1189.

CHEMICAL DESIGNATIONS: CAS: 110-49-6; DOT: 1189; mf: $C_5H_{10}O_3$; fw: 118.13; RTECS: KL 5950000.

PROPERTIES: Colorless liq with an ether-like odor. Mp: -65°, bp: 144-145°, *d*: 1.0090 @ 19/19°, fl p: 43.9°, lel: 1.1%, uel: 8.2%, S_o: sol in alcohol and ether, S_w: misc, vap d: 4.83 g/L @ 25°, 4.08 (air = 1), vp: 2 mm @ 20°.

TRANSFORMATION PRODUCTS: Anticipated hydrolysis products include methyl cellosolve and acetic acid.

EXPOSURE LIMITS: NIOSH REL: IDLH 4,000 ppm; OSHA PEL: TWA 25 ppm; ACGIH TLV: TWA 5 ppm.

SYMPTOMS OF EXPOSURE: Kid and brain damage, eye irrit.

USES: Solvent for cellulose acetate, nitrocellulose, various gums, resins, waxes, oils; textile printing; lacquers; dopes; textile printing; photographic film.

METHYL CHLORIDE

SYNONYMS: Artic; **Chloromethane**; Monochloromethane; RCRA waste number U045; UN 1063.

CHEMICAL DESIGNATIONS: CAS: 74-87-3; DOT: 1063; mf: CH_3Cl; fw: 50.48; RTECS: PA 6300000.

PROPERTIES: Liquified compressed gas, colorless, odorless or sweet ethereal odor. Mp: -97.1°, bp: -24.2°, *d*: 0.9159 @ 20/4°, fl p: -50°, lel: 7.6%, uel: 19%, K_H: 10^{-2} atm·m^3/mol @ 25°, IP: 11.26 eV, log K_{oc}: 1.40 (calcd), log K_{ow}: 0.90, 0.91, S_o: misc with chloroform, ether and glacial acetic acid; sol in ethanol, S_w: 7,400 mg/L @ 25°, vap d: 2.06 g/L @ 25°, 1.74 (air = 1), vp: 3,789 mm @ 20°.

TRANSFORMATION PRODUCTS: Enzymatic degradation of methyl chloride was reported to yield formaldehyde. Hydrolyzes in water forming methanol and hydrochloric acid (est $t_{1/2}$ = 0.93 yr @ 25° and pH 7). Reported photooxidation products via hydroxyl radicals include formyl chloride, CO, hydrogen chloride and phosgene. In the presence of water, formyl chloride hydrolyzes to hydrochloric acid and CO whereas phosgene hydrolyzes to hydrogen chloride and CO.

EXPOSURE LIMITS: NIOSH REL: lowest feasible limit; OSHA PEL: TWA 100 ppm, C 200 ppm, 5-min 3-h peak 300 ppm; ACGIH TLV: TWA 50 ppm, STEL 100 ppm.

SYMPTOMS OF EXPOSURE: Dizz, nau, vomit; vis dist; stagger; slur speech; convuls, coma; liv and kid damage; frost.

USES: Coolant and refrigerant; herbicide and fumigant; organic synthesis-methylating agent; manufacture of silicone polymers pharmaceuticals, tetraethyl lead, synthetic rubber, methyl cellulose, agricultural chemicals and non-flammable films; preparation of methylene chloride, carbon tetrachloride, chloroform; low

METHYL FORMATE

SYNONYMS: Formic acid methyl ester; Methyl methanoate; UN 1243.

CHEMICAL DESIGNATIONS: CAS: 107-31-3; DOT: 1243; mf: $C_2H_4O_2$; fw: 60.05; RTECS: LQ 8925000.

PROPERTIES: Colorless liq with a pleasant odor. Mp: -99°, bp: 31.5°, *d*: 0.9742 @ 20/4°, fl p: -18.9°, lel: 4.5%, uel: 23.0%, K_H: 2.23 x 10^{-4} atm·m^3/mol @ 25°, IP: 10.815 eV, log K_{ow}: -0.18 (calcd), S_o: sol in ether and misc with alcohol, S_w: 304 g/L @ 20°, vap d: 2.45 g/L @ 25°, 2.07 (air = 1), vp: 476 mm @ 20°.

TRANSFORMATION PRODUCTS: Methyl formate, formed from the irradiation of dimethyl ether in the presence of chlorine, degraded to CO_2, water and small amounts of formic acid. Continued irradiation degraded formic acid to CO_2, water and hydrogen chloride. Anticipated hydrolysis products include methyl alcohol and formic acid.

EXPOSURE LIMITS: NIOSH REL: IDLH 5,000 ppm; OSHA PEL: TWA 100 ppm.

SYMPTOMS OF EXPOSURE: Eye, nose irrit; chest oppression, dysp; vis dist; CNS depres.

USES: Fumigant and larvacide for tobacco, cereals, dried fruits; cellulose acetate solvent; military poison gases; organic synthesis.

3-METHYLHEPTANE

SYNONYM: 5-Methylheptane.

CHEMICAL DESIGNATIONS: CAS: 589-81-1; DOT: none assigned; mf: C_8H_{18}; fw: 114.23; RTECS: none assigned.

PROPERTIES: Colorless liq. Mp: -120.5°, bp: 119°, *d*: 0.7058 @ 20/4°, K_H: 3.70 atm·m^3/mol @ 25° (calcd), log K_{ow}: 3.97 (calcd), S_o: sol in acetone, alcohol, benzene, chloroform and ether, S_w: 0.792 ppm @ 25°, vap d: 4.67 g/L @ 25°, 3.94 (air = 1), vp: 19.5 mm @ 25°.

USES: Calibration; gasoline component; organic synthesis.

5-METHYL-3-HEPTANONE

SYNONYMS: Amyl ethyl ketone; EAK; Ethyl amyl ketone; Ethyl *sec*-amyl ketone; 3-Methyl-5-heptanone.

CHEMICAL DESIGNATIONS: CAS: 541-85-5; DOT: 2271; mf: $C_8H_{16}O$; fw: 128.21; RTECS: MJ 7350000.

PROPERTIES: Colorless liq with a fruity odor. Mp: -56.7°, bp: 158.9°, *d*: 0.820–0.824 @ 20/20°, fl p: 43.3°, K_H: 1.30 x 10^{-4} atm·m^3/mol @ 20° (calcd), log K_{ow}: 1.96 (calcd), S_w: 0.26 wt % @ 20°, vap d: 5.24 g/L @ 25°, 4.43 (air = 1), vp: 2 mm @ 20°.

EXPOSURE LIMITS: NIOSH REL: IDLH 3,000 ppm; OSHA PEL: TWA 25 ppm.

SYMPTOMS OF EXPOSURE: Irrit eyes, muc memb; head; narco, coma; derm.

USES: Solvent for vinyl resins, nitrocellulose-alkyd and nitrocellulose-maleic acid resins.

2-METHYLHEXANE

SYNONYMS: Ethylisobutylmethane; Isoheptane.

CHEMICAL DESIGNATIONS: CAS: 591-76-4; DOT: none assigned; mf: C_7H_{16}; fw: 100.20; RTECS: none assigned.

PROPERTIES: Colorless liq. Mp: -118.3°, bp: 90°, *d*: 0.6787 @ 20/4°, fl p: < -17.7°, K_H: 3.42 atm·m^3/mol @ 25° (calcd), log K_{ow}: 3.30, S_o: sol in acetone, alcohol, benzene, chloroform, lignoin and ether, S_w: 3.8 mg/L @ 23°, vap d: 4.10 g/L @ 25°, 3.46 (air = 1), vp: 65.9 mm @ 25°.

USE: Organic synthesis.

3-METHYLHEXANE

SYNONYM: 4-Methylhexane.

CHEMICAL DESIGNATIONS: CAS: 589-34-4; DOT: none assigned; mf: C_7H_{16}; fw: 100.20; RTECS: none assigned.

PROPERTIES: Colorless liq. Mp: -119°, bp: 92°, *d*: 0.6860 @ 20/4°, fl p: -3.9°, K_H: 1.55-1.64 atm·m^3/mol @ 25° (calcd), log K_{ow}: 3.41, S_o: sol in acetone, alcohol, benzene, chloroform, lignoin and ether, S_w: 2.90 mg/L @ 23°, vap d: 4.10 g/L @ 25°, 3.46 (air = 1), vp: 61.6 mm @ 25°.

USES: Oil extender solvent; gasoline component; organic synthesis.

METHYLHYDRAZINE

SYNONYMS: Hydrazomethane; 1-Methylhydrazine; MMH; Monomethylhydrazine.

CHEMICAL DESIGNATIONS: CAS: 60-34-4; DOT: 1244; mf: CH_6N_2; fw: 46.07; RTECS: MV 5600000.

PROPERTIES: Fuming, colorless liq with an ammonia-like odor. Mp: -52.4°, -20.9°, bp: 87.5°, *d*: 0.874 @ 25/4°, fl p: -8.3°, lel: 2.5%, uel: 92%, IP: 8.00 eV, 7.67 eV, S_o: sol in alcohol and ether, S_w: misc, vap d: 1.88 g/L @ 25°, 1.59 (air = 1), vp: 36 mm @ 20°.

TRANSFORMATION PRODUCTS: It was suggested that the rapid disappearance of methylhydrazine in sterile and nonsterile soil (Arrendondo fine sand) under aerobic conditions was due to chemical oxidation. Though the oxidation product was not identified, it biodegraded to CO_2 in the nonsterile soil. The oxidation product did not degrade further in the sterile soil.

EXPOSURE LIMITS: NIOSH REL: 2-h C 0.08 mg/m^3; OSHA PEL: C 0.2 ppm.

SYMPTOMS OF EXPOSURE: Irrit eyes, vomit, diarr, resp irrit, tremors, atax, anoxia, cyan, convuls.

USES: Rocket fuel; solvent; intermediate; organic synthesis.

METHYL IODIDE

SYNONYMS: Halon 10,001; **Iodomethane**; RCRA waste number U138; UN 2644.

CHEMICAL DESIGNATIONS: CAS: 74-88-4; DOT: 2644; mf: CH_3I; fw: 141.94; RTECS: PA 9450000.

PROPERTIES: Colorless liq which may become yellow, red or brown on exposure to light and moisture. Mp: -66.4°, bp: 42.4°, *d*: 2.279 @ 20/4°, K_H: 5.48 x 10^{-3} atm·m^3/mol @ 25°, IP: 9.54 eV, log K_{oc}: 1.36 (calcd), log K_{ow}: 1.51, 1.69, S_o: sol in acetone and benzene; misc with alcohol and ether, S_w: 14 g/L @ 20°, vap d: 5.80 g/L @ 25°, 4.90 (air = 1), vp: 375 mm @ 20°.

TRANSFORMATION PRODUCTS: Anticipated products from the reaction of methyl iodide with O_3 or hydroxyl radicals in the atmosphere are formaldehyde, iodoformaldehyde, CO and I•. With hydroxyl radicals, CH_2, methyl radical, HOI and water are possible reaction products. Hydrolyzes in water forming methyl alcohol and hydriodic acid (est $t_{1/2}$ = 110 days @ 25° and pH 7). May react with chlorides in seawater to form methyl chloride.

EXPOSURE LIMITS: OSHA PEL: TWA 5 ppm; ACGIH TLV: TWA 2 ppm.

SYMPTOMS OF EXPOSURE: Nau, vomit, verti, atax, slur speech, drow, derm, eye irrit, blister skin.

USES: Microscopy; medicine; testing for pyridine; methylating agent in organic synthesis.

METHYL ISOCYANATE

SYNONYMS: Isocyanic acid methyl ester; **Isocyanatomethane**; RCRA waste number P064; TL 1450; UN 2480.

CHEMICAL DESIGNATIONS: CAS: 624-83-9; DOT: 2480; mf: C_2H_3NO; fw: 57.05; RTECS: NQ 9450000.

PROPERTIES: Colorless liq with a sharp, penetrating odor. Mp: -45°, bp: 39.1-40.1°, *d*: 0.9230 @ 27/4°, fl p: < -17.8° (OC), lel: 5.3%, uel: 26%, K_H: 3.89 x 10^{-4} atm·m^3/mol @ 20° (calcd), S_w: 6.7 wt % @ 20°, vap d: 2.33 g/L @ 25°, 1.97 (air = 1), vp: 348 mm @ 20°.

EXPOSURE LIMITS: NIOSH REL: IDLH 20 ppm; OSHA PEL: TWA 0.02 ppm.

SYMPTOMS OF EXPOSURE: Irrit eyes, nose, throat; cough, secretions, chest pain, dysp; asth; eye, skin inj.

USE: Chemical intermediate.

METHYL MERCAPTAN

SYNONYMS: Mercaptomethane; **Methanethiol**; Methyl sulfhydrate; Methyl thioalcohol; RCRA waste number U153; Thiomethanol; Thiomethyl alcohol; UN 1064.

CHEMICAL DESIGNATIONS: CAS: 74-93-1; DOT: 1064; mf: CH_4S; fw: 48.10; RTECS: PB 4375000.

PROPERTIES: Colorless gas with a garlic-like or rotten cabbage odor. Mp: -123°, bp: 6-7.6°, *d*: 0.8665 @ 20/4°, fl p: -17.8° (OC), lel: 3.9%, uel: 21.8%, pK_a: 10.70 @ 25°, K_H: 3.01 x 10^{-3} atm·m^3/mol @ 25°, IP: 9.440 eV, S_o: sol in alcohol, ether and petroleum naphtha, S_w: 23.30 g/L @ 20°, vap d: 1.97 g/L @ 25°, 1.66 (air = 1), vp: 1,516 mm @ 25°.

TRANSFORMATION PRODUCTS: Sunlight irradiation of a methyl mercaptan-NO_x mixture in an outdoor chamber yielded formaldehyde, sulfur dioxide, nitric acid, methyl nitrate, methanesulfonic acid and an inorganic sulfate. In the presence of NO, methyl mercaptan reacted with hydroxyl radicals to give methyl sulfenic acid and methyl thionitrite.

EXPOSURE LIMITS: NIOSH REL: 15-min C 0.5 ppm, IDLH 400 ppm; OSHA PEL: 15-min C 10 ppm; ACGIH TLV: TWA 0.5 ppm.

SYMPTOMS OF EXPOSURE: Narco, cyan, convul, pulm irrit, resp para, head, nau.

USES: Synthesis of methionine; intermediate in the manufacture of pesticides, fungicides, jet fuels, plastics; catalyst.

METHYL METHACRYLATE

SYNONYMS: Diakon; Methacrylic acid methyl ester; Methyl-α-methylacrylate; Methyl methacrylate monomer; Methyl-2-methyl-2-propenoate; MME; "Monocite" methacrylate monomer; NA 1,247; NCI-C50680; **2-Propenoic acid 2-methyl methyl ester**; RCRA waste number U162; UN 1247.

CHEMICAL DESIGNATIONS: CAS: 80-62-6; DOT: 1247; mf: $C_5H_8O_2$; fw: 100.12; RTECS: OZ 5075000.

PROPERTIES: Colorless liq with a penetrating, fruity odor. Mp: -48°, bp: 100-101°, *d*: 0.9440 @ 20/4°, fl p: 10° (OC), lel: 1.7%, uel: 8.2%, K_H: 2.46 x 10^{-4} atm·m^3/mol @ 20° (calcd), log K_{ow}: 1.33 (calcd), S_o: sol in acetone, alcohol, ether, 2-butanone, tetrahydrofuran, esters, aromatic and chlorinated hydrocarbons, S_w: 1.5 wt % @ 20°, vap d: 4.09 g/L @ 25°, 3.46 (air = 1), vp: 28 mm @ 20°.

TRANSFORMATION PRODUCTS: Polymerizes easily.

EXPOSURE LIMITS: NIOSH REL: IDLH 4,000 ppm; OSHA PEL: TWA 100 ppm.

SYMPTOMS OF EXPOSURE: Irrit eyes, nose, throat; derm; narco.

USES: Manufacturing methacrylate resins and plastics; impregnation of concrete.

2-METHYLNAPHTHALENE

SYNONYM: β-Methylnaphthalene.

CHEMICAL DESIGNATIONS: CAS: 91-57-6; DOT: none assigned; mf: $C_{11}H_{10}$;

fw: 142.20; RTECS: QJ 9635000.

PROPERTIES: Solid. Mp: 34.6°, bp: 241.052°, *d*: 1.0058 @ 20/4°, fl p: 97°, pK_a: > 15, K_H: 3.18 x 10^{-4} atm·m^3/mol @ 25°, IP: 7.955 eV, 8.48 eV, log K_{oc}: 3.40-3.93, log K_{ow}: 3.86, 4.11, S_o: sol in most solvents, S_w: 24.6 mg/L @ 25°, vp: 5.4 x 10^{-2} mm @ 25° (est).

TRANSFORMATION PRODUCTS: An aq soln containing chlorine dioxide in the dark for 3.5 days at room temperature oxidized 2-methylnaphthalene into the following: 1-chloro-2-methylnaphthalene, 3-chloro-2-methylnaphthalene, 1,3-dichloro-2-methylnaphthalene, 3-hydroxymethylnaphthalene, 2-naphthaldehyde, 2-naphthoic acid and 2-methyl-1,4-naphthoquinone. 2-Naphthoic acid was reported to be the biooxidation product of 2-methylnaphthalene by *Nocardia* sp. in soil using *n*-hexadecane as the substrate.

USES: Organic synthesis; insecticides; jet fuel component; minor component in gasoline. Derived from industrial and experimental coal gasification operations where the maximum concn detected in gas, liq and coal tar streams were 2.1, 0.22 and 10 mg/m^3, respectively.

4-METHYLOCTANE

SYNONYM: 5-Methyloctane.

CHEMICAL DESIGNATIONS: CAS: 2216-34-4; DOT: none assigned; mf: C_9H_{20}; fw: 128.26; RTECS: none assigned.

PROPERTIES: Liq. Mp: -113.2°, bp: 142.4°, *d*: 0.7199 @ 20/4°, K_H: 10.27 atm·m^3/mol @ 25° (calcd), log K_{ow}: 4.69 (calcd), S_o: sol in acetone, alcohol, benzene and ether, S_w: 0.115 ppm @ 25°, vap d: 5.24 g/L @ 25°, 4.43 (air = 1), vp: 7 mm @ 25°.

USES: Gasoline component; organic synthesis.

2-METHYLPENTANE

SYNONYMS: Dimethylpropylmethane; Isohexane; UN 2462.

CHEMICAL DESIGNATIONS: CAS: 107-83-5; DOT: 2462; mf: C_6H_{14}; fw: 86.18; RTECS: SA 2995000.

PROPERTIES: Colorless liq. Mp: -153.7°, bp: 60.3°, *d*: 0.6532 @ 20/4°, fl p: -23.3°, lel: 1.2%, uel: 7.0%, K_H: 1.732 atm·m^3/mol @ 25°, IP: 10.12 eV, log K_{ow}: 2.77, S_o: sol in acetone, alcohol, benzene, chloroform, ether, and many hydrocarbons, S_w: 14.0 mg/L @ 23°, vap d: 3.52 g/L @ 25°, 2.98 (air = 1), vp: 211.8 mm @ 25°.

TRANSFORMATION PRODUCTS: Synthetic air containing gaseous nitrous acid and exposed to artificial sunlight (λ = 300-450 nm) photooxidized 2-methylpentane into acetone, propionaldehyde, peroxyacetal nitrate, peroxypropionyl nitrate and possibly 2 isomers of hexyl nitrate and propyl nitrate.

USES: Solvent; gasoline component; organic synthesis.

3-METHYLPENTANE

SYNONYMS: Diethylmethylmethane; UN 2462.

CHEMICAL DESIGNATIONS: CAS: 96-14-0; DOT: 2462; mf: C_6H_{14}; fw: 86.18; RTECS: SA 2995500.

PROPERTIES: Colorless liq. Mp: -117.8°, bp: 63.3°, *d*: 0.6645 @ 20/4°, fl p: < -6.6°, lel: 1.2%, uel: 7.0%, K_H: 1.693 atm·m^3/mol @ 25°, IP: 10.08 eV, log K_{ow}: 2.88, S_o: sol in acetone, alcohol, benzene and ether, S_w: 10.5 mg/L @ 23°, vap d: 3.52 g/L @ 25°, 2.98 (air = 1), vp: 189.8 mm @ 25°.

USES: Solvent; gasoline component; organic synthesis.

4-METHYL-2-PENTANONE

SYNONYMS: Hexanone; Hexone; Isobutyl methyl ketone; Isopropylacetone; Methyl isobutyl ketone; 2-Methyl-4-pentanone; MIBK; MIK; RCRA waste number U161; Shell MIBK; UN 1245.

CHEMICAL DESIGNATIONS: CAS: 108-10-1; DOT: 1245; mf: $C_6H_{12}O$; fw: 100.16; RTECS: SA 9275000.

PROPERTIES: Clear, colorless, watery liq with a mild pleasant odor. Mp: -84.7°, bp: 116.8°, *d*: 0.7978 @ 20/4°, fl p: 22.8°, lel: 1.4%, uel: 7.5%, K_H: 1.49 x 10^{-5} atm·m^3/mol @ 25° (calcd), IP: 9.30 eV, log K_{oc}: 0.79 (est), log K_{ow}: 1.09, S_o: sol in acetone, ethanol, benzene, chloroform, ether and many other solvents, S_w: 17 g/L @ 20°, vap d: 4.09 g/L @ 25°, 3.46 (air = 1), vp: 14.5-16 mm @ 20°.

TRANSFORMATION PRODUCTS: Synthetic air containing gaseous nitrous acid and exposed to artificial sunlight (λ = 300-450 nm) photooxidized 4-methyl-2-pentanone into acetone, peroxyacetal nitrate and methyl nitrate. In a subsequent experiment, the hydroxylinitiated photooxidation of 4-methyl-2-pentanone in a smog chamber produced acetone (90% yield) and peroxyacetal nitrate. No other products were found. Irradiation of 4-methyl-2-pentanone (λ = 3130 Å) yielded acetone, propyldiene and free radicals.

EXPOSURE LIMITS: NIOSH REL: 10-h TWA 50 ppm, IDLH 3,000 ppm; OSHA PEL: TWA 100 ppm; ACGIH TLV: TWA 50 ppm, STEL 75 ppm.

SYMPTOMS OF EXPOSURE: Irrit eyes, muc memb; head; narcosis, coma; derm.

USES: Denaturant for ethyl alcohol; solvent for paints, varnishes nitrocellulose lacquers; preparation of methyl amyl alcohol; extraction of uranium from fission products; organic synthesis.

2-METHYL-1-PENTENE

SYNONYMS: 2-Methylpentene; 1-Methyl-1-propylethene; 1-Methyl-1-propylethylene.

CHEMICAL DESIGNATIONS: CAS: 763-29-1; DOT: none assigned; mf: C_6H_{12}; fw: 84.16; RTECS: SB 2230000.

PROPERTIES: Colorless liq. Mp: -135.7°, bp: 60.7°, *d*: 0.6799 @ 20/4°, fl p: < -7°, K_H: 2.77 x 10^{-1} atm·m^3/mol @ 25° (calcd), log K_{ow}: 2.54 (calcd), S_o: sol in alcohol, benzene, chloroform and petroleum, S_w: 78 mg/L @ 20°, vap d: 3.44 g/L @ 25°, 2.91 (air = 1), vp: 195.4 mm @ 25°.

USES: Flavors; perfumes; medicines; dyes; oils; resins; organic synthesis.

4-METHYL-1-PENTENE

SYNONYMS: 1-Isopropyl-2-methylethene; 1-Isopropyl-2-methylethylene.

CHEMICAL DESIGNATIONS: CAS: 691-37-2; DOT: none assigned; mf: C_6H_{12}; fw: 84.16; RTECS: none assigned.

PROPERTIES: Colorless liq. Mp: -153.6°, bp: 53.9°, *d*: 0.6642 @ 20/4°, fl p: -31.6°, K_H: 6.15 x 10^{-1} atm·m^3/mol @ 25°, log K_{ow}: 2.70 (calcd), S_o: sol in alcohol, benzene, chloroform and petroleum, S_w: 48 mg/L @ 20°, vap d: 3.44 g/L @ 25°, 2.91 (air = 1), vp: 270.8 mm @ 25°.

USES: Manufacture of plastics used in automobiles, laboratory ware and electronic components; organic synthesis.

1-METHYLPHENANTHRENE

SYNONYM: α-Methylphenanthrene.

CHEMICAL DESIGNATIONS: CAS: 832-69-9; DOT: none assigned; mf: $C_{15}H_{12}$; fw: 192.26; RTECS: none assigned.

PROPERTIES: Solid. Mp: 123°, bp: 358.6°, *d*: 1.161 (calcd), log K_{oc}: 4.56 (calcd), log K_{ow}: 5.27 (calcd), S_o: sol in alcohol, S_w: 193 ppb @ 19.2°.

USES: Chemical research; organic synthesis.

2-METHYLPHENOL

SYNONYMS: 2-Cresol; *o*-Cresol; *o*-Cresylic acid; 1-Hydroxy-2-methylbenzene; 2-Hydroxytoluene; *o*-Hydroxytoluene; 2-Methylhydroxybenzene; *o*-Methylhydroxybenzene; *o*-Methylphenol; *o*-Methylphenylol; Orthocresol; *o*-Oxytoluene; RCRA waste number U052; 2-Toluol; *o*-Toluol; UN 2076.

CHEMICAL DESIGNATIONS: CAS: 95-48-7; DOT: 2076; mf: C_7H_8O; fw: 108.14; RTECS: GO 6300000.

PROPERTIES: Colorless solid or liq with a phenolic odor; darkens on exposure to air. Mp: 30.9°, bp: 191.0°, *d*: 1.0273 @ 20/4°, fl p: 81°, lel: 1.35%, pK_a: 10.26 @ 25°, K_H: 1.23 x 10^{-6} atm·m^3/mol @ 25°, IP: 8.98 eV, log K_{oc}: 1.34, log K_{ow}: 1.93-1.99, S_o: misc with ethanol, benzene, ether and glycerol, S_w: 24.5 g/L @ 20°, vp: 0.24 mm @ 25°.

TRANSFORMATION PRODUCTS: Bacterial degradation of 2-methylphenol may introduce a hydroxyl group to produce *m*-methylcatechol. In phenol acclimated activated sludge, metabolites identified included 3-methylcatechol, 4-

methylresorcinol and methylhydroquinone. Other metabolites identified included α-ketobutyric acid, dihydroxybenzaldehyde and trihydroxytoluene. Groundwater contaminated with phenol and other phenols degraded in a methanogenic aquifer to methane and CO_2. These results could not duplicated in the laboratory utilizing an anaerobic digester. Chloroperoxidase, a fungal enzyme isolated from *Caldariomyces fumago*, reacted with 2-methylphenol forming 2-methyl-4-chlorophenol (38% yield) and 2-methyl-6-chlorophenol. Ozonation of an aq soln containing 2-methylphenol (200-600 mg/L) yielded formic, acetic, propionic, glyoxylic, oxalic and salicylic acid. In a smog chamber experiment, 2-methylphenol reacted with NO_x to form nitrocresols, dinitrocresols and hydroxynitrocresols. Sunlight irradiation of 2-methylphenol and NO_x in air yielded the following gas-phase products: acetaldehyde, formaldehyde, pyruvic acid, peroxyacetylnitrate, nitrocresols and trace levels of nitric acid and methyl nitrate. Particulate phase products were also identified and these included 2-hydroxy-3-nitrotoluene, 2-hydroxy-5-nitrotoluene, 2-hydroxy-3,5-dinitrotoluene and tentatively identified nitrocresol isomers. Anticipated products from the reaction of 2-methylphenol with O_3 or hydroxyl radicals in the atmosphere are hydroxynitrotoluenes and ring cleavage compounds.

EXPOSURE LIMITS: NIOSH REL: 10-h TWA 2.3 ppm, IDLH 250 ppm; OSHA PEL: TWA 5 ppm.

USES: Disinfectant; phenolic resins; tricresyl phosphate; ore flotation; textile scouring agent; organic intermediate; manufacture of salicylaldehyde, coumarin and herbicides; surfactant; synthetic food flavors (*para* isomer only); food antioxidant; dye, perfume, plastics and resins manufacturing.

4-METHYLPHENOL

SYNONYMS: 4-Cresol; *p*-Cresol; *p*-Cresylic acid; 1-Hydroxy-4-methylbenzene; *p*-Hydroxytoluene; 4-Hydroxytoluene; *p*-Kresol; 1-Methyl-4-hydroxybenzene; 4-Methylhydroxybenzene; *p*-Methylhydroxybenzene; *p*-Methylphenol; 4-Oxytoluene; *p*-Oxytoluene; Paracresol; Paramethylphenol; RCRA waste number U052; 4-Toluol; *p*-Toluol; *p*-Tolyl alcohol; UN 2076.

CHEMICAL DESIGNATIONS: CAS: 106-44-5; DOT: 2076; mf: C_7H_8O; fw: 108.14; RTECS: GO 6475000.

PROPERTIES: Colorless solid with a phenolic odor. Mp: 34.8°, bp: 201.9°, *d*: 1.0178 @ 20/4°, fl p: 86°, lel: 1.06%, uel: 1.4%, pK_a: 10.26 @ 25°, K_H: 7.92 x 10^{-7} atm·m^3/mol @ 25°, IP: 8.97 eV, log K_{oc}: 1.69-3.53, log K_{ow}: 1.67-3.01, S_o: misc with ethanol, benzene, ether and glycerol, S_w: 23 g/L @ 25°, vp: 4 x 10^{-2} mm @ 20°.

TRANSFORMATION PRODUCTS: Protocatechuic acid (3,4-dihydroxybenzoic acid) is the central metabolite in the bacterial degradation of 4-methylphenol. Intermediate byproducts included *p*-hydroxybenzyl alcohol, *p*-hydroxybenzaldehyde and *p*-hydroxybenzoic acid. In addition, 4-methylphenol may undergo hydroxylation to form *p*-methylcatechol. Chloroperoxidase, a fungal enzyme isolated from *Caldariomyces fumago*, reacted with 4-methylphenol forming 4-

methyl-2-chlorophenol. Photooxidation products reported include 2,2′-dihydroxy-4,4′-dimethylbiphenyl, 2-hydroxy-3,4′-dimethylbiphenyl ether and 4-methylcatechol. Anticipated products from the reaction of 4-methylphenol with O_3 or hydroxyl radicals in the atmosphere are hydroxynitrotoluene and ring cleavage compounds. A species of *Pseudomonas*, isolated from creosote-contaminated soil, degraded 4-methylphenol into *p*-hydroxybenzaldehyde and *p*-hydroxybenzoate. Both metabolites were then converted into protocatechuate. Under methanogenic conditions, inocula from a municipal sewage treatment plant digester degraded 4-methylphenol to phenol prior to being mineralized to CO_2 and methane.

EXPOSURE LIMITS: NIOSH REL: TWA 250 ppm, 10-h TWA 2.3 ppm; OSHA PEL: TWA 5 ppm.

USES: Disinfectant; phenolic resins; tricresyl phosphate; ore flotation; textile scouring agent; organic intermediate; manufacture of salicylaldehyde, coumarin and herbicides; surfactant; synthetic food flavors.

2-METHYLPROPANE

SYNONYMS: Isobutane; Liquified petroleum gas; Trimethylmethane; UN 1075; UN 1969.

CHEMICAL DESIGNATIONS: CAS: 75-28-5; DOT: 1969mf: C_4H_{10}; fw: 58.12; RTECS: TZ 4300000.

PROPERTIES: Colorless gas with a faint odor. Mp: -159.4°, -145°, bp: -11.633°, *d*: 0.549 @ 20/4°, fl p: -83°, lel: 1.8%, uel: 8.4%, K_H: 1.171 atm·m^3/mol @ 25°, IP: 10.57 eV, log K_{ow}: 2.29 (calcd), S_o: sol in alcohol, chloroform and ether, S_w: 49 mg/L @ 20°, vap d: 2.38 g/L @ 25°, 2.01 (air = 1), vp: 2 atm @ 7.5°.

USES: Gasoline component; organic synthesis.

2-METHYLPROPENE

SYNONYMS: γ-Butylene; *unsym*-Dimethylethylene; Isobutene; Isobutylene; Methylpropene; **2-Methyl-1-propene**; 2-Methylpropylene.

CHEMICAL DESIGNATIONS: CAS: 115-11-7; DOT: 1055/1075; mf: C_4H_8; fw: 56.11; RTECS: UD 0890000.

PROPERTIES: Volatile liq or gas with a coal gas odor. Mp: -140.3°, bp: -6.900°, *d*: 0.5942 @ 20/4°, fl p: -76°, lel: 1.8%, uel: 9.6%, K_H: 2.1 x 10^{-1} atm·m^3/mol @ 25°, IP: 9.23 eV, log K_{ow}: 1.99 (calcd), S_o: very sol in alcohol, ether and sulfuric acid, S_w: 263 mg/L @ 20°, vap d: 2.29 g/L @ 25°, 1.94 (air = 1), vp: 2,270 mm @ 25°.

TRANSFORMATION PRODUCTS: Polymerizes readily. Products identified from the photoirradiation of 1-butene with NO_2 in air are 2-butanone, 2-methylpropanal, acetone, CO, CO_2, methanol, methyl nitrate and nitric acid.

USES: Production of isooctane, butyl rubber, polyisobutene resins, high octane aviation fuels, *tert*-butyl chloride, *tert*-butyl methacrylates; copolymer resins with acrylonitrile, butadiene and other unsaturated hydrocarbons; organic synthesis.

α-METHYLSTYRENE

SYNONYMS: AMS; Isopropenylbenzene; **(1-Methylethenyl)benzene**; 1-Methyl-1-phenylethylene; 2-Phenylpropene; β-Phenylpropene; 2-Phenylpropylene; β-Phenylpropylene.

CHEMICAL DESIGNATIONS: CAS: 98-83-9; DOT: 2303; mf: C_9H_{10}; fw: 118.18; RTECS: WL 5250000.

PROPERTIES: Colorless liq with a characteristic odor. Mp: -23.21°, bp: 163-164°, *d*: 0.9082 @ 20/4°, fl p: 53.9°, lel: 1.9%, uel: 6.1%, IP: 8.35 eV, S_o: sol in benzene and chloroform; misc with alcohol and ether, vap d: 4.83 g/L @ 25°, 4.08 (air = 1), vp: 1.9 mm @ 20°.

TRANSFORMATION PRODUCTS: Polymerizes in the presence of heat or catalysts.

EXPOSURE LIMITS: NIOSH REL: IDLH 5,000 ppm; OSHA PEL: TWA 100 ppm; ACGIH TLV: 50 ppm.

SYMPTOMS OF EXPOSURE: Irrit eyes, nose, throat; drow; derm.

USES: Manufacture of polyesters.

MEVINPHOS

SYNONYMS: Apavinphos; 2-Butenoic acid 3-[(dimethoxyphosphinyl)oxy]methyl ester; 2-Carbomethoxy-1-methylvinyl dimethyl phosphate; α-Carbomethoxy-1-methylvinyl dimethyl phosphate; 2-Carbomethoxy-1-propen-2-yl dimethyl phosphate; CMDP; Compound 2046; **3-[(Dimethoxyphosphinyl)oxy]-2-butenoic acid methyl ester**; *O,O*-Dimethyl-*O*-(2-carbomethoxy-1-methylvinyl)phosphate; Dimethyl-1-carbomethoxy-1-propen-2-yl phosphate; *O,O*-Dimethyl 1-carbomethoxy-1-propen-2-yl phosphate; Dimethyl 2-methoxycarbonyl-1-methylvinyl phosphate; Dimethyl methoxycarbonylpropenyl phosphate; Dimethyl (1-methoxycarboxypropen-2-yl)phosphate; *O,O*-Dimethyl *O*-(1-methyl-2-carboxyvinyl)-phosphate; Dimethyl phosphate of methyl-3-hydroxy-*cis*-crotonate; Duraphos; ENT 22,324; Fosdrin; Gesfid; Gestid; 3-Hydroxycrotonic acid methyl ester dimethyl phosphate; Meniphos; Menite; 2-Methoxycarbonyl-1-methylvinyl dimethyl phosphate; *cis*-2-Methoxycarbonyl-1-methylvinyl dimethyl phosphate; 1-Methoxycarbonyl-1-propen-2-yl dimethyl phosphate; Methyl 3-(dimethoxyphosphinyloxy)crotonate; NA 2,783; OS 2,046; PD 5; Phosdrin; *cis*-Phosdrin; Phosfene; Phosphoric acid (1-methoxycarboxypropen-2-yl) dimethyl ester.

CHEMICAL DESIGNATIONS: CAS: 7786-34-7; DOT: 2783; mf: $C_7H_{13}O_6P$; fw: 224.16; RTECS: GQ 5250000.

PROPERTIES: Colorless to yellow liq. Mp: -56.1°, bp: 106-107.5° @ 1 mm, *d*: 1.25 @ 20/4°, fl p: 79.4° (OC), S_o: misc with acetone, benzene, carbon tetrachloride, chloroform, ethanol, isopropanol, toluene and xylene; sol in carbon disulfide and kerosene (50 g/L), S_w: misc, vap d: 9.16 g/L @ 25°, 7.74 (air = 1), vp: 2.2×10^{-3} mm @ 20°.

TRANSFORMATION PRODUCTS: Though no products were identified, the reported hydrolysis $t_{1/2}$ ranged from 3 to 35 days.

EXPOSURE LIMITS: NIOSH REL: IDLH 40 mg/m^3; OSHA PEL: TWA 0.1 mg/m^3.

SYMPTOMS OF EXPOSURE: Miosis; rhin; head; chest, wheez, lar spas; salv, cyan; anor, nau, abdom cramp, diarr, para, atax, convul, low BP; card; irrit skin, eyes.

USES: Insecticide and acaricide.

MORPHOLINE

SYNONYMS: Diethyleneimide oxide; Diethyleneimid oxide; Diethylene oximide; Diethylenimide oxide; 1-Oxa-4-azacyclohexane; Tetrahydro-1,4-isoxazine; Tetrahydro-1,4-oxazine; Tetrahydro-2*H*-1,4-oxazine; UN 2054.

CHEMICAL DESIGNATIONS: CAS: 110-91-8; DOT: 2054; mf: C_4H_9NO; fw: 87.12; RTECS: QD 6475000.

PROPERTIES: Colorless liq with a weak ammonia-like odor. Mp: -4.7°, bp: 128.3°, *d*: 1.0005 @ 20/4°, fl p: 35°, lel: 1.4%, uel: 11%, pK_a: 8.33 @ 25°, log K_{ow}: -1.08, S_o: misc with in acetone, benzene, castor oil, ethanol, ether, ethylene glycol, 2-hexanone, linseed oil, methanol, pine oil and turpentine, S_w: misc, vap d: 3.56 g/L @ 25°, 3.01 (air = 1), vp: 8.0 mm @ 20°.

TRANSFORMATION PRODUCTS: In an aq soln, chloramine reacted with morpholine to form reaction of NO_2 (1-99 ppm) and morpholine yielded *N*-nitromorpholine and *N*-nitromorpholine.

EXPOSURE LIMITS: NIOSH REL: IDLH 8,000 ppm; OSHA PEL: TWA 20 ppm.

SYMPTOMS OF EXPOSURE: Vis dist; nose irrit; cough; resp irrit; eye, skin irrit.

USES: Solvent for waxes, casein, dyes and resins; rubber accelerator; solvent; optical brightener for detergents; corrosion inhibitor; additive to boiler water; preservation of book paper; organic synthesis.

NALED

SYNONYMS: Arthodibrom; Bromchlophos; Bromex; Dibrom; 1,2-Dibromo-2,2-dichloroethyldimethyl phosphate; Dimethyl 1,2-dibromo-2,2-dichloroethyl phosphate; *O,O*-Dimethyl-*O*-(1,2-dibromo-2,2-dichloroethyl)phosphate; *O,O*-Dimethyl *O*-(2,2-dichloro-1,2-dibromoethyl)phosphate; ENT 24,988; Hibrom; NA 2,783; Ortho 4,355; Orthodibrom; Orthodibromo; **Phosphoric acid 1,2-dibromo-2,2-dichloroethyl dimethyl ester**; RE-4,355.

CHEMICAL DESIGNATIONS: CAS: 300-76-5; DOT: 2783; mf: $C_4H_7Br_2Cl_2O_4P$; fw: 380.79; RTECS: TB 9450000.

PROPERTIES: Colorless to pale yellow liq or solid with a pungent odor. Mp: 26.5-27.5°, bp: 110° @ 0.5 mm, *d*: 1.96 @ 25/4°, S_o: Freely sol in ketone, alcohols, aromatic and chlorinated hydrocarbons but sparingly sol in petroleum solvents and mineral oils, vp: 2 x 10^{-3} @ 20°.

TRANSFORMATION PRODUCTS: Completely hydrolyzed in water within 2 days.

EXPOSURE LIMITS: NIOSH REL: IDLH 1,800 mg/m^3; OSHA PEL: TWA 3 mg/m^3.

USES: Insecticide; acaricide.

NAPHTHALENE

SYNONYMS: Camphor tar; Mighty 150; Mighty RD1; Moth balls; Moth flakes; Naphthalin; Naphthaline; Naphthene; NCI-C52904; RCRA waste number U165; Tar camphor; UN 1334; White tar.

CHEMICAL DESIGNATIONS: CAS: 91-20-3; DOT: 1334; mf: $C_{10}H_8$; fw: 128.18; RTECS: QJ 0525000.

PROPERTIES: White, cryst flakes with an odor resembling coal-tar. Mp: 80.5°, bp: 217.942°, *d*: 1.162 @ 20/4°, fl p: 79°, lel: 0.9%, uel: 5.9%, pK_a: > 15, K_H: 7.34 x 10^{-4} atm·m^3/mol @ 25°, IP: 8.14 eV, log K_{oc}: 2.72-3.52, log K_{ow}: 3.20-4.70, S_o: sol in acetone, methanol or ethanol (1 g/13 mL), benzene or toluene (1 g/3.5 mL), olive oil or turpentine (1 g/ 8 mL), chloroform or carbon tetrachloride (1 g/2 mL), carbon disulfide (1 g/2 mL), S_w: 30 mg/L @ 25°, vp: 5.4 x 10^{-2} mm @ 20°.

TRANSFORMATION PRODUCTS: Under certain conditions, *Pseudomonas* sp. oxidized naphthalene to *cis*-1,2-dihydro-1,2-dihydroxynaphthalene. This metabolite may be further oxidized by *Pseudomonas putida* to CO_2 and water. Under aerobic conditions, *Cuninghamella elegans* biodegraded naphthalene to α-naphthol, β-naphthol, *trans*-1,2-dihydroxy-1,2-dihydronaphthalene, 4-hydroxy-1-tetralene and 1,4-naphthoquinone. Also under aerobic conditions, *Agnenellum*, *Oscillatoria* and *Anabaena* reportedly biodegraded naphthalene into 1-naphthol, *cis*-1,2-dihydroxyl-1,2-dihydronaphthalene and 4-hydroxy-1-tetralene. Cultures of *Bacillus* sp. oxidized naphthalene to (+)-*trans*-1,2-dihydro-1,2-dihydroxy-naphthalene. In the presence of reduced nicotinamide adeninedinucleotide phosphate ($NADPH_2$) and ferrous ions, a cell extract oxidized naphthalene to *trans*-naphthalenediol. Hydroxylation by pure microbial cultures yielded an unidentified phenol, 1- and 2-hydroxynaphthalene. An aq soln containing chlorine dioxide in the dark for 3.5 days oxidized naphthalene to chloronaphthalene, 1,4-dichloronaphthalene and methyl esters of phthalic acid. In the presence bromide ions and a chlorinating agent (sodium hypochlorite), major products identified at various reaction times and pH's included 1-bromonaphthalene, dibromonaphthalene and 2-bromo-1,4-naphthoquinone. Minor products identified include monochloronaphthalene, dibromonaphthalene, bromochloronaphthalene, bromonaphthol, dibromonaphthol, 2-bromonaphthoquinone, dichloronaphthalene and chlorodibromonaphthalene. The gas-phase reaction of N_2O_5 and naphthalene in an environmental chamber at room temperature resulted in the formation of 1- and 2-nitronaphthalene with yields of ≈ 18 and ≈ 7.5%, respectively. The reaction of naphthalene with NO_x to form nitronaphthalene was reported to occur in urban air from St. Louis, MO. Irradiation of naphthalene and NO_2 using a high pressure mercury lamp (λ > 290 nm) yielded the following principal products: 1- and 2-

hydroxynaphthalene, 1-hydroxy-2-nitronaphthalene, 1-nitronaphthalene, 2,3-dinitronaphthalene, phthalic anhydride, 1,3-, 1,5- and 1,8-dinitronaphthalene. A CO_2 yield of 30.0% was achieved when naphthalene adsorbed on silica gel was irradiated with light ($\lambda > 290$ nm) for 17 h. In activated sludge, 9.0% of the applied amount mineralized to CO_2 after 5 days. It was suggested that the chlorination of naphthalene in tap water accounted for the presence of chloro- and dichloronaphthalenes.

EXPOSURE LIMITS: NIOSH REL: IDLH 500 ppm; OSHA PEL: TWA 10 ppm; ACGIH TLV: STEL 15 ppm.

SYMPTOMS OF EXPOSURE: Eye irrit; head; conf, excitement; mal;, nau, vomit, abdom pain; irrit bladder; profuse sweat; jaun; hema; hemog; renal shut; derm.

USES: Intermediate for phthalic anhydride, naphthol, 1,4-napththoquinone, 1,4-dihydronaphthalene, 1,2,3,4-tetrahydronaphthalene (tetralin), decahydronaphthalene (decalin), 1-nitronaphthalene, halogenated naphthalenes, naphthyl, naphthol derivatives, dyes; mothballs manufacturing; preparation of pesticides, fungicides, dyes, detergents and wetting agents, synthetic resins, celluloids and lubricants; synthetic tanning; preservative; textile chemicals; emulsion breakers; scintillation counters; smokeless powders.

1-NAPHTHYLAMINE

SYNONYMS: 1-Aminonaphthalene; C.I. azoic diazo component 114; Fast garnet B base; Fast garnet base B; **1-Naphthalenamine**; Naphthalidam; Naphthalidine; α-Naphthylamine; RCRA waste number U167; UN 2077.

CHEMICAL DESIGNATIONS: CAS: 134-32-7; DOT: 2077; mf: $C_{10}H_9N$; fw: 143.19; RTECS: QM 1400000.

PROPERTIES: Colorless cryst with an unpleasant odor. Becomes purplish-red on color on exposure to air. Mp: 50°, bp: 300.8° (subl), *d*: 1.1229 @ 25/25°, fl p: 157.2, pK_a: 3.92 @ 25°, K_H: 1.27 x 10^{-10} atm·m^3/mol @ 25°, log K_{oc}: 3.51, log K_{ow}: 2.07, S_o: sol in alcohol and ether, S_w: 1,700 mg/L, vp: 6.5 x 10^{-5} mm @ 20-30°.

TRANSFORMATION PRODUCTS: 1-Naphthylamine added to 3 different soils was incubated in the dark @ 23° under a CO_2-free atmosphere. After 308 days, 16.6 to 30.7% of the added 1-naphthylamine to soil biodegraded to CO_2.

SYMPTOMS OF EXPOSURE: Derm; hemorragic cystitis, dysp, atax, methemoglobinemia; hema; dys.

USES: Manufacturing dyes and dye intermediates; agricultural chemicals.

2-NAPHTHYLAMINE

SYNONYMS: 2-Aminonaphthalene; C.I. 37,270; Fast scarlet base B; 2-Naphthalamine; **2-Naphthalenamine**; β-Naphthylamine; 6-Naphthylamine; 2-Naphthylamine mustard; RCRA waste number U168; UN 1650; USAF CB-22.

CHEMICAL DESIGNATIONS: CAS: 91-59-8; DOT: 1650; mf: $C_{10}H_9N$; fw: 143.19; RTECS: QM 2100000.

PROPERTIES: White to reddish leaflets. Becomes purplish-red in color on exposure to air. Mp: 113°, bp: 306.1°, *d*: 1.0614 @ 98/4°, fl p: 157°, pK_a: 4.11 @ 25°, K_H: 2.01 x 10^{-9} atm·m^3/mol @ 25°, log K_{oc}: 2.11 (calcd), log K_{ow}: 2.07, S_o: sol in alcohol and ether, S_w: 586 mg/L @ 20-30°, vp: 2.56 x 10^{-4} mm @ 20-30°.

SYMPTOMS OF EXPOSURE: Derm; hemorragic cystitis, dysp, atax, methemoglobinemia; hema; dys.

USES: Manufacture of dyes.

NITRAPYRIN

SYNONYMS: **2-Chloro-6-(trichloromethyl)pyridine**; Dowco-163; N-serve; N-serve nitrogen stabilizer.

CHEMICAL DESIGNATIONS: CAS: 1929-82-4; DOT: none assigned; mf: $C_6H_3Cl_4N$; fw: 230.90; RTECS: US 7525000.

PROPERTIES: Solid. Mp: 62-63°, *d*: 1.744 (calcd), K_H: 2.13 x 10^{-3} atm·m^3/mol (calcd), log K_{oc}: 2.24-2.68, log K_{ow}: 3.02, 3.41, S_o: sol in acetone, alcohol, benzene, chloroform, ether and lignoin, S_w: 40 mg/L, vp: 0.0028 mm @ 20°.

TRANSFORMATION PRODUCTS: 6-Chloropicolinic acid and CO_2 were reported as biodegradation products. Photolysis of nitrapyrin in water yielded 6-chloropicolinic acid, 6-hydroxypicolinic acid and an unidentified polar material.

USES: Nitrification inhibitor in ammonium fertilizers.

2-NITROANILINE

SYNONYMS: 1-Amino-2-nitrobenzene; Azoene fast orange GR base; Azoene fast orange GR salt; Azofix orange GR salt; Azogene fast orange GR; Azoic diazo component 6; Brentamine fast orange GR base; Brentamine fast orange GR salt; C.I. 37,025; C.I. azoic diazo component 6; Devol orange B; Devol orange salt B; Diazo fast orange GR; Fast orange base GR; Fast orange base GR salt; Fast orange base JR; Fast orange GR base; Fast orange O base; Fast orange O salt; Fast orange salt JR; Hiltonil fast orange GR base; Hiltosal fast orange GR salt; Hindasol orange GR salt; Natasol fast orange GR salt; *o*-Nitraniline; *o*-Nitroaniline; **2-Nitrobenzenamine**; ONA; Orange base CIBA II; Orange base IRGA II; Orange GRS salt; Orange salt CIBA II; Orange salt IRGA II; Orthonitroaniline; UN 1661.

CHEMICAL DESIGNATIONS: CAS: 88-74-4; DOT: 1661; mf: $C_6H_6N_2O_2$; fw: 138.13; RTECS: BY 6650000.

PROPERTIES: Orange-yellow cryst with a musty odor. Mp: 71.5°, bp: 284.1°, *d*: 1.44 @ 20/4°, fl p: 168°, pK_a: -0.26 @ 25°, K_H: 9.72 x 10^{-5} atm·m^3/mol @ 25° (calcd), IP: 8.66 eV, log K_{oc}: 1.23-1.62 (calcd), log K_{ow}: 1.44-1.83, S_o: sol in acetone, ether, benzene (208 g/kg @ 25°), chloroform (11.7 g/kg @ 0°) and ethanol (278.7 g/kg @ 25°), S_w: 1,260 mg/L @ 25°, vp: 8.1 mm @ 25°.

TRANSFORMATION PRODUCTS: Under aerobic and anaerobic conditions using a sewage inoculum, 2-nitroaniline degraded to 2-methylbenzimidazole and 2-nitroacetanilide. A *Pseudomonas* sp. strain P6, isolated from a Matapeake silt loam, did not grow on 2-nitroaniline as the sole source of carbon. However, in the presence of 4-nitroaniline, ≈ 50% of the applied 2-nitroaniline metabolized to nonvolatile products which could not be identified by HPLC. 2-Nitroaniline was degraded by tomatoe cell suspension cultures (*Lycopericon lycopersicum*). Transformation products identified included 2-nitroanilino-β-D-glucopyranoside, β-(2-amino-3-nitrophenyl)glucopyranoside and β-(4-amino-3-nitrophenyl)glucopyranoside.

USE: Organic synthesis.

3-NITROANILINE

SYNONYMS: Amarthol fast orange R base; 1-Amino-3-nitrobenzene; 3-Aminonitrobenzene; *m*-Aminonitrobenzene; Azobase MNA; C.I. 37,030; C.I. azoic diazo component 7; Daito orange base R; Devol orange R; Diazo fast orange R; Fast orange base R; Fast orange M base; Fast orange MM base; Fast orange R base; Fast orange R salt; Hiltonil fast orange R base; MNA; Naphtoelan orange R base; Nitranilin; *m*-Nitraniline; 3-Nitroaminobenzene; *m*-Nitroaminobenzene; *m*-Nitroaniline; **3-Nitrobenzenamine**; *m*-Nitrobenzenamine; *m*-Nitrophenylamine; Orange base IRGA I; UN 1661.

CHEMICAL DESIGNATIONS: CAS: 99-09-2; DOT: 1661; mf: $C_6H_6N_2O_2$; fw: 138.13; RTECS: BY 6825000.

PROPERTIES: Yellow, rhombic cryst. Mp: 114°, bp: 306.4°, *d*: 0.9011 @ 25/4°, fl p: 306° (calcd), pK_a: 2.46 @ 25°, K_H: 1.93 x 10^{-5} atm·m^3/mol @ 25° (calcd), IP: 8.80 eV, log K_{oc}: 1.26 (calcd), log K_{ow}: 1.37, S_o: sol in acetone, ether, benzene (27.18 g/kg @ 25°), chloroform (32.16 g/kg @ 25°) and ethanol (77.78 g/kg @ 25°), S_w: 1,100 mg/L @ 20°, vp: 9.56 x 10^{-5} mm @ 25°.

TRANSFORMATION PRODUCTS: A bacterial culture isolated from the Oconee River in North Georgia degraded 3-nitroaniline to 4-nitrocatechol. A *Pseudomonas* sp. strain P6, isolated from a Matapeake silt loam, did not grow on 3-nitroaniline as the sole source of carbon. However, in the presence of 4-nitroaniline, all of the applied 3-nitroaniline metabolized completely to CO_2.

USE: Organic synthesis.

4-NITROANILINE

SYNONYMS: 1-Amino-4-nitrobenzene; 4-Aminonitrobenzene; *p*-Aminonitrobenzene; Azoamine red ZH; Azofix Red GG salt; Azoic diazo compound 37; C.I. 37,035; C.I. azoic diazo component 37; C.I. developer 17; Developer P; Devol red GG; Diazo fast red GG; Fast red base GG; Fast red base 2J; Fast red 2G base; Fast red 2G salt; Fast red GG base; Fast red GG salt; Fast red MP base; Fast red P base; Fast red P salt; Fast red salt GG; Fast red salt 2J; IG base; Naphtolean red GG base; NCI-C60786; 4-Nitraniline; *p*-Nitraniline; Nitrazol 2F extra; *p*-Nitroaniline;

4-Nitrobenzenamine; *p*-Nitrobenzenamine; *p*-Nitrophenylamine; PNA; RCRA waste number P077; Red 2G base; Shinnippon fast red GG base; UN 1661.

CHEMICAL DESIGNATIONS: CAS: 100-01-6; DOT: 1661; mf: $C_6H_6N_2O_2$; fw: 138.13; RTECS: BY 7000000.

PROPERTIES: Bright yellow powder. Mp: 148-149°, bp: 331.7°, *d*: 1.424 @ 20/4°, fl p: 165°, pK_a: 0.99 @ 25°, K_H: 1.14 x 10^{-8} atm·m^3/mol @ 25° (calcd), IP: 8.85 eV, log K_{oc}: 1.08 (est), log K_{ow}: 1.39, S_o: sol in acetone, ether, benzene (5.794 g/kg @ 25°), chloroform (9.29 g/kg @ 25°) and ethanol (60.48 g/kg @ 25°), S_w: 800 mg/L @ 18.5°, vp: 1.5 x 10^{-3} mm @ 20°.

TRANSFORMATION PRODUCTS: A *Pseudomonas* sp. strain P6, isolated from a Matapeake silt loam, was grown using a yeast extract. After 8 days, 4-nitroaniline degraded completely to CO_2.

EXPOSURE LIMITS: NIOSH REL: TWA 3 mg/m^3, IDLH 300 mg/m^3; OSHA PEL: TWA 1 ppm.

SYMPTOMS OF EXPOSURE: Cyan, atax, tachycard, tachypnea, dysp, irrity, vomit, diarr, convuls, resp ar, anem, methemoglobinemia.

USES: Intermediate for dyes and antioxidants; inhibits gum formation in gasoline; corrosion inhibiter; organic synthesis (preparation of *p*-phenylenediamine).

NITROBENZENE

SYNONYMS: Essence of mirbane; Essence of myrbane; Mirbane oil; NCI-C60082; Nitrobenzol; Oil of bitter almonds; Oil of mirbane; Oil of myrbane; RCRA waste number U169; UN 1662.

CHEMICAL DESIGNATIONS: CAS: 98-95-3; DOT: 1662; mf: $C_6H_5NO_2$; fw: 123.11; RTECS: DA 6475000.

PROPERTIES: Clear, light yellow to brown, oily liq with an almond or shoe polish odor. Mp: 5.7°, bp: 210.8°, *d*: 1.2037 @ 20/4°, fl p: 88°, lel: 1.8%, K_H: 2.45 x 10^{-5} atm·m^3/mol @ 25°, pK_a: > 15, IP: 9.92 eV, log K_{oc}: 1.49-2.36, log K_{ow}: 1.70-1.88, S_o: sol in acetone, ethanol, benzene and ether, S_w: 1.9 g/L @ 20°, vap d: 5.03 g/L @ 25°, 4.25 (air = 1), vp: 0.15 mm @ 20.0°.

TRANSFORMATION PRODUCTS: Irradiation of nitrobenzene in the vapor phase produced nitrosobenzene and 4-nitrophenol. Titanium dioxide suspended in an aq soln and irradiated with UV light (λ = 365 nm) converted nitrobenzene to CO_2 at a significant rate. A CO_2 yield of 6.7% was achieved when nitrobenzene adsorbed on silica gel was irradiated with light (λ > 290 nm) for 17 h. In activated sludge, 0.4% of the applied nitrobenzene mineralized to CO_2 after 5 days. An aq soln containing nitrobenzene (5 x 10^{-4} M) and H_2O_2 (100 μM) was irradiated with UV light (λ = 285-360 nm). After 18 h, 2% of the substrate was converted into *o*-, *m*- and *p*-nitrophenols having an isomer distribution of 50, 29.5 and 20.5%, respectively. Under anaerobic conditions using a sewage inoculum, nitrobenzene degarded to aniline.

EXPOSURE LIMITS: NIOSH REL: IDLH 200 ppm; OSHA PEL: TWA 1 ppm.

SYMPTOMS OF EXPOSURE: Anoxia, irrit eyes, derm, anem, dizz, nau, vomit, dysp.

USES: Solvent for cellulose ethers; modifying esterification of cellulose acetate; ingredient of metal polishes and shoe polishes; manufacture of aniline, benzidine, quinoline, azobenzene, drugs, photographic chemicals.

4-NITROBIPHENYL

SYNONYMS: **4-Nitro-1,1′-biphenyl**; 4-Nitrodiphenyl; *p*-Nitrobiphenyl; 4-Phenylnitrobenzene; *p*-Phenylnitrobenzene; PNB.

CHEMICAL DESIGNATIONS: CAS: 92-93-3; DOT: none assigned; mf: $C_{12}H_9NO_2$; fw: 199.21; RTECS: DV 5600000.

PROPERTIES: White to yellow cryst. Mp: 114°, bp: 340°, fl p: 143°, S_o: sol in acetic acid, benzene, chloroform and ether.

SYMPTOMS OF EXPOSURE: Head, lethargy, dizz, dysp, atax, weak, methemoglobinemia, urinary buring, acute hemorr cystitis.

USES: Organic synthesis.

NITROETHANE

SYNONYM: UN 2842.

CHEMICAL DESIGNATIONS: CAS: 79-24-3; DOT: 2842; mf: $C_2H_5NO_2$; fw: 75.07; RTECS: KI 5600000.

PROPERTIES: Colorless liq with a fruity odor. Mp: -50°, bp: 115°, *d*: 1.0448 @ 25/4°, fl p: 27.8°, lel: 3.4%, pK_a: 8.44 @ 25°, K_H: 4.66 x 10^{-5} atm·m^3/mol @ 25°, IP: 10.88 eV, log K_{ow}: 0.18, S_o: sol in acetone; misc with alcohol, chloroform and ether, S_w: 4.5 wt % @ 20°, vap d: 3.07 g/L @ 25°, 2.59 (air = 1), vp: 15.6 mm @ 20°.

EXPOSURE LIMITS: NIOSH REL: IDLH 1,000 ppm; OSHA PEL: TWA 100 ppm.

SYMPTOMS OF EXPOSURE: Derm.

USES: Solvent for nitrocellulose; cellulose acetate; cellulose acetobutyrate; cellulose acetopropionate, waxes, fats, dyestuffs, vinyl and alkyd resins; experimental propellant; fuel additive; organic synthesis (Friedel-Crafts reactions).

NITROMETHANE

SYNONYMS: Nitrocarbol; UN 1261.

CHEMICAL DESIGNATIONS: CAS: 75-52-5; DOT: 1261; mf: CH_3NO_2; fw: 61.04; RTECS: PA 9800000.

PROPERTIES: Colorless liq with a strong disagreeable odor. Mp: -29°, -17°, bp: 100.8°, *d*: 1.1371 @ 20/4°, fl p: 35°, lel: 7.3%, pK_a: 10.21 @ 25°, K_H: 2.86 x 10^{-5} atm·m^3/mol, IP: 11.08 eV, log K_{ow}: -0.33 to 0.08, S_o: sol in acetone, alcohol, ether and dimethylformamide, S_w: 9.5 wt % @ 20°, vap d: 2.49 g/L @ 25°, 2.11 (air = 1), vp: 27.8 mm @ 20°.

EXPOSURE LIMITS: NIOSH REL: IDLH 1,000 ppm; OSHA PEL: TWA 100 ppm.

SYMPTOMS OF EXPOSURE: Derm.

USES: Rocket fuel; coatings industry; solvent for cellulosic compounds, polymers, waxes, fats; gasoline additive; organic synthesis.

2-NITROPHENOL

SYNONYMS: 2-Hydroxynitrobenzene; *o*-Hydroxynitrobenzene; 2-Nitro-1-hydroxybenzene; *o*-Nitrophenol; ONP; UN 1663.

CHEMICAL DESIGNATIONS: CAS: 88-75-5; DOT: 1663; mf: $C_6H_5NO_3$; fw: 139.11; RTECS: SM 2100000.

PROPERTIES: Pale yellow cryst with an aromatic odor. Mp: 44-46°, bp: 216°, *d*: 1.495 @ 20/4°, fl p: 73.5°, pK_a: 7.23 @ 25°, K_H: 3.5 x 10^{-6} atm·m^3/mol, log K_{oc}: 2.06, log K_{ow}: 1.73-1.79, S_o: sol in acetone, ether, chloroform, carbon disulfide, alcohol (340 g/L @ RT), benzene (1,472 g/kg @ 20°, 3,597 g/kg @ 30°) and ethanol (460 g/kg @ 25°), S_w: 2.1 mg/L @ 20°, vp: 0.20 mm @ 25°.

TRANSFORMATION PRODUCTS: A microorganism, *Pseudomonas putida*, isolated from soil degraded *o*-nitrophenol to nitrite. Degradation by enzymatic mechanisms produced nitrite and catechol. Catechol subsequently degraded to β-ketoadipic acid. Oxidation by Fenton's reagent (H_2O_2 and Fe^{3+}) produced nitrohydroquinone and 3-nitrocatechol.

USES: Indicator; preparation of *o*-nitroanisole and other organic compounds.

4-NITROPHENOL

SYNONYMS: 4-Hydroxynitrobenzene; *p*-Hydroxynitrobenzene; NCI-C55992; 4-Nitro-1-hydroxybenzene; *p*-Nitrophenol; PNP; RCRA waste number U170; UN 1663.

CHEMICAL DESIGNATIONS: CAS: 100-02-7; DOT: 1663; mf: $C_6H_5NO_3$; fw: 139.11; RTECS: SM 2275000.

PROPERTIES: Colorless to pale yellow, odorless cryst. Mp: 114°, bp: 279° (decomp), *d*: 1.479 @ 20/4°, pK_a: 7.15 @ 25°, K_H: 3.0 x 10^{-5} atm·m^3/mol @ 20° (calcd), IP: 9.52 eV, log K_{oc}: 1.74-2.73, log K_{ow}: 1.85-2.04, S_o: sol in acetone, ether, chloroform, benzene (9.2 g/kg @ 20°), ethanol (1,895 g/kg @ 25°) and toluene (227 g/kg @ 70°), S_w: 16 g/L @ 25°, vp: 10^{-4} mm @ 20°.

TRANSFORMATION PRODUCTS: Under anaerobic conditions, 4-nitrophenol may undergo nitro-reduction to produce *p*-aminophenol. Estuarine sediment samples collected from the Mississippi River near Leeville, LA were used to study the mineralization of *p*-nitrophenol under aerobic and anaerobic conditions. The rate of mineralization to CO_2 was found to be faster under aerobic conditions (1.04 x 10^{-3} μg/day/g dry sediment) than under anaerobic conditions (2.95 x 10^{-5} μg/day/g dry sediment). An aq soln containing 200 ppm *p*-nitrophenol exposed to sunlight for 1-2 mo yielded hydroquinone, 4-nitrocatechol and an unidentified polymeric substance. Under artificial sunlight, river water containing 2-5 ppm 4-nitrophenol photodegraded to produce trace amounts of 4-aminophenol. Wet oxidation of 4-nitrophenol @ 320° yielded formic and acetic acids. In activated sludge, *p*-nitrophenol biodegraded to hydroquinone with smaller quantities of oxyhydroquinone. A CO_2 yield of 39.5% was achieved when 4-nitrophenol adsorbed on silica gel was irradiated with light ($\lambda > 290$ nm) for 17 h. In activated sludge, 0.5% mineralized to CO_2 after 5 days. Wet oxidation of *p*-nitrophenol at an elevated pressure and temperature gave the following products: acetone, acetaldehyde, formic, acetic, maleic, oxalic and succinic acids.

USES: Fungicide for leather; production of parathion; preparation of *p*-nitrophenyl acetate and other organic compounds.

1-NITROPROPANE

SYNONYM: UN 2608.

CHEMICAL DESIGNATIONS: CAS: 108-03-2; DOT: 2608; mf: $C_3H_7NO_2$; fw: 89.09; RTECS: TZ 5075000.

PROPERTIES: Colorless liq with a mild, fruity odor. Mp: -108°, bp: 130-131°, *d*: 0.9934 @ 25/4°, fl p: 34°, lel: 2.2%, K_H: 8.68 x 10^{-5} atm·m^3/mol @ 25°, IP: 10.81 eV, log K_{ow}: 0.65, 0.87, S_o: sol in alcohol, chloroform and ether; misc with many organic solvents, S_w: 1.4 wt % @ 20°, vap d: 3.64 g/L @ 25°, 3.08 (air = 1), vp: 7.5 mm @ 20°.

EXPOSURE LIMITS: NIOSH REL: IDLH 2,300 ppm; OSHA PEL: TWA 25 ppm.

SYMPTOMS OF EXPOSURE: Eye irrit, head, nau, vomit, diarr.

USES: Solvent for cellulose acetate, lacquers, vinyl resins, fats, oils, dyes, synthetic rubbers; chemical intermediate; propellant; gasoline additive.

2-NITROPROPANE

SYNONYMS: Dimethylnitromethane; Isonitropropane; Nipar S-20 solvent; Nipar S-30 solvent; Nitroisopropane; 2-NP; RCRA waste number U171; UN 2608.

CHEMICAL DESIGNATIONS: CAS: 79-46-9; DOT: 2608; mf: $C_3H_7NO_2$; fw: 89.09; RTECS: TZ 5250000.

PROPERTIES: Colorless liq with a mild, fruity odor. Mp: -93°, bp: 120°, *d*: 0.9876 @ 20/4°, fl p: 24°, lel: 2.6%, K_H: 1.23 x 10^{-4} atm·m^3/mol @ 25°, IP: 10.71 eV, S_o: sol in chloroform; misc with many organic solvents, S_w: 1.7 wt % @ 20°, vap d: 3.64 g/L @ 25°, 3.08 (air = 1), vp: 12.9 mm @ 20°.

TRANSFORMATION PRODUCTS: Anticipated products from the reaction of 2-nitropropane with O_3 or hydroxyl radicals in the atmosphere are formaldehyde and acetaldehyde.

EXPOSURE LIMITS: NIOSH REL: lowest feasible limit; OSHA PEL: TWA 25 ppm; ACGIH TLV: TWA 10 ppm.

SYMPTOMS OF EXPOSURE: Head, anor, nau, vomit, diarr, irrit resp sys.

USES: Solvent for cellulose acetate, lacquers, vinyl resins, fats, oils, dyes, synthetic rubbers; chemical intermediate; propellant; gasoline additive.

N-NITROSODIMETHYLAMINE

SYNONYMS: Dimethylnitrosamine; *N*-Dimethylnitrosamine; *N,N*-Dimethylnitrosamine; Dimethylnitrosomine; DMN; DMNA; ***N*-Methyl-*N*-nitrosomethanamine**; NDMA; Nitrous dimethylamide; RCRA waste number P082.

CHEMICAL DESIGNATIONS: CAS: 62-75-9; DOT: 1955; mf: $C_2H_6N_2O$; fw: 74.09; RTECS: IQ 0525000.

PROPERTIES: Yellow liq with a faint characteristic odor. Bp: 154°, *d*: 1.0059 @ 20/4°, fl p: 61°, K_H: 0.143 atm·m^3/mol @ 25° (calcd), log K_{oc}: 1.41 (calcd), log K_{ow}: 0.06, S_o: sol in solvents including ethanol and ether, S_w: misc, vp: 2.7 mm @ 20°.

TRANSFORMATION PRODUCTS: A Teflon bag containing air and *N*-nitrosodimethylamine was subjected to sunlight on 2 different days. On a cloudy day, half of the *N*-nitrosodimethylamine was photolyzed in 1 h. On a sunny day, half of the *N*-nitrosodimethylamine was photolyzed in 30 min. Photolysis products included NO, CO, formaldehyde and an unidentified compound. Irradiation of an O_3-rich atmosphere containing *N*-nitrosodimethylamine yielded the following photolysis products: dimethylnitramine, nitromethane, formaldehyde, CO, NO_2, N_2O_5 and nitric acid. Two of 7 microorganisms, *Escherichia coli* and *Pseudomonas fluorescens*, were capable of slowly degrading *N*-nitrosodimethylamine to dimethylamine.

SYMPTOMS OF EXPOSURE: Nau, vomit, diarr, abdom cramps; head; fvr; enl liv; jaun; reduced function of liv, kid and lungs.

USES: Rubber accelerator; solvent in fiber and plastic industry; rocket fuels; lubricants; condensers to increase dielectric constant; industrial solvent; antioxidant; nematocide; softener of copolymers; research chemical; plasticizer in acrylonitrile polymers; inhibit nitrification in soil; chemical intermediate for 1,1-dimethylhydrazine.

N-NITROSODIPHENYLAMINE

SYNONYMS: Benzenamine; Curetard A; Delac J; Diphenylnitrosamine; Diphenyl-*N*-nitrosamine; *N,N*-Diphenylnitrosamine; Naugard TJB; NCI-C02880; NDPA; NDPhA; Nitrosodiphenylamine; *N*-Nitroso-*n*-phenylamine; *N*-Nitroso-*n*-phenylbenzenamine; Nitrous diphenylamide; Redax; Retarder J; TJB; Vulcalent A; Vulcatard; Vulcatard A; Vultrol.

CHEMICAL DESIGNATIONS: CAS: 86-30-6; DOT: none assigned; mf: $C_{12}H_{10}N_2O$; fw: 198.22; RTECS: JJ 9800000.

PROPERTIES: Green platy cryst or dark blue cryst. Mp: 66.5°, 144°, K_H: 2.33 x 10^{-8} atm·m^3/mol @ 25° (calcd), log K_{oc}: 2.76 (est), log K_{ow}: 3.13, S_o: sol in ethanol, benzene, ether, chloroform and slightly sol in petroleum ether, S_w: 35.1 mg/L @ 25°, vp: 0.1 mm @ 25° (est).

TRANSFORMATION PRODUCTS: Above 85°, technical grades may decompose to NO_x.

USES: Chemical intermediate for *N*-phenyl-*p*-phenylenediamine; rubber processing (vulcanization retarder).

N-NITROSODI-*n*-PROPYLAMINE

SYNONYMS: Dipropylnitrosamine; Di-*n*-propylnitrosamine; DPN; DPNA; NDPA; *N*-Nitrosodipropylamine; ***N*-Nitroso-*n*-propyl-1-propanamine**; RCRA waste number U111.

CHEMICAL DESIGNATIONS: CAS: 621-64-7; DOT: none assigned; mf: $C_6H_{14}N_2O$; fw: 130.19; RTECS: JL 9700000.

PROPERTIES: Yellow to gold colored liq. Bp: 205.9°, *d*: 0.9160 @ 20/4°, log K_{oc}: 1.01 (est), log K_{ow}: 1.31 (calcd), S_o: sol in ethanol and ether, S_w: 9,900 mg/L @ 25°.

USE: Research chemical.

2-NITROTOLUENE

SYNONYMS: **1-Methyl-2-nitrobenzene**; 2-Methylnitrobenzene; *o*-Methylnitrobenzene; *o*-Nitrotoluene; ONT; UN 1664.

CHEMICAL DESIGNATIONS: CAS: 88-72-2; DOT: 1664; mf: $C_7H_7NO_2$; fw: 137.14; RTECS: XT 3150000.

PROPERTIES: Yellowish liq with an aromatic odor. Mp: -9.5° (needles), -2.5° (cryst), bp: 221.7°, *d*: 1.1629 @ 20/4°, fl p: 106.1°, lel: 2.2%, K_H: 4.51 x 10^{-5} atm·m^3/mol @ 20° (calcd), log K_{ow}: 2.30, S_o: sol in alcohol, ether, benzene and petroleum ether, S_w: 652 mg/kg @ 30°, vap d: 5.61 g/L @ 25°, 4.73 (air = 1), vp: 0.15 mm @ 20°.

EXPOSURE LIMITS: NIOSH REL: IDLH 200 ppm; OSHA PEL: TWA 5 ppm; ACGIH TLV: TWA 2 ppm.

SYMPTOMS OF EXPOSURE: Anoxia, cyan, head, weak, dizz, atax, dysp, tacar, nau, vomit.

USES: Manufacture of dyes, nitrobenzoic acids, toluidines, etc.

3-NITROTOLUENE

SYNONYMS: **1-Methyl-3-nitrobenzene**; 3-Methylnitrobenzene; *m*-Methylnitrobenzene; MNT; *m*-Nitrotoluene; 3-Nitrotoluol; UN 1664.

CHEMICAL DESIGNATIONS: CAS: 99-08-1; DOT: 1664; mf: $C_7H_7NO_2$; fw: 137.14; RTECS: XT 2975000.

PROPERTIES: Yellowish liq with an aromatic odor. Mp: 16°, bp: 232.6°, *d*:

1.1571 @ 20/4°, fl p: 101.1°, lel: 1.6%, K_H: 5.41 x 10^{-5} atm·m^3/mol @ 20° (calcd), IP: 9.65 eV, log K_{ow}: 2.40-2.45, S_o: sol in alcohol, benzene and ether, S_w: 498 mg/kg @ 30°, vap d: 5.61 g/L @ 25°, 4.73 (air = 1), vp: 0.15 mm @ 20°.

TRANSFORMATION PRODUCTS: Under anaerobic conditions using a sewage inoculum, 3-nitrotoluene and 4-nitrotoluene both degraded to toluidine.

EXPOSURE LIMITS: NIOSH REL: IDLH 200 ppm; OSHA PEL: TWA 5 ppm; ACGIH TLV: 2 ppm.

SYMPTOMS OF EXPOSURE: Anoxia, cyan, head, weak, dizz, atax, dysp, tacar, nau, vomit.

USES: Manufacture of dyes, nitrobenzoic acids, toluidines, etc; organic synthesis.

4-NITROTOLUENE

SYNONYMS: **1-Methyl-4-nitrobenzene**; 4-Methylnitrobenzene; *p*-Methylnitrobenzene; NCI-C60537; *p*-Nitrotoluene; 4-Nitrotoluol; PNT; UN 1664.

CHEMICAL DESIGNATIONS: CAS: 99-99-0; DOT: 1664; mf: $C_7H_7NO_2$; fw: 137.14; RTECS: XT 3325000.

PROPERTIES: Yellowish cryst with a weak aromatic odor. Mp: 54.5°, bp: 238.3°, *d*: 1.286 @ 20°, fl p: 106.1°, lel: 1.6%, K_H: 5.0 x 10^{-5} atm·m^3/mol @ 25° (calcd), IP: 9.82 eV, log K_{ow}: 2.37, 2.42, S_o: sol in acetone, alcohol, benzene, ether and chloroform, S_w: 0.005 wt % @ 20°, vp: 0.12 mm @ 20°.

TRANSFORMATION PRODUCTS: Under anaerobic conditions using a sewage inoculum, 3- and 4-nitrotoluene both degraded to toluidine.

EXPOSURE LIMITS: NIOSH REL: IDLH 200 ppm; OSHA PEL: TWA 5 ppm; ACGIH TLV: TWA 2 ppm.

SYMPTOMS OF EXPOSURE: Anoxia, cyan, head, weak, dizz, atax, dysp, tacar, nau, vomit.

USES: Manufacture of dyes, nitrobenzoic acids, toluidines, etc.

n-NONANE

SYNONYMS: **Nonane**; Nonyl hydride; Shellsol 140; UN 1920.

CHEMICAL DESIGNATIONS: CAS: 111-84-2; DOT: 1920; mf: C_9H_{20}; fw: 128.26; RTECS: RA 6115000.

PROPERTIES: Colorless liq. Mp: -51°, bp: 150.8°, *d*: 0.7176 @ 20/4°, fl p: 30°, lel: 0.8%, uel: 2.9%, K_H: 5.95 atm·m^3/mol @ 25° (calcd), log K_{ow}: 4.67 (calcd), S_o: sol in acetone, alcohol, benzene, chloroform and ether, S_w: 0.122 ppm @ 25°, vap d: 5.24 g/L @ 25°, 4.43 (air = 1), vp: 3.22 mm @ 20°.

TRANSFORMATION PRODUCTS: *n*-Nonane may biodegrade in 2 ways. The first is the formation of nonyl hydroperoxide which decomposes to *n*-nonanol followed by oxidation to nonanoic acid. The other pathway involves dehydrogenation to 1-nonene which may react with water giving *n*-nonanol.

USES: Solvent; standardized hydrocarbon; manufacturing paraffin products;

biodegradable detergents; jet fuel research; rubber industry; paper processing industry; distillation chaser; organic synthesis.

OCTACHLORONAPHTHALENE

SYNONYM: Halowax 1051.

CHEMICAL DESIGNATIONS: CAS: 2234-13-1; DOT: none assigned; mf: $C_{10}Cl_8$; fw: 403.73; RTECS: QK 0250000.

PROPERTIES: Light yellow solid with an aromatic odor. Mp: 197-198°, bp: ≈ 410°, S_o: sol in benzene, chloroform and lignoin, vp: < 1 mm @ 20°.

EXPOSURE LIMITS: NIOSH REL: IDLH 200 mg/m^3; OSHA PEL: TWA 0.1 mg/m^3.

SYMPTOMS OF EXPOSURE: Acneform derm, liv damage, jaun.

USES: Chemical research; organic synthesis.

n-OCTANE

SYNONYMS: **Octane**; UN 1262.

CHEMICAL DESIGNATIONS: CAS: 111-65-9; DOT: 1262; mf: C_8H_{18}; fw: 114.23; RTECS: RG 8400000.

PROPERTIES: Clear liq. Mp: -56.8°, bp: 125.7°, *d*: 0.7025 @ 20/4°, fl p: 13.3°, lel: 1%, uel: 4.7%, K_H: 3.225 atm·m^3/mol @ 25°, log K_{ow}: 5.18, S_o: sol in acetone, alcohol, chloroform, ether and petroleum hydrocarbons; misc with benzene, gasoline, petroleum ether, S_w: 0.431 ppm @ 25°, vap d: 4.67 g/L @ 25°, 3.94 (air = 1), vp: 11 mm @ 20°.

TRANSFORMATION PRODUCTS: *n*-Octane may biodegrade in 2 ways. The first is the formation of octyl hydroperoxide which decomposes to *n*-octanol followed by oxidation to octanoic acid. The other pathway involves dehydrogenation to 1-octene which may react with water giving *n*-octanol.

EXPOSURE LIMITS: NIOSH REL: 10-h TWA 75 ppm, 15-min C 385 ppm, IDLH 5,000 ppm; OSHA PEL: TWA 500 ppm; ACGIH TLV: TWA 300 ppm.

SYMPTOMS OF EXPOSURE: Irrit eyes, nose; drow; derm; chem pneu.

USES: Solvent; rubber and paper industries; calibrations; azeotropic distillations; organic synthesis.

1-OCTENE

SYNONYMS: 1-Caprylene; 1-Octylene.

CHEMICAL DESIGNATIONS: CAS: 111-66-0; DOT: none assigned; mf: C_8H_{16}; fw: 112.22; RTECS: none assigned.

PROPERTIES: Colorless liq. Mp: -101.7°, bp: 121.3°, *d*: 0.7149 @ 20/4°, fl p: 21° (OC), K_H: 9.52 x 10^{-1} atm·m^3/mol @ 25°, log K_{ow}: 2.79, S_o: sol in acetone, benzene and chloroform; misc with alcohol and ether, S_w: 2.7 mg/L @ 20°, vap d:

4.59 g/L @ 25°, 3.87 (air = 1), vp: 17.4 mm @ 25°.

TRANSFORMATION PRODUCTS: Biooxidation of 1-octene may occur yielding 7-octen-1-ol which may further oxidize to give 7-octenoic acid.

USES: Plasticizer; surfactants; organic synthesis.

OXALIC ACID

SYNONYMS: Ethanedioic acid; Ethanedionic acid; NCI-C55209.

CHEMICAL DESIGNATIONS: CAS: 144-62-7; DOT: 2449; mf: $C_2H_2O_4$; fw: 90.04; RTECS: RO 2450000.

PROPERTIES: Colorless rhombic cryst. Mp: 182-189.5° (anhydrous), 101.5° (hydrated), bp: 157°, *d*: 1.653 @ 18/4°, pK_a: @ 25°: pK_1 = 1.27, pK_2 = 4.27, K_H: 1.43 x 10^{-10} atm·m^3/mol @ pH 4, log K_{oc}: 0.89 (calcd), log K_{ow}: -0.43, -0.81, S_o: sol in alcohol (40 g/L), ether (10 g/L), glycerol (181.8 g/L), S_w: 95 g/L @ 15°, vp: 3 x 10^{-4} mm @ 30°.

TRANSFORMATION PRODUCTS: Above 189.5°, decomposes to CO_2, CO, formic acid and water. Ozonolysis of oxalic acid in distilled water @ 25° under acidic conditions (pH 6.3) yielded CO_2. Absorbs moisture in air to form the dihydrate.

EXPOSURE LIMITS: NIOSH REL: IDLH 500 mg/m^3; OSHA PEL: TWA 1 mg/m^3.

SYMPTOMS OF EXPOSURE: Lac, vis dist, conj; head; cough; dysp; derm; skin burns; kid damage.

USES: Calico printing and dyeing, analytical reagent; bleaching agent; removing ink or rust stains, paint or varnish; stripping agent for permanent press resins; reducing agent; metal polishes; ceramics and pigments; cleaning wood; laboratory reagent; purifying methanol; cleanser in the metallurgical industry; in paper industry and photography; producing glucose from starch; in process engraving; leather tanning; rubber manufacturing industry; automobile radiator cleanser; purifying agent and intermediate for many compounds; catalyst; rare earth processing; organic synthesis.

PARATHION

SYNONYMS: AAT; AATP; AC 3,422; ACC 3,422; Alkron; Alleron; American Cyanamide 3,422; Aphamite; B 404; Bay E-605; Bladan; Bladan F; Compound 3422; Corothion; Corthion; Corthione; Danthion; DDP; *O,O*-Diethyl-*O*-4-nitrophenyl phosphorothioate; *O,O*-Diethyl *O*-*p*-nitrophenyl phosphorothioate; Diethyl-4-nitrophenyl phosphorothionate; Diethyl-*p*-nitrophenyl thionophosphate; *O,O*-Diethyl-*O*-4-nitrophenyl thionophosphate; *O,O*-Diethyl-*O*-*p*-nitrophenyl thionophosphate; Diethyl-*p*-nitrophenyl thiophosphate; *O,O*-Diethyl-*O*-*p*-nitrophenyl thiophosphate; Diethylparathion; DNTP; DPP; Drexel parathion 8E; E 605; Ecatox; Ekatin WF & WF ULV; Ekatox; ENT 15,108; Ethlon; Ethyl parathion; Etilon; Folidol; Folidol E605; Folidol E & E 605; Fosfermo; Fosferno;

Fosfex; Fosfive; Fosova; Fostern; Fostox; Gearphos; Genithion; Kolphos; Kypthion; Lethalaire G-54; Lirothion; Murfos; NA 2,783; NCI-C00226; Niran; Niran E-4; Nitrostigmine; Nitrostygmine; Niuif-100; Nourithion; Oleofos 20; Oleoparaphene; Oleoparathion; Orthophos; Pac; Panthion; Paradust; Paraflow; Paramar; Paramar 50; Paraphos; Paraspray; Parathene; Parathionethyl; Parawet; Penphos; Pestox plus; Pethion; Phoskil; Phosphemol; Phosphenol; **Phosphorothioic acid *O,O*-diethyl *O*-(4-nitrophenyl) ester**; Phosphostigmine; RB; RCRA waste number P089; Rhodiasol; Rhodiatox; Rhodiatrox; Selephos; Sixty-three special E.C. insecticide; SNP; Soprathion; Stabilized ethyl parathion; Stathion; Strathion; Sulphos; Super rodiatox; T-47; Thiofos; Tiophos; Thiophos 3,422; Tox 47; Vapophos; Vitrex.

CHEMICAL DESIGNATIONS: CAS: 56-38-2; DOT: 2783; mf: $C_{10}H_{14}NO_5PS$; fw: 291.27; RTECS: TF 4550000.

PROPERTIES: Pale yellow to brown liq with a garlic-like odor. Mp: 6.1°, bp: 375°, *d*: 1.26 @ 25/4°, fl p: 174°, K_H: 8.56 x 10^{-8} atm·m^3/mol @ 25°, log K_{oc}: 2.50-4.20, log K_{ow}: 2.15-3.93, S_o: sol in acetone, alcohol, benzene and ether; freely sol in alcohols, aromatic hydrocarbons, esters, ethers and ketones, S_w: 12.9 mg/L @ 20°, vap d: 11.91 g/L @ 25°, 10.06 (air = 1), vp: 4 x 10^{-4} mm @ 20°.

TRANSFORMATION PRODUCTS: Diethyl-*O*-thiophosphoric acid and *p*-nitrophenol were reported as biodegradation products. *p*-Nitrophenol and paraoxon were formed from the irradiation of parathion in water by a low-pressure mercury lamp. When parathion was released in the atmosphere on a sunny day, it was rapidly converted to the photochemical paraoxon (est $t_{1/2}$ = 2 min). The reaction involving the oxidation of parathion to paraoxon is catalyzed in the presence of UV light, O_3, soil dust, or clay minerals. Parathion was reported to biologically hydrolyze to *p*-nitrophenol in different soils under flooded conditions. *p*-Nitrophenol also was identified as a hydrolysis product. The reported hydrolysis half-lives @ pH 7.4 @ 20 and 37.5° were 130 and 26.8 days, respectively. When equilibrated with a prereduced pokkali soil, parathion instantaneously degraded to aminoparathion. The quick rate of reaction was reportedly due to soil enzymes and/or other heat labile substances. Aminoparathion also was formed when parathion (500 ppm) was incubated in a flooded alluvial soil. The amount of parathion remaining after 6 and 12 days were 43.0 and 0.09%, respectively. Reported ozonation products of parathion in drinking water include sulfuric acid, paraoxon, 2,4-dinitrophenol, picric and phosphoric acids.

EXPOSURE LIMITS: NIOSH REL: 10-h TWA 0.05 mg/m^3, IDLH 20 mg/m^3; OSHA PEL: TWA 0.1 mg/m^3.

SYMPTOMS OF EXPOSURE: Miosis; rhin; head; chest wheez; lar spas; salv; cyan; anor, abdom cramp; diarr; sweat; musc fasc; weak; para, atax; convuls; low BP; card; derm.

USES: Insecticide; acaricide

PCB-1016

SYNONYMS: Arochlor 1016; **Aroclor 1016**; Chlorodiphenyl (41% Cl).

CHEMICAL DESIGNATIONS: CAS: 12674-11-2; DOT: 2315; mf: not definitive; fw: 257.9 (avg); RTECS: TQ 1351000.

PROPERTIES: Oily light yellow liq or white powder with a weak odor. Bp: distills @ 325-356°, *d*: 1.33 @ 25/4°, K_H: 750 atm/mol fraction, log K_{oc}: 4.70 (est), log K_{ow}: 4.38, 5.88, S_o: sol in most solvents, S_w: 49 μg/L @ 24°, vp: 9.03 x 10^{-4} mm @ 25°.

TRANSFORMATION PRODUCTS: Reported degradation products by the microorganism *Alcaligenes* BM-2 for a mixture of polychlorinated biphenyls include monohydroxychlorobiphenyl, 2-hydroxy-6-oxochlorophenylhexa-2,4-dieonic acid, chlorobenzoic acid, chlorobenzoylpropionic acid, chlorophenylacetic acid and 3-chlorophenyl-2-chloropropenic acid.

USES: Insulator fluid for electric condensers and as an additive in very high pressure lubricants.

PCB-1221

SYNONYMS: Arochlor 1221; **Aroclor 1221**; Chlorodiphenyl (21% Cl).

CHEMICAL DESIGNATIONS: CAS: 11104-28-2; DOT: 2315; mf: not definitive; fw: 192 (avg); RTECS: TQ 1352000.

PROPERTIES: Oily, colorless to light yellow liq with a weak odor. Mp: 1°, bp: distills @ 275-320°, *d*: 1.15 @ 25/4°, fl p: 141-150°, K_H: 3.24 x 10^{-4} atm·m^3/mol (calcd), log K_{oc}: 2.44 (est), log K_{ow}: 2.8 (est), S_o: sol in most solvents, S_w: 590 μg/L @ 24°, vp: 6.7 x 10^{-3} mm @ 25° (est).

TRANSFORMATION PRODUCTS: Reported degradation products by the microorganism *Alcaligenes* BM-2 for a mixture of polychlorinated biphenyls include monohydroxychlorobiphenyl, 2-hydroxy-6-oxochlorophenylhexa-2,4-dieonic acid, chlorobenzoic acid, chlorobenzoylpropionic acid, chlorophenylacetic acid and 3-chlorophenyl-2-chloropropenic acid. In sewage wastewater, *Pseudomonas* sp. 7509 degraded PCB-1221 into a yellow compound tentatively identified as chlorinated derivatives of α-hydroxymuconic acid.

USES: In polyvinyl acetate to improve fiber-tear properties; plasticizer for polystyrene; in epoxy resins to improve adhesion and resistance to chemical attack; as an insulator fluid for electric condensers and as an additive in very high pressure lubricants.

PCB-1232

SYNONYMS: Arochlor 1232; **Aroclor 1232**; Chlorodiphenyl (32% Cl).

CHEMICAL DESIGNATIONS: CAS: 11141-16-5; DOT: 2315; mf: not definitive; fw: 221 (avg); RTECS: TQ 1354000.

PROPERTIES: Oily, almost colorless to light yellow liq with a weak odor. Mp: -35.5° (pour point), bp: distills @ 290-325°, *d*: 1.24 @ 25/4°, fl p: 152-154°, K_H: 8.64 x 10^{-4} atm·m^3/mol (calcd), log K_{oc}: 2.83 (est), log K_{ow}: 3.2 (est), S_o: sol in most solvents, S_w: 1.45 mg/L @ 25° (est), vap d: ≈ 9.03 g/L @ 25°, 7.63 (air = 1), vp: 4.6 x 10^{-3} mm @ 25° (est).

TRANSFORMATION PRODUCTS: PCB-1232 in a 90% acetonitrile/water soln containing 0.2-0.3 M sodium borohydride and irradiated with UV light (λ = 254 nm) reacted to yield dechlorinated biphenyls. Without sodium borohydride, the reaction proceeded more slowly. Reported degradation products by the microorganism *Alcaligenes* BM-2 for a mixture of polychlorinated biphenyls include monohydroxychlorobiphenyl, 2-hydroxy-6-oxochlorophenylhexa-2,4-dieonic acid, chlorobenzoic acid, chlorobenzoylpropionic acid, chlorophenylacetic acid and 3-chlorophenyl-2-chloropropenic acid.

USES: In polyvinyl acetate to improve fiber-tear properties; as an insulator fluid for electric condensers and as an additive in very high pressure lubricants.

PCB-1242

SYNONYMS: Arochlor 1242; **Aroclor 1242**; Chlorodiphenyl (42% Cl).

CHEMICAL DESIGNATIONS: CAS: 53469-21-9; DOT: 2315; mf: not definitive; fw: 261 (avg); RTECS: TQ 1356000.

PROPERTIES: Almost colorless to light yellow oily liq with a weak odor. Mp: -19°, bp: distills @ 325-366°, *d:* 1.392 @ 15/4°, fl p: 176-180°, K_H: 2.8 x 10^{-4} atm·m^3/mol @ 20°, log K_{oc}: 3.71 (est), log K_{ow}: 4.11, S_o: sol in most solvents, S_w: 200 μg/L @ 20°, vap d: ≈ 10.67 g/L @ 25°, 9.01 (air = 1), vp: 2.5 x 10^{-4} mm @ 20°.

TRANSFORMATION PRODUCTS: A strain of *Alcaligenes eutrophus* degraded 81% of the congeners by dechlorination under anaerobic conditions. A bacterial culture isolated from Hamilton Harbour, Ontario was capable of degrading a commercial mixture of PCB-1242. The metabolites identified by GC/MS included isohexane, isooctane, ethylbenzene, isoheptane, isopropylbenzene, *n*-propylbenzene, isobutylbenzene, *n*-butylbenzene and isononane. A strain of *Pseudomonas*, isolated from activated sludge and grown with biphenyl as the sole carbon source, degraded 2,4′-dichlorobiphenyl yielding the following compounds: monochlorobenzoic acids, 2 monohydroxydichlorobiphenyls and the yellow compound hydroxyoxo(chlorophenyl)chlorohexadienoic acid. Irradiation of the mixture containing these compounds led to the formation of 2 monochloroacetophenones and the disappearance of the yellow compound. Similar compounds were found when 2,4′-dichlorobiphenyl was replaced with PCB-1242 mixture. PCB-1242 in a 90% acetonitrile/water soln containing 0.2-0.3 M sodium borohydride and irradiated with UV light (λ = 254 nm) reacted to yield dechlorinated biphenyls. Without sodium borohydride, the reaction proceeded much more slowly. When PCB-1242 contaminated sand was treated with a poly(ethylene glycol)/potassium hydroxide mixture at room temperature, 27% reacted after 2 wk forming aryl poly(ethylene glycols). Reported degradation products by the microorganism *Alcaligenes* BM-2 for a mixture of polychlorinated biphenyls include monohydroxychlorobiphenyl, 2-hydroxy-6-oxochlorophenylhexa-2,4-dieonic acid, chlorobenzoic acid, chlorobenzoylpropionic acid, chlorophenylacetic acid and 3-chlorophenyl-2-chloropropenic acid.

EXPOSURE LIMITS: NIOSH REL: 10-h TWA 1.0 μg/m^3; OSHA PEL: TWA 1 mg/m^3.

SYMPTOMS OF EXPOSURE: Irrit eyes, chloracne, liv damage.

USES: Dielectric liquids; heat-transfer liquid widely used in transformers; swelling agents for transmission seals; ingredient in lubricants, oils and greases; plasticizers for cellulosics, vinyl and chlorinated rubbers; in polyvinyl acetate to improve fiber-tear properties.

PCB-1248

SYNONYMS: Arochlor 1248; **Aroclor 1248**; Chlorodiphenyl (48% Cl).

CHEMICAL DESIGNATIONS: CAS: 12672-29-6; DOT: 2315; mf: not definitive; fw: 288 (avg); RTECS: TQ 1358000.

PROPERTIES: Oily, light yellow liq with a weak odor. Mp: -7° (pour point), bp: distills @ 340-375°, *d*: 1.41 @ 25/4°, fl p: 193-196°, K_H: 3.5 x 10^{-3} atm·m^3/mol, log K_{oc}: 5.64 (est), log K_{ow}: 6.11, S_o: sol in most solvents, S_w: 50 μg/L @ 20°, vp: 4.94 x 10^{-4} mm @ 25°.

TRANSFORMATION PRODUCTS: Heating PCB-1248 in oxygen @ 270-300° for 1 wk resulted in the formation of mono-, di-, tri-, tetra- and pentachlorodibenzofurans (PCDFs). Above 330°, PCDFs decompose. Reported degradation products by the microorganism *Alcaligenes* BM-2 for a mixture of polychlorinated biphenyls include monohydroxychlorobiphenyl, 2-hydroxy-6-oxochlorophenylhexa-2,4-dieonic acid, chlorobenzoic acid, chlorobenzoylpropionic acid, chlorophenylacetic acid and 3-chlorophenyl-2-chloropropenic acid.

USES: In epoxy resins to improve adhesion and resistance to chemical attack; insulator fluid for electric condensers; additive in very high pressure lubricants.

PCB-1254

SYNONYMS: Arochlor 1254; **Aroclor 1254**; Chlorodiphenyl (54% Cl); NCI-C02664.

CHEMICAL DESIGNATIONS: CAS: 11097-69-1; DOT: 2315; mf: not definitive; fw: 327 (avg); RTECS: TQ 1360000.

PROPERTIES: Light yellow, viscous liq with a weak odor. Mp: 10°, bp: distills @ 365-390°, *d*: 1.505 @ 15.5/4°, fl p: 222°, K_H: 1.9 x 10^{-4} atm·m^3/mol @ 20°, log K_{oc}: 4.40-5.61, log K_{ow}: 5.61, 6.47, S_o: sol in most solvents, S_w: 50 μg/L @ 20°, vap d: ≈ 13.36 g/L @ 25°, 11.29 (air = 1), vp: 6 x 10^{-5} mm @ 20°.

TRANSFORMATION PRODUCTS: A strain of *Alcaligenes eutrophus* degraded 35% of the congeners by dechlorination under anaerobic conditions. Indigenous microbes in the Center Hill Reservoir, TN biooxidized 2-chlorobiphenyl (a congener present in trace quantities) into chlorobenzoic acid and [illegible]lorobenzoylformic acid. Biooxidation of the PCB mixture containing 54 wt % [illegible]rine was not observed. Using methanol, ethanol or 2-propanol in the presence [illegible]ickel chloride and sodium borohydride, dechlorination resulted in the [illegible]ion of biphenyls with smaller quantities of mono- and dichlorobiphenyls. [illegible]4 in a 90% acetonitrile/water soln containing 0.2-0.3 M sodium borohydride and irradiated with UV light (λ = 254 nm) reacted to yield dechlorinated biphenyls. After 16 h, no chlorinated biphenyls were detected. Without sodium borohydride, only 25% of PCB-1254 was destroyed after 16 h. When PCB-1254 contaminated sand was treated with a poly(ethylene glycol)/potassium hydroxide mixture at room temperature, 81% reacted after 6 days forming aryl poly(ethylene glycols). When PCB-1254 in a methanol/water soln containing sodium methyl siliconate was irradiated with UV light (λ = 300 nm) for 5 h, most of the congeners dechlorinated to biphenyl. Reported degradation products by the microorganism *Alcaligenes* BM-2 for a mixture of polychlorinated biphenyls include monohydroxychlorobiphenyl, 2-hydroxy-6-oxochlorophenylhexa-2,4-dieonic acid, chlorobenzoic acid, chlorobenzoylpropionic acid, chlorophenylacetic acid and 3-chlorophenyl-2-chloropropenic acid.

EXPOSURE LIMITS: NIOSH REL: 10-h TWA 1 $\mu g/m^3$; OSHA PEL: TWA 0.5 mg/m^3.

SYMPTOMS OF EXPOSURE: Irrit eyes, skin; acne-form derm; jaun; dark urine.

USES: Secondary plasticizer for polyvinyl chloride; co-polymers of styrene-butadiene and chlorinated rubber to improve chemical resistance to attack.

PCB-1260

SYNONYMS: Arochlor 1260; **Aroclor 1260**; Chlorodiphenyl (60% Cl); Clophen A60; Kanechlor; Phenoclor DP6.

CHEMICAL DESIGNATIONS: CAS: 11096-82-5; DOT: 2315; mf: not definitive; fw: 370 (avg); RTECS: TQ 1362000.

PROPERTIES: Light yellow sticky, soft resin with a weak odor. Mp: 31°, bp: distills @ 385-420°, *d*: 1.566 @ 15.5/4°, K_H: 1.7 x 10^{-4} atm·m^3/mol @ 20°, log K_{oc}: 6.42, log K_{ow}: 6.91, S_o: sol in most solvents, S_w: 14.4 μg/L @ 20°, vp: 6.31 x 10^{-6} mm @ 20°.

TRANSFORMATION PRODUCTS: PCB-1260 in a 90% acetonitrile/water soln containing 0.2-0.3 M sodium borohydride and irradiated with UV light (λ = 254 nm) reacted to yield dechlorinated biphenyls. After 2 h, about 75% of the cogeners were destroyed. Without sodium borohydride, only 10% of the cogeners had reacted. Products identified by GC include biphenyl, 2-, 3- and 4-chlorobiphenyl, 6 dichlorobiphenyls, 3 trichlorobiphenyls, 1-phenyl-1,4-cyclohexadiene and 1-phenyl-3-cyclohexene. When PCB-1260 contaminated sand was treated with a poly(ethylene glycol)/potassium hydroxide mixture at room temperature, more than 99% reacted after 2 days forming aryl poly(ethylene glycols). Reported degradation products by the microorganism *Alcaligenes* BM-2 for a mixture of polychlorinated biphenyls include monohydroxychlorobiphenyl, 2-hydroxy-6-oxochlorophenylhexa-2,4-dieonic acid, chlorobenzoic acid, chlorobenzoylpropionic acid, chlorophenylacetic acid and 3-chlorophenyl-2-chloropropenic acid.

USES: Secondary plasticizer for polyvinyl chloride; in polyester resins increase strength of fiberglass; varnish formulations to improve water and - resistance; as an insulator fluid for electric condensers and as an additive i high pressure lubricants.

PENTACHLOROBENZENE

SYNONYMS: QCB; RCRA waste number U183.

CHEMICAL DESIGNATIONS: CAS: 608-93-5; DOT: none assigned; mf: C_6HCl_5; fw: 250.34; RTECS: DA 6640000.

PROPERTIES: Needles from alcohol. Mp: 86°, bp: 277°, *d*: 1.8342 @ 16.5/4°, K_H: 0.0071 atm·m^3/mol @ 20°, log K_{oc}: 6.3, log K_{ow}: 4.88-5.75, S_o: very sol in ether, S_w: 5.32 μM/L @ 25°, vp: 6.0 x 10^{-3} mm @ 20-30°.

TRANSFORMATION PRODUCTS: UV irradiation (λ = 2537 Å) of pentachlorobenzene in hexane soln for 3 h produced a 50% yield of 1,2,4,5-tetrachlorobenzene and a 13% yield of 1,2,3,5-tetrachlorobenzene. Irradiation (λ ≥ 285 nm) of pentachlorobenzene (1.1-1.2 mM/L) in an acetonitrile-water mixture containing acetone (concn = 0.553 mM/L) as a sensitizer gave the following products (% yield): 1,2,3,4-tetrachlorobenzene (6.6), 1,2,3,5-tetrachlorobenzene (52.8), 1,2,4,5-tetrachlorobenzene (15.1), 1,2,4-trichlorobenzene (1.9), 1,3,5-trichlorobenzene (5.3), 1,3-dichlorobenzene (0.9), 2,2′,3,3′,4,4′,5,6,6′-nonachlorobiphenyl (2.08), 2,2′,3,3′,4,4′,5,5′,6-nonachlorobiphenyl (0.34), 2,2′,3,3′,4,5,5′,6,6′-nonachlorobiphenyl (trace), 1 octachlorobiphenyl (0.53) and 1 heptachlorobiphenyl (0.49). Without acetone, the identified photolysis products (% yield) included 1,2,3,4-tetrachlorobenzene (3.7), 1,2,3,5-tetrachlorobenzene (13.5), 1,2,4,5-tetrachlorobenzene (2.8), 1,2,4-trichlorobenzene (12.7), 1,3,5-trichlorobenzene (1.0) and 1,4-dichlorobenzene (6.7). A CO_2 yield of 2.0% was achieved when pentachlorobenzene adsorbed on silica gel was irradiated with light (λ > 290 nm) for 17 h. In activated sludge, < 0.1% mineralized to CO_2 after 5 days.

USES: Chemical research; organic synthesis.

PENTACHLOROETHANE

SYNONYMS: Ethane pentachloride; NCI-C53894; Pentalin; RCRA waste number U184; UN 1669.

CHEMICAL DESIGNATIONS: CAS: 76-01-7; DOT: 1669; mf: C_2HCl_5; fw: 202.28; RTECS: KI 6300000.

PROPERTIES: Colorless liq with a chloroform-like odor. Mp: -29°, bp: 162°, *d*: 1.6796 @ 20/4°, K_H: 2.45 x 10^{-3} atm·m^3/mol @ 25°, log K_{oc}: 3.28, log K_{ow}: 2.89, S_o: misc with alcohol and ether, S_w: 500 mg/L @ 20°, vap d: 8.27 g/L @ 25°, 6.98 (air = 1), vp: 3.4 mm @ 20°.

TRANSFORMATION PRODUCT: At various pH's, pentachloroethane hydrolyzed to tetrachloroethylene ($t_{1/2}$ = 3.6 days @ 25° and pH 7).

USES: Solvent for oil and grease in metal cleaning.

PENTACHLOROPHENOL

SYNONYMS: Acutox; Chem-penta; Chem-tol; Chlorophen; Cryptogil OL; Dowcide 7; Dowicide 7; Dowicide EC-7; Dowicide G; Dow pentachlorophenol

DP-2 antimicrobial; Durotox; EP 30; Fungifen; Fungol; Glazd penta; Grundier arbezol; Lauxtol; Lauxtol A; Liroprem; Monsanto penta; Moosuran; NCI-C54933; NCI-C55378; NCI-C56655; PCP; Penchlorol; Penta; Pentachlorofenol; Pentachlorofenolo; Pentachlorophenate; Pentachlorphenol; 2,3,4,5,6-Pentachlorophenol; Pentacon; Penta-kil; Pentasol; Penwar; Peratox; Permacide; Permaguard; Permasan; Permatox DP-2; Permatox Penta; Permite; Priltox; RCRA waste number U242; Santobrite; Santophen; Santophen 20; Sinituho; Term-i-trol; Thompson's wood fix; Weedone; Witophen P.

CHEMICAL DESIGNATIONS: CAS: 87-86-5; DOT: 2020; mf: C_6HCl_5O; fw: 266.34; RTECS: SM 6300000.

PROPERTIES: White to dark-colored flakes or beads with a characteristic odor. Mp: 174-191°, bp: 310°, *d*: 1.978 @ 22/4°, pK_a: 4.74, K_H: 2.1-3.4 x 10^{-7} atm·m^3/mol, log K_{oc}: 2.47-4.40, log K_{ow}: 3.32-5.86, S_o: very sol in acetone, carbitol, cellosolve, ethanol and ether; sol in hot benzene; slightly sol in solvents and ligroin, S_w: 14 mg/L @ 20° @ pH 5, vp: 1.7 x 10^{-4} mm @ 20°.

TRANSFORMATION PRODUCTS: Under aerobic conditions, microbes in estuarine water partially dechlorinated pentachlorophenol to trichlorophenol. In distilled water, pentachlorophenol photolyzed to tetrachlorophenols, trichlorophenols, chlorinated dihydroxybenzenes and dichloromaleic acid. Under anaerobic conditions, pentachlorophenol may undergo sequential dehalogenation to produce tetra-, tri-, di- and *m*-chlorophenol. In aerobic and anaerobic soils, pentachloroanisole was the major metabolite along with 2,3,6-trichlorophenol, 2,3,4,5- and 2,3,5,6-tetrachlorophenol. The disapearance of pentachlorophenol was studied in 4 aquaria with and without mud under aerobic and anaerobic conditions. Potential biological and/or chemical products identified include pentachloroanisole, 2,3,4,5-, 2,3,4,6- and 2,3,5,6-tetrachlorophenol. Wood treated with pure pentachlorophenol did not photolyze under natural sunlight or laboratory induced UV radiation. However, in the presence of an antimicrobial (Dowcide EC-7), pure pentachlorophenol degraded to chlorinated dibenzo-*p*-dioxin. Wood containing composited technical grade pentachlorophenol yielded similar results. Pentachlorophenol degraded in anaerobic sludge to 3,4,5-trichlorophenol which was further reduced to 3,5-dichlorophenol. Metabolites identified in soil beneath a sawmill environment using pentachlorophenol as a wood preservative include pentachloroanisole, 2,3,4,6-tetrachloroanisole, tetrachlorocatechol, tetrachlorohydroquinone, 3,4,5-trichlorocatechol, 2,3,6-trichlorohydroquinone, 3,4,6-trichlorocatechol and 2,3,4,6-tetrachlorophenol. An aq soln containing pentachlorophenol and exposed to sunlight or laboratory UV light yielded tetrachlorocatechol, tetrachlororesorcinol and tetrachlorohydroquinone. These compounds were air-oxidized to chloranil, hydroxyquinones and 2,3-dichloromaleic acid (DCM). Other compounds identified include a cyclic dichlorodiketone, tetra- and trichlorophenols. UV irradiation (λ = 2537 Å) of pentachlorophenol in hexane soln for 32 hours produced a 30% yield of 2,3,5,6-tetrachlorophenol and about a 10% yield of a compound tentatively identified as an isomeric tetrachlorophenol. Wet oxidation of pentachlorophenol @ 320° yielded formic and acetic acids. A CO_2 yield of 50.0% was achieved when pentachlorophenol adsorbed on silica gel was irradiated with light (λ > 290 nm) for 17 h. In activated sludge, only 0.2% mineralized to CO_2 after 5 days. When an aq soln containing pentachlorophenol (4.5 x 10^{-5} M) and a suspension of titanium dioxide (2 g/L) was irradiated with UV light, CO_2 and hydrochloric acid formed in quantatative amounts ($t_{1/2}$ = 9 min @ 45-50°).

EXPOSURE LIMITS: NIOSH REL: IDLH 150 mg/m^3; OSHA PEL: TWA 0.5 mg/m^3.

SYMPTOMS OF EXPOSURE: Irrit eyes, nose, throat; sneez, cough; weak, anor, low-wgt; sweat; head; dizz; nau, vomit; dysp; chest pain; derm.

USES: Manufacture of insecticides, algicides, herbicides, fungicides and bactericides; wood preservative.

1,4-PENTADIENE

SYNONYMS: None.

CHEMICAL DESIGNATIONS: CAS: 591-93-5; DOT: none assigned; mf: C_5H_8; fw: 68.12; RTECS: none assigned.

PROPERTIES: Liq or gas. Mp: -148.3°, bp: 26°, *d*: 0.6608 @ 20/4°, fl p: < 0°, K_H: 1.20 x 10^{-1} atm·m^3/mol @ 25°, log K_{ow}: 1.48, S_o: sol in acetone, alcohol, benzene and ether, S_w: 558 ppm @ 25°, vap d: 2.78 g/L @ 25°, 2.35 (air = 1), vp: 734.6 mm @ 25°.

USES: Chemical research; organic synthesis.

n-PENTANE

SYNONYMS: Amyl hydride; Dimethylmethane; **Pentane**; UN 1265.

CHEMICAL DESIGNATIONS: CAS: 109-66-0; DOT: 1265; mf: C_5H_{12}; fw: 72.15; RTECS: RZ 9450000.

PROPERTIES: Colorless liq. Mp: -130°, bp: 36.1°, *d*: 0.6262 @ 20/4°, fl p: -49.4°, lel: 1.5%, uel: 7.8%, K_H: 1.255 atm·m^3/mol @ 25°, IP: 10.35 eV, log K_{ow}: 3.23-3.62, S_o: sol in acetone, benzene and chloroform; misc with alcohol, ether and many organic solvents, S_w: 39.5 ppm @ 25°, vap d: 2.95 g/L @ 25°, 2.49 (air = 1), vp: 426 mm @ 20°.

TRANSFORMATION PRODUCTS: Synthetic air containing gaseous nitrous acid and exposed to artificial sunlight (λ = 300-450 nm) photooxidized *n*-pentane into methyl nitrate, pentyl nitrate, peroxyacetal nitrate and peroxypropionyl nitrate. *n*-Pentane may biodegrade in 2 ways. The first is the formation of pentyl hydroperoxide which decomposes to *n*-pentanol followed by oxidation to pentanoic acid. The other pathway involves dehydrogenation to 1-pentene which may react with water giving *n*-pentanol.

EXPOSURE LIMITS: NIOSH REL: 10-h TWA 120 ppm, IDLH 15,000 ppm; OSHA PEL: TWA 1,000 ppm; ACGIH TLV: 600 ppm.

SYMPTOMS OF EXPOSURE: Drow; irrit eyes, nose; derm; chem pneu.

USES: Solvent recovery and extraction; blowing agent for plastic foams; low temperature thermometers; natural gas processing plants; production of olefin, hydrogen, ammonia; fuel production; pesticide; manufacture of artificial ice; organic synthesis.

2-PENTANONE

SYNONYMS: Ethyl acetone; Methyl propyl ketone; Methyl *n*-propyl ketone; MPK; Propyl methyl ketone; UN 1249.

CHEMICAL DESIGNATIONS: CAS: 107-87-9; DOT: 1249; mf: $C_5H_{10}O$; fw: 86.13; RTECS: SA 7875000.

PROPERTIES: Colorless liq with a characteristic odor. Mp: -77.8°, bp: 102°, *d*: 0.8089 @ 20/4°, fl p: 7.2°, lel: 1.5%, uel: 8.2%, K_H: 6.44 x 10^{-5} atm·m^3/mol @ 25°, IP: 9.37 eV, log K_{ow}: 0.91, S_o: misc with alcohol and ether, S_w: 43,065 mg/L @ 20°, vap d: 3.52 g/L @ 25°, 2.97 (air = 1), vp: 12 mm @ 20°.

EXPOSURE LIMITS: NIOSH REL: 10-h TWA 150 ppm, IDLH 5,000 ppm; OSHA PEL: TWA 200 ppm.

SYMPTOMS OF EXPOSURE: Irrit eyes, muc memb; head; derm; narcosis, coma.

USES: Solvent; substitute for 3-pentanone; flavoring.

1-PENTENE

SYNONYMS: α-*n*-Amylene; Propylethylene.

CHEMICAL DESIGNATIONS: CAS: 109-67-1; DOT: none assigned; mf: C_5H_{10}; fw: 70.13; RTECS: none assigned.

PROPERTIES: Colorless liq. Mp: -165°, -138°, bp: 30°, *d*: 0.6405 @ 20/4°, fl p: -17.7° (OC), lel: 1.5%, uel: 8.7%, K_H: 4.06 x 10^{-1} atm·m^3/mol @ 25°, IP: 9.50 eV, log K_{ow}: 2.26 (calcd), S_o: misc with acetone, alcohol, benzene, ether and many hydrocarbons, S_w: 148 ppm @ 25°, vap d: 2.87 g/L @ 25°, 2.42 (air = 1), vp: 637.7 mm @ 25°.

TRANSFORMATION PRODUCTS: Biooxidation of 1-pentene may occur yielding 4-penten-1-ol which may further oxidize to give 4-pentenoic acid. Washed cell suspensions of bacteria belonging to the genera *Mycobacterium*, *Nocardia*, *Xanthobacter* and *Pseudomonas* and growing on selected alkenes metabolized 1-pentene to 1,2-epoxypentane. *Mycobacterium* sp., growing on ethene, hydrolyzed 1,2-epoxypropane to 1,2-propanediol.

USES: Blending agent for high octane motor fuel; organic synthesis.

cis-2-PENTENE

SYNONYMS: *cis*-Pentene-2; **(Z)-2-Pentene.**

CHEMICAL DESIGNATIONS: CAS: 627-20-3; DOT: none assigned; mf: C_5H_{10}; fw: 70.13; RTECS: none assigned.

PROPERTIES: Colorless liq. Mp: -180 to -178°, -151.4°, bp: 36.9°, *d*: 0.6556 @ 20/4°, fl p: < -20°, K_H: 2.25 x 10^{-1} atm·m^3/mol @ 25° (calcd), IP: 9.13 eV, log K_{ow}: 2.15 (calcd), S_o: sol in alcohol, benzene and ether, S_w: 203 ppm @ 25°, vap d: 2.87 g/L @ 25°, 2.42 (air = 1), vp: 494.6 mm @ 25°.

USES: Polymerization inhibitor; organic synthesis.

trans-2-PENTENE

SYNONYMS: **(*E*)-2-Pentene**; *trans*-Pentene-2.

CHEMICAL DESIGNATIONS: CAS: 646-04-8; DOT: none assigned; mf: C_5H_{10}; fw: 70.13; RTECS: none assigned.

PROPERTIES: Colorless liq. Mp: -136°, bp: 36.3°, *d*: 0.6482 @ 20/4°, fl p: < -20°, K_H: 2.34 x 10^{-1} atm·m^3/mol @ 25°, IP: 9.13 eV, log K_{ow}: 2.15 (calcd), S_o: sol in alcohol, benzene and ether, S_w: 203 mg/L @ 20°, vap d: 2.87 g/L @ 25°, 2.42 (air = 1), vp: 505.5 mm @ 25°.

USES: Polymerization inhibitor; organic synthesis.

PENTYLCYCLOPENTANE

SYNONYMS: None.

CHEMICAL DESIGNATIONS: CAS: 3741-00-2; DOT: none assigned; mf: $C_{10}H_{20}$; fw: 140.28; RTECS: none assigned.

PROPERTIES: Liq. Bp: 180.6°, log K_{ow}: 4.90 (calcd), S_w: 0.115 ppm @ 25°, vap d: 5.73 g/L @ 25°, 4.84 (air = 1).

USES: Organic synthesis; gasoline component.

PHENANTHRENE

SYNONYMS: Phenanthren; Phenantrin.

CHEMICAL DESIGNATIONS: CAS: 85-01-8; DOT: none assigned; mf: $C_{14}H_{10}$; fw: 178.24; RTECS: SF 7175000.

PROPERTIES: Colorless, monoclinic cryst. Mp: 100.5°, bp: 340°, pK_a: > 15, *d*: 1.179 @ 25/4°, fl p: 171° (OC), K_H: 2.35 x 10^{-5} atm·m^3/mol @ 25°, IP: 8.22 eV, log K_{oc}: 3.72-4.59, log K_{ow}: 4.16-4.57, S_o: sol in acetone, ethanol, acetic acid, benzene, carbon disulfide, absolute alcohol (1 g/25 mL), toluene or carbon tetrachloride (1 g/2.4 mL) and anhydrous ether (1 g/3.3 mL), S_w: 0.994-1.29 mg/L @ 25°, vp: 2.1 x 10^{-4} mm @ 20°.

TRANSFORMATION PRODUCTS: Catechol is the central metabolite in the bacterial degradation of phenanthrene. Intermediate byproducts include 1-hydroxy-2-napthoic acid, 1,2-dihydroxynaphthalene and salicylic acid. It was reported that *Beijerinckia*, under aerobic conditions, degraded phenanthrene to *cis*-3,4-dihydroxy-3,4-dihydrophenanthracene. A CO_2 yield of 24.2% was achieved when phenanthrene adsorbed on silica gel was irradiated with light (λ > 290 nm) for 17 h. In activated sludge, 39.6% mineralized to CO_2. In a 2-wk experiment, ^{14}C-labelled phenanthrene applied to soil-water suspensions under aerobic and anaerobic conditions gave $^{14}CO_2$ yields of 7.2 and 6.3%, respectively. The aq chlorination of phenanthrene @ pH < 4 produced phenanthrene-9,10-dione and 9-chlorophenanthrene. at high pH (> 8.8), phenanthrene-9,10-oxide, phenanthrene-9,10-dione and 9,10-dihydrophenanthrenediol were identified as major products. It was suggested that the chlorination of phenanthrene in tap water accounted for the presence of chloro- and dichlorophenanthrenes.

EXPOSURE LIMITS: No individual standards have been set, however, as a constituent in coal tar pitch volatiles, the following exposure limits have been established: NIOSH REL: 10-h TWA 0.1 mg/m^3 (cyclohexane-extractable fraction); OSHA PEL: TWA 0.2 mg/m^3 (benzene-soluble fraction); ACGIH TLV: TWA 0.2 mg/m^3 (benzene solubles).

USES: Explosives; dyestuffs; biochemical research; synthesis of drugs; preparation of 9,10-phenanthrenequinone, 9,10-dihydrophenanthrene, 9-bromophenanthrene, 9,10-dibromo-9,10-dihydrophenanthrene and many other organic compounds.

PHENOL

SYNONYMS: Baker's P and S liquid and ointment; Benzenol; Carbolic acid; Hydroxybenzene; Monohydroxybenzene; NA 2,821; NCI-C50124; Oxybenzene; Phenic acid; Phenyl hydrate; Phenyl hydroxide; Phenylic acid; Phenylic alcohol; RCRA waste number U188; UN 1671; UN 2312; UN 2821.

CHEMICAL DESIGNATIONS: CAS: 108-95-2; DOT: 1671; mf: C_6H_6O; fw: 94.11; RTECS: SJ 3325000.

PROPERTIES: White cryst or light pink liq which slowly turns brown on exposure to air. Sweet tarry odor. Mp: 43°, bp: 181.7°, *d*: 1.0576 @ 20/4°, fl p: 79°, lel: 1.7%, uel: 8.6%, pK_a: 9.99 @ 25°, K_H: 3.97 x 10^{-7} $atm{\cdot}m^3/mol$ @ 25°, IP: 8.47 eV, 8.51 eV, log K_{oc}: 1.43-3.46, log K_{ow}: 1.39-1.48, S_o: sol in carbon disulfide and chloroform; very sol in ether; misc with carbon tetrachloride, hot benzene and alcohol, S_w: 93 g/L @ 25°, vp: 0.2 mm @ 20°.

TRANSFORMATION PRODUCTS: In an aq, oxygenated soln exposed to artificial light (λ = 234 nm), phenol photolyzed to hydroquinone, catechol, 2,2′-, 2,4′- and 4,4′-dihydroxybiphenyl. Titanium dioxide suspended in an aq soln and irradiated with UV light (λ = 365 nm) converted phenol to CO_2 at a significant rate. When an aq soln containing potassium nitrate (10^{-2} M) and phenol (10^{-3} M) was irradiated with UV light (λ = 290-350 nm) up to a conversion of 10%, the following products formed: hydroxyhydroquinone, hydroquinone, resorcinol, hydroxybenzoquinone, benzoquinone, catechol, nitrosophenol, 4-nitrocatechol, nitrohydroquinone, 2- and 4-nitrophenol. In an environmental chamber, nitrogen trioxide (10,000 ppb) reacted quickly with phenol (concn 200-1400 ppb) to form phenoxy radicals and nitric acid. The phenoxy radicals may react with oxygen and NO_2 to form quinones and nitrohydroxy derivatives, respectively. Irradiation of phenol with UV light (λ = 254 nm) in the presence of oxygen yielded substituted biphenyls, hydroquinone and catechol. Anticipated products from the reaction of phenol with O_3 or hydroxyl radicals in the atmosphere are dihydroxybenzenes, nitrophenols and ring cleavage products. Groundwater contaminated with phenol and other phenols degraded in a methanogenic aquifer. Similar results were obtained in the laboratory utilizing an anaerobic digester. Methane and CO_2 were reported as degradation products. Under methanogenic conditions, inocula from a municipal sewage treatment plant digester degraded phenol to CO_2 and methane.

In a methanogenic enrichment culture, phenol anaerobically biodegraded to CO_2 and methane (yield 70%). Chloroperoxidase, a fungal enzyme isolated from *Caldariomyces fumago*, reacted with phenol forming 2- and 4-chlorophenol, the latter in a 25% yield. Ozonization of phenol in water resulted in the formation of many oxidation products. The identified products in the order of degradation are catechol, hydroquinone, *o*-quinone, *cis,cis*-muconic acid, maleic (or fumaric) and oxalic acids. In addition, glyoxylic, formic and acetic acids also were reported as ozonization products prior to further oxidation to CO_2. Ozonation of an aq soln of phenol subjected to UV light (120 watt low pressure mercury lamp) gave glyoxal, glyoxylic, oxalic and formic acids as major products. Minor products included catechol, hydroquinone, muconic, fumaric and maleic acids. Wet oxidation of phenol @ 320° yielded formic and acetic acids. Chlorination of water containing bromide ions converted phenol to 2,4,6-tribromophenol. Bromodichlorophenol, dibromochlorophenol and tribromophenol have also been reported to form from the chlorination of natural water under simulated conditions. A CO_2 yield of 32.5% was achieved when phenol adsorbed on silica gel was irradiated with light (λ > 290 nm) for 17 h. In activated sludge, 41.4% mineralized to CO_2 after 5 days. Wet oxidation of phenol at elevated pressure and temperature gave the following products: acetone, acetaldehyde, formic, acetic, maleic, oxalic and succinic acids. Chlorine dioxide reacted with phenol in an aq soln forming *p*-benzoquinone and hypochlorous acid. Reacts with sodium and potassium hydroxide to form sodium and potassium phenolate, respectively.

EXPOSURE LIMITS: NIOSH REL: 10-h TWA 20 mg/m^3, 15-min C 60 mg/m^3, IDLH 250 ppm; OSHA PEL: TWA 5 ppm; ACGIH TLV: TWA 5 ppm.

SYMPTOMS OF EXPOSURE: Irrit eyes, nose, throat, anor, low-wgt; weak, musc ache, pain; dark urine; cyan; liv and kid damage; skin burn; derm; ochronosis; tremor, convuls; twitch.

USES: Antiseptic and disinfectant; pharmaceuticals; dyes; indicators; slimicide; phenolic resins; epoxy resins (bisphenol-A); nylon-6 (caprolactum); 2,4-D; solvent for refining lubricating oils; preparation of adipic acid, salicylic acid, phenolphthalein, pentachlorophenol, acetophenetidin, picric acid, anisole, phenoxyacetic acid, phenyl benzoate, *o*-phenolsulfonic acid, *p*-phenolsulfonic acid, *o*-nitrophenol, *p*-nitrophenol, 2,4,6-tribromophenol, *p*-bromophenol, *p-tert*-butylphenol, salicyladehyde and many other organic compounds; germicidal paints; laboratory reagent.

p-PHENYLENEDIAMINE

SYNONYMS: 4-Aminoaniline; *p*-Aminoaniline; BASF ursol D; **1,4-Benzenediamine**; *p*-Benzenediamine; Benzofur D; C.I. 76,060; C.I. developer 13; C.I. oxidation base 10; Developer 13; Developer PF; 1,4-Diaminobenzene; *p*-Diaminobenzene; Durafur black R; Fouramine D; Fourrine 1; Fourrine D; Fur black 41,867; Fur brown 41,866; Furro D; Fur yellow; Futramine D; Nako H; Orsin; Oxidation base 10; Para; Pelagol D; Pelagol DR; Pelagol grey D; Peltol D; 1,4-Phenylenediamine; PPD; Renal PF; Santoflex IC; Tertral D; UN 1673; Ursol D;

USAF EK-394; Vulkanox 4020; Zoba black D.

CHEMICAL DESIGNATIONS: CAS: 106-50-3; DOT: 1673; mf: $C_6H_8N_2$; fw: 108.14; RTECS: SS 8050000.

PROPERTIES: White, red or brown cryst. Mp: 140°, bp: 267°, fl p: 155.6°, lel: 0.025 g/L (dust), pK_a: @ 25°: pK_1 = 3.29, pK_2 = 6.08, IP: 7.58 eV, S_o: sol in alcohol, chloroform and ether, S_w: 38 g/L @ 24°.

TRANSFORMATION PRODUCTS: A CO_2 yield of 53.7% was achieved when phenylenediamine (presumably an isomeric mixture) adsorbed on silica gel was irradiated with light (λ > 290 nm) for 17 h. In activated sludge, 3.8% mineralized to CO_2 after 5 days.

EXPOSURE LIMITS: NIOSH REL: IDLH 25 mg/m^3; OSHA PEL: TWA 0.1 mg/m^3.

SYMPTOMS OF EXPOSURE: Irrit pharyphotonx, larynx; bron asth; sens derm.

USES: Manufacturing azo dyes, intermediates for antioxidants and accelerators for rubber; photochemical measurements; laboratory reagent; dyeing hair and fur.

PHENYL ETHER

SYNONYMS: Biphenyl ether; Biphenyl oxide; Diphenyl ether; Diphenyl oxide; Geranium crystals; **1,1′-Oxybisbenzene**; Phenoxybenzene.

CHEMICAL DESIGNATIONS: CAS: 101-84-8; DOT: none assigned; mf: $C_{12}H_{10}O$; fw: 170.21; RTECS: KN 8970000.

PROPERTIES: Colorless solid or liq with a geranium-like odor. Mp: 27.2°, bp: 257.9°, *d*: 1.0748 @ 20/4°, fl p: 115°, lel: 0.8%, uel: 1.5%, K_H: 2.13 x 10^{-4} atm·m^3/mol @ 20° (calcd), IP: 8.82 eV, 12.90 eV, log K_{ow}: 3.79-4.36, S_o: sol in acetic acid, alcohol, benzene and ether, S_w: 21 ppm @ 25°, vp: 0.02 mm @ 20°.

EXPOSURE LIMITS: OSHA PEL: TWA 1 ppm.

SYMPTOMS OF EXPOSURE: Nau; irrit eyes, nose, skin.

USES: Heat transfer liquid; perfuming soaps; resins for laminated electrical insulation; organic synthesis.

PHENYLHYDRAZINE

SYNONYMS: Hydrazine-benzene; Hydrazinobenzene; UN 2562.

CHEMICAL DESIGNATIONS: CAS: 100-63-0; DOT: 2572; mf: $C_6H_8N_2$; fw: 108.14; RTECS: MV 8925000.

PROPERTIES: Yellow monoclinic cryst or oil. Turns reddish-brown on exposure to air. Mp: 19.8°, bp: 243° (decomp), *d*: 1.0986 @ 20/4°, fl p: 88.9°, pK_a: 5.20 @ 25°, IP: 7.66 eV, log K_{ow}: 1.25, S_o: sol in acetone; misc with alcohol, benzene, chloroform and ether, vap d: 4.42 g/L @ 25°, 3.73 (air = 1), vp: < 0.1 mm @ 20°.

EXPOSURE LIMITS: NIOSH REL: 2-h C 0.14 ppm; OSHA PEL: TWA 5 ppm; ACGIH TLV: TWA 5 ppm.

SYMPTOMS OF EXPOSURE: Skin sens, hemolytic anem, dysp, cyan, juan; kid damage.

USES: Reagent for aldehydes, ketones, sugars; manufacturing dyes, nitron and antipyrine.

PHTHALIC ANHYDRIDE

SYNONYMS: 1,2-Benzenedicarboxylic acid anhydride; 1,3-Dioxophthalan; ESEN; 1,3-Dihydro-1,3-dioxoisobenzofuran; **1,3-Isobenzofurandione**; NCI-C03601; Phthalandione; 1,3-Phthalandione; Phthalic acid anhydride; RCRA waste number U190; Retarder AK; Retarder ESEN; Retarder PD; UN 2214.

CHEMICAL DESIGNATIONS: CAS: 85-44-9; DOT: 2214; mf: $C_8H_4O_3$; fw: 148.12; RTECS: TI 3150000.

PROPERTIES: White cryst with a characteristic odor. Mp: 131.6°, bp: 283.9°, *d*: 1.527 @ 4°, fl p: 151.1°, lel: 1.7%, uel: 10.4%, K_H: 6.29 x 10^{-9} atm·m^3/mol @ 20° (calcd), log K_{oc}: 1.90 (calcd), log K_{ow}: -0.62, S_o: sol in alcohol and in 125 parts carbon disulfide; slightly sol in ether, S_w: 0.62 wt % @ 20°, vp: 2 x 10^{-4} mm @ 20°.

TRANSFORMATION PRODUCTS: Reacts with water to form phthalic acid. Pyrolysis of phthalic anhydride in the presence of polyvinyl chloride @ 600° yielded biphenyl, fluorene, benzophenone, 9-fluorenone, *o*-terphenyl, 9-phenylfluorene and 3 unidentified compounds.

EXPOSURE LIMITS: NIOSH REL: IDLH 1,670 ppm; OSHA PEL: TWA 2 ppm; ACGIH TLV: 1 ppm.

SYMPTOMS OF EXPOSURE: Conj, nas ulc bleed, upper resp irrit, bron, asth, derm.

USES: Manufacturing phthalates, phthaleins, benzoic acid, synthetic indigo, pharmaceuticals, insecticides, chlorinated products and artificial resins.

PICRIC ACID

SYNONYMS: Carbazotic acid; C.I. 10,305; 2-Hydroxy-1,3,5-trinitrobenzene; Lyddite; Melinite; Nitroxanthic acid; Pertite; Phenol trinitrate; Picronitric acid; Shimose; 1,3,5-Trinitrophenol; **2,4,6-Trinitrophenol**; UN 0154.

CHEMICAL DESIGNATIONS: CAS: 88-89-1; DOT: 1344; mf: $C_6H_3N_3O_7$; fw: 229.11; RTECS: TJ 7875000.

PROPERTIES: Colorless to yellow liq or cryst. Mp: 122-123°, bp: subl and explodes above 300°, *d*: 1.763, fl p: 150°, pK_a: 0.29 @ 25°, K_H: < 2.15 x 10^{-5} atm·m^3/mol @ 20° (calcd), log K_{ow}: 1.34, 2.03, S_o: sol in acetic acid, acetone, alcohol (83.3 g/L), benzene (100 g/L), chloroform (28.57 g/L) and ether (15.38 g/L), S_w: 14 g/L @ 20°, vp: < 1 mm @ 20°.

EXPOSURE LIMITS: NIOSH REL: IDLH 100 mg/m^3; OSHA PEL: TWA 0.1 mg/m^3.

SYMPTOMS OF EXPOSURE: Irrit eyes; sens derm; yellow-stain hair, teeth; weak, myalgia, anuria, polyuria, bitter taste, GI; hepatitis; hema, album, neph.

USES: Explosives, matches; electric batteries; in leather industry; manufacturing colored glass; etching copper; textile mordant; reagent.

PINDONE

SYNONYMS: Chemrat; **2-(2,2-Dimethyl-1-oxopropyl)-1*H*-indene-1,3(2*H*)-dione**; Pivacin; Pival; 2-Pivaloylindane-1,3-dione; 2-Pivaloyl-1,3-indanedione; Pivalyl; Pivalyl indandione; 2-Pivalyl-1,3-indandione; Pivalyl valone; Tri-Ban; UN 2472.

CHEMICAL DESIGNATIONS: CAS: 83-26-1; DOT: 2472; mf: $C_{14}H_{14}O_3$; fw: 230.25; RTECS: NK 6300000.

PROPERTIES: Bright yellow cryst. Mp: 108-110°, log K_{oc}: 2.95 (calcd), log K_{ow}: 3.18 (calcd), S_w: 18 ppm @ 25°.

EXPOSURE LIMITS: NIOSH REL: IDLH 200 mg/m^3; OSHA PEL: TWA 0.1 mg/m^3.

SYMPTOMS OF EXPOSURE: Nosebleed, excess bleeding minor cuts, bruises; smokey urine, black tarry stools; pain abdom, back.

USES: Insecticide; rodenticide; pharmaceutical intermediate.

PROPANE

SYNONYMS: Dimethylmethane; Propyl hydride; UN 1075; UN 1978.

CHEMICAL DESIGNATIONS: CAS: 74-98-6; DOT: 1978; mf: C_3H_8; fw: 44.10; RTECS: TX 2275000.

PROPERTIES: Colorless gas with a natural gas odor. Mp: -189.7°, bp: -42.1°, *d*: 0.5843 @ -45/4°, fl p: -105°, lel: 2.2%, uel: 9.5%, pK_a: ≈ 44, K_H: 7.06 x 10^{-1} $atm{\cdot}m^3/mol$ @ 25°, IP: 11.07 eV, log K_{ow}: 2.36, S_o: sol in alcohol (790 vol % @ 16.6° and 754 mm), benzene (1,452 vol % @ 21.5° and 757 mm), chloroform (1,299 vol % @ 21.6°, 757 mm), ether (926 vol % @ 16.6° and 757 mm) and turpentine (1,587 vol % @ 17.7° and 757 mm), S_w: 62.4 ppm @ 25°, vap d: 1.8324 g/L @ 25°; 1.52 (air = 1), vp: 8.6 atm @ 20°.

TRANSFORMATION PRODUCTS: Synthetic air containing gaseous nitrous acid and exposed to artificial sunlight (λ = 300-450 nm) photooxidized propane to acetone with a yield of 56%. In the presence of methane, *Pseudomonas methanica* degraded propane to *n*-propanol, propionic acid and acetone. The presence of CO_2 was required for "*Nocardia paraffinicum*" to degrade propane to propionic acid. Propane may biodegrade in 2 pathways. The first is the formation of propyl hydroperoxide which decomposes to propanol followed by oxidation to propanoic acid. The other pathway involves dehydrogenation to propene which may react with water giving propanol. Incomplete combustion of propane in the presence of excess hydrogen chloride resulted in a high number of different chlorinated compounds including but not limited to alkanes, alkenes, monoaromatics, alicyclic hydrocarbons and polynuclear aromatic hydrocarbons. Without hydrogen chloride, 13 non-chlorinated polynuclear aromatic hydrocarbons were formed. Complete combustion yields CO_2 and water.

EXPOSURE LIMITS: NIOSH REL: IDLH 20,000 ppm; OSHA PEL: TWA 1,000 ppm.

SYMPTOMS OF EXPOSURE: Dizz, disorientation, excitation, frost.

USES: Organic synthesis; refrigerant; fuel gas; manufacture of ethylene; solvent; extractant; aerosol propellant; mixture for bubble chambers.

β-PROPIOLACTONE

SYNONYMS: Betaprone; BPL; Hydracrylic acid β-lactone; 3-Hydroxypropionic acid lactone; β-Lactone; NSC 21,626; **2-Oxetanone**; Propanolide; Propiolactone; 1,3-Propiolactone; 3-Propiolactone; β-Propionolactone; 3-Propionolactone; β-Proprolactone.

CHEMICAL DESIGNATIONS: CAS: 57-57-8; DOT: none assigned; mf: C_3H_6O; fw: 72.06; RTECS: RQ 7350000.

PROPERTIES: Colorless liq with a sweet odor. Mp: -33.4°, bp: 162° (decomp), *d*: 1.1460 @ 20/5°, fl p: 70°, lel: 2.9%, K_H: 7.63 x 10^{-7} atm·m^3/mol @ 25° (calcd), S_o: misc with acetone, alcohol, chloroform and ether, S_w: 37 vol % @ 25°, vap d: 2.95 g/L @ 25°, 2.49 (air = 1), vp: 3.4 mm @ 25°.

TRANSFORMATION PRODUCTS: Slowly hydrolyzes to hydracrylic acid. In a reactor heated to 250° and a pressure of 12 mm, β-propiolactone decomposed to give equal amounts of ethylene and CO_2.

EXPOSURE LIMITS: ACGIH TLV: TWA 0.5 ppm.

SYMPTOMS OF EXPOSURE: Skin irrit, blistering, burns; corn opac; frequent urination, dys; hema.

USES: Organic synthesis; disinfectant; vapor sterilant.

n-PROPYL ACETATE

SYNONYMS: Acetic acid propyl ester; Acetic acid *n*-propyl ester; 1-Acetoxypropane; Propyl acetate; 1-Propyl acetate; UN 1276.

CHEMICAL DESIGNATIONS: CAS: 109-60-4; DOT: 1276; mf: $C_5H_{10}O_2$; fw: 102.12; RTECS: AJ 3675000.

PROPERTIES: Clear, colorless liq with a pleasant, pear-like odor. Mp: -95°, bp: 101.6°, *d*: 0.8878 @ 20/4°, fl p: 13°, lel: 2%, uel: 8%, K_H: 1.99 x 10^{-4} atm·m^3/mol @ 25°, IP: 10.04 eV, log K_{ow}: 1.24, S_o: misc with alcohol, ether, hydrocarbons and ketones, S_w: 18.9 g/L @ 20°, vap d: 4.17 g/L @ 25°, 3.53 (air = 1), vp: 25 mm @ 20°.

TRANSFORMATION PRODUCTS: Slowly hydrolyzes in water forming acetic acid and propyl alcohol.

EXPOSURE LIMITS: NIOSH REL: IDLH 8,000 ppm; OSHA PEL: TWA 200 ppm; ACGIH TLV: TWA 200 ppm, STEL 250 ppm.

SYMPTOMS OF EXPOSURE: Irrit eyes, nose, throat; narco; derm.

USES: Manufacture of flavors and perfumes; solvent for plastics, cellulose products and resins; lacquers, paints; natural and synthetic resins; lab reagent; organic synthesis.

n-PROPYL ALCOHOL

SYNONYMS: Ethyl carbinol; 1-Hydroxypropane; Optal; Osmosol extra; Propanol;

1-Propanol; *n*-Propanol; Propyl alcohol; 1-Propyl alcohol; Propylic alcohol; UN 1274.

CHEMICAL DESIGNATIONS: CAS: 71-23-8; DOT: 1274; mf: C_3H_8O; fw: 60.10; RTECS: UH 8225000.

PROPERTIES: Colorless liq with an alcohol-like odor. Mp: -126.5°, bp: 97.4°, *d*: 0.8035 @ 20/4°, fl p: 15°, lel: 2%, uel: 14%, K_H: 6.74 x 10^{-6} atm·m^3/mol @ 25°, IP: 10.1 eV, 10.22 eV, log K_{ow}: 0.25, 0.34, S_o: sol in acetone and benzene; misc with alcohol and ether, S_w: a saturated soln in equilibrium with its own vapor had a concn of 250.4 g/L @ 25°, vap d: 2.46 g/L @ 25°, 2.07 (air = 1), vp: 14.5 mm @ 20°.

EXPOSURE LIMITS: NIOSH REL: IDLH 4,000 ppm; OSHA PEL: TWA 200 ppm.

SYMPTOMS OF EXPOSURE: Mild irrit eyes, nose, throat; dry cracking skin; drow, head, atax; GI pain; abdom cramps; nau, vomit, diarr.

USES: Solvent for cellulose esters and resins; in manufacture of printing inks, nail polishes, polymerization and spinning of acrylonitrile, dyeing wool, polyvinyl chloride adhesives, esters, waxes, vegetable oils; brake fluids; solvent degreasing; antiseptic; organic synthesis.

n-PROPYLBENZENE

SYNONYMS: Isocumene; 1-Phenylpropane; **Propylbenzene**; UN 2364.

CHEMICAL DESIGNATIONS: CAS: 103-65-1; DOT: 2364; mf: C_9H_{12}; fw: 120.19; RTECS: DA 8750000.

PROPERTIES: Colorless liq. Mp: -99.5°, bp: 159.2°, *d*: 0.8620 @ 20/4°, fl p: 30°, lel: 0.8%, uel: 6%, K_H: 1.0 x 10^{-2} atm·m^3/mol @ 25°, IP: 8.73 eV, 9.14 eV, log K_{oc}: 2.87, log K_{ow}: 3.57-3.72, S_o: sol in acetone and benzene; misc with alcohol and ether, S_w: 55-120 mg/L @ 25°, vap d: 4.91 g/L @ 25°, 4.15 (air = 1), vp: 2.5 mm @ 20°.

TRANSFORMATION PRODUCTS: A *Nocardia* sp., growing on *n*-octadecane, biodegraded *n*-propyl-benzene to phenyl acetic acid.

USES: In textile dyeing and printing; solvent for cellulose acetate; manufacturing methylstyrene.

PROPYLCYCLOPENTANE

SYNONYM: 1-Cyclopentylpropane.

CHEMICAL DESIGNATIONS: CAS: 2040-96-2; DOT: none assigned; mf: C_8H_{16}; fw: 112.22; RTECS: none assigned.

PROPERTIES: Colorless liq with an ether-like odor. Mp: -117.3°, bp: 131°, *d*: 0.7763 @ 20/4°, K_H: 8.90 x 10^{-1} atm·m^3/mol @ 25° (calcd), log K_{ow}: 3.63 (calcd), S_o: sol in acetone, alcohol, benzene and ether, S_w: 2.04 ppm @ 25°, vap d: 4.59 g/L @ 25°, 3.87 (air = 1), vp: 12.3 mm @ 25°.

USES: Organic synthesis; gasoline component.

PROPYLENE OXIDE

SYNONYMS: Epoxypropane; 1,2-Epoxypropane; Methyl ethylene oxide; **Methyloxirane**; NCI-C50099; Propene oxide; 1,2-Propylene oxide; UN 1280.

CHEMICAL DESIGNATIONS: CAS: 75-56-9; DOT: 1280; mf: C_3H_6O; fw: 58.08; RTECS: TZ 2975000.

PROPERTIES: Colorless liq with an ethereal odor. Mp: -112.2°, bp: 34.3°, *d*: 0.859 @ 0/4°, fl p: -37.2°, lel: 2.1%, uel: 37%, K_H: 8.34 x 10^{-5} atm·m^3/mol @ 20° (calcd), IP: 10.22 eV, log K_{ow}: 0.08, S_o: misc with alcohol and ether, S_w: 405 g/L @ 20°, vap d: 2.37 g/L @ 25°, 2.01 (air = 1), vp: 442 mm @ 20°.

TRANSFORMATION PRODUCTS: Anticipated products from the reaction of propylene oxide with O_3 or hydroxyl radicals in the atmosphere are formaldehyde, pyruvic acid, $CH_3C(O)OCHO$ and HC(O)OCHO. Hydrolyzes in water forming 1,2-propanediol ($t_{1/2}$ = 14.6 days @ 25° and pH 7).

EXPOSURE LIMITS: NIOSH REL: IDLH 2,000 ppm; OSHA PEL: TWA 100 ppm; ACGIH TLV: TWA 20 ppm.

SYMPTOMS OF EXPOSURE: Irrit eyes, upper resp, lungs; skin irrit, blister, burns.

USES: Preparation of propylene and dipropylene glycols, lubricants, oil demulsifiers, surfactants, isopropanol amines, polyols for urethane foams; solvent; soil sterilant; fumigant.

n-PROPYL NITRATE

SYNONYMS: **Nitric acid propyl ester**; Propyl nitrate; UN 1865.

CHEMICAL DESIGNATIONS: CAS: 627-13-4; DOT: 1865; mf: $C_3H_7NO_3$; fw: 105.09; RTECS: UK 0350000.

PROPERTIES: Colorless to light yellow liq with an ether-like odor. Mp: < -101.1°, bp: 110° @ 762 mm, *d*: 1.0538 @ 20/4°, fl p: 20°, lel: 2%, uel: 100%, S_o: sol in alcohol and ether, vap d: 4.30 g/L @ 25°, 3.63 (air = 1), vp: 18 mm @ 20°.

EXPOSURE LIMITS: NIOSH REL: IDLH 2,000 ppm; OSHA PEL: TWA 25 ppm; ACGIH TLV: TWA 25 ppm.

USES: Rocket fuel formulations; organic synthesis.

PROPYNE

SYNONYMS: Allylene; Methyl acetylene; Propine; **1-Propyne**.

CHEMICAL DESIGNATIONS: CAS: 74-99-7; DOT: none assigned; mf: C_3H_4; fw: 40.06; RTECS: UK 4250000.

PROPERTIES: Colorless gas with a sweet odor. Mp: -101.5°, bp: -23.2°, *d*: 0.678 @ -27/4°, lel: 1.7%, uel: 11.7%, K_H: 1.1 x 10^{-1} atm·m^3/mol @ 25°, IP: 10.36 eV, log K_{ow}: 1.61 (calcd), S_o: sol in alcohol, benzene and chloroform, S_w: 3,640 mg/l @ 20°, vap d: 1.64 g/L @ 25°, 1.38 (air = 1), vp: 5.2 atm @ 20°.

TRANSFORMATION PRODUCTS: When passed through a cold soln containing hydrobromite ions, 1-bromo-1-propyne was formed.

EXPOSURE LIMITS: NIOSH REL: IDLH 17,000 ppm; OSHA PEL: TWA 1,000 ppm.

USES: Chemical intermediate; specialty fuel.

PYRENE

SYNONYMS: Benzo[*def*]phenanthrene; β-Pyrene; β-Pyrine.

CHEMICAL DESIGNATIONS: CAS: 129-00-0; DOT: none assigned; mf: $C_{16}H_{10}$; fw: 202.26; RTECS: UR 2450000.

PROPERTIES: Colorless solid (tetracene impurities impart a yellow color). Solns impart a slight blue fluorescence. Mp: 156°, bp: 393°, *d*: 1.271 @ 23/4°, pK_a: > 15, K_H: 1.87 x 10^{-5} atm·m^3/mol, IP: 7.50-7.70 eV, log K_{oc}: 4.66-5.23, log K_{ow}: 4.88-5.52, S_o: sol in most solvents, S_w: 135 μg/L @ 25°, vp: 2.5 x 10^{-6} mm @ 25°.

TRANSFORMATION PRODUCTS: Adsorption onto garden soil for 10 days @ 32° and irradiated with UV radiation produced 1,1′-bipyrene, 1,6-pyrenedione, 1,8-pyrenedione and 3 unidentified compounds. Microbial degradation by *Mycobacterium* sp. yielded the following ring-fission products: 4-phenanthroic acid, 4-hydroxyperinaphthenone, cinnamic and phthalic acids. The compounds pyrenol and the *cis*- and *trans*-4,5-dihydrodiols of pyrene were identified as ring-oxidation products. Silica gel coated with pyrene and suspended in an aq soln containing nitrite ion and subjected to UV radiation yielded the tentatively identified product 1-nitropyrene. 1-Nitropyrene coated on glass surfaces and exposed to natural sunlight resulted in the formation of hydroxypyrene, possibly pyrene dione and dihydroxy pyrene and other unidentified compounds. At room temperature, concentrated sulfuric acid will react with pyrene to form a mixture of disulfonic acids. In addition, an atmosphere containing 10% sulfur dioxide transformed pyrene into many sulfur compounds including pyrene-1-sulfonic acid and pyrenedisulfonic acid. 2-Nitropyrene was the sole product formed from the gas-phase reaction of pyrene with hydroxyl radicals in a NO_x atmosphere. Pyrene adsorbed on glass fiber filters reacted rapidly with N_2O_5 to form 1-nitropyrene with a 60-70% yield. 1-Nitropyrene also formed when pyrene deposited on glass filter paper containing sodium nitrite was irradiated with UV light at room temperature. This compound was reported to have formed from the reaction of pyrene with NO_x in urban air from St. Louis, MO. When pyrene adsorbed from the vapor phase onto coal fly ash, silica and alumina was exposed to NO_2, no reaction occurred. However, in the presence of nitric acid, nitrated compounds were produced. Ozonation of water containing pyrene (10-200 μg/L) yielded short chain aliphatic compounds as the major products.

EXPOSURE LIMITS: No individual standards have been set, however, as a constituent in coal tar pitch volatiles, the following exposure limits have been established: NIOSH REL: 10-h TWA 0.1 mg/m^3 (cyclohexane-extractable fraction); OSHA PEL: TWA 0.2 mg/m^3 (benzene-soluble fraction); ACGIH TLV: TWA 0.2 mg/m^3 (benzene solubles).

USE: Research chemical. Derived from industrial and experimental coal gasification operations where maximum concn detected in gas and coal tar streams were 9.2 and 24 mg/m^3, respectively.

PYRIDINE

SYNONYMS: Azabenzene; Azine; NCI-C55301; RCRA waste number U196; UN 1282.

CHEMICAL DESIGNATIONS: CAS: 110-86-1; DOT: 1282; mf: C_5H_5N; fw: 79.10; RTECS: UR 8400000.

PROPERTIES: Colorless liq with a sharp, penetrating odor. Mp: -42°, bp: 115.5°, *d*: 0.9819 @ 20/4°, fl p: 20°, lel: 1.8%, uel: 12.4%, pK_a: 5.19, 5.21 @ 25°, K_H: 1.2 x 10^{-5} atm·m^3/mol @ 25°, IP: 9.27 eV, 9.32 eV, log K_{ow}: 0.64-1.28, S_o: sol in acetone, alcohol, benzene and ether; misc with alcohol, ether, petroleum ether, oils and many other organic liqs, S_w: a saturated soln in equilibrium with its own vapor had a concn of 233.4 g/L @ 25°, vap d: 3.23 g/L @ 25°, 2.73 (air = 1), vp: 14 mm @ 20°.

TRANSFORMATION PRODUCTS: Irradiation of an aq soln @ 50° for 24 h resulted in a 23.06% yield of CO_2. The gas-phase reaction of O_3 with pyridine in synthetic air @ 23° yielded a nitrated salt having the formula: $[C_6H_5NH]^+NO_3^-$. The titanium dioxide-mediated photocatalytic oxidation of pyridine in the presence of hydroxyl radicals yielded ammonium and nitrate ions. When an aq soln containing pyridine and titanium dioxide was illuminated by UV light, ammonium and nitrate ions formed as the major products.

EXPOSURE LIMITS: NIOSH REL: IDLH 3,600 ppm; OSHA PEL: TWA 5 ppm; ACGIH TLV: 5 ppm.

SYMPTOMS OF EXPOSURE: Head; ner; dizz; insom; nau, anor; frequent urination; eye irrit; derm; liv and kid damage.

USES: Organic synthesis (vitamins and drugs); analytical chemistry; solvent for anhydrous mineral salts; denaturant for alcohol; antifreeze mixtures; textile dyeing; waterproofing; fungicides; rubber chemicals.

p-QUINONE

SYNONYMS: 1,4-Benzoquine; Benzoquinone; 1,4-Benzoquinone; *p*-Benzoquinone; Chinone; Cyclohexadienedione; 1,4-Cyclohexadienedione; **2,5-Cyclohexadiene-1,4-dione**; 1,4-Cyclohexadiene dioxide; 1,4-Dioxybenzene; NCI-C55845; Quinone; 4-Quinone; RCRA waste number U197; UN 2587; USAF P-220.

CHEMICAL DESIGNATIONS: CAS: 106-51-4; DOT: 2587; mf: $C_6H_4O_2$; fw: 108.10; RTECS: DK 2625000.

PROPERTIES: Light yellow cryst with an acrid odor. Mp: 115-117°, bp: subl, *d*: 1.318 @ 20/4°, K_H: 9.48 x 10^{-7} atm·m^3/mol @ 20° (calcd), IP: 9.67 eV, log K_{ow}: 0.20, S_o: sol in alcohol, ether and hot petroleum ether, S_w: 1.5 wt % @ 20°, vp: 0.1 mm @ 20°.

EXPOSURE LIMITS: NIOSH REL: IDLH 75 ppm; OSHA PEL: TWA 0.1 ppm.

SYMPTOMS OF EXPOSURE: Eye and skin irrit, conj kera; ulceration of skin.

USES: Oxidizing agent; manufacturing hydroquinone and dyes; in photography; strengthening animal fibers; tanning hides; analytical reagent; making gelatin insoluble; fungicides.

RONNEL

SYNONYMS: Dermaphos; Dimethyl trichlorophenyl thiophosphate; *O,O*-Dimethyl-*O*-2,4,5-trichlorophenyl phosphorothioate; *O,O*-Dimethyl *O*-(2,4,5-trichlorophenyl)thiophosphate; Dow ET 14; Dow ET 57; Ectoral; ENT 23,284; ET 14; ET 57; Etrolene; Fenchlorfos; Fenchlorophos; Fenchchlorphos; Karlan; Korlan; Korlane; Nanchor; Nanker; Nankor; **Phosphorothioic acid *O,O*-dimethyl *O*-(2,4,5-trichlorophenyl)ester**; Trichlorometafos; Trolen; Trolene; Viozene.

CHEMICAL DESIGNATIONS: CAS: 299-84-3; DOT: 2922; mf: $C_8H_8Cl_3O_3PS$; fw: 321.57; RTECS: TG 0525000.

PROPERTIES: White to light brown, waxy solid. Mp: 41°, bp: 97° @ 0.01 mm, *d*: 1.48 @ 25/4°, K_H: 8.46 x 10^{-6} atm·m^3/mol @ 25° (calcd), log K_{oc}: 2.76 (calcd), log K_{ow}: 4.67-5.068, S_o: sol in acetone, carbon tetrachloride, ether, kerosene, methylene chloride and toluene, S_w: 40 mg/L @ 25°, vp: 5.29 x 10^{-5} mm @ 20°.

TRANSFORMATION PRODUCTS: Though no products were identified, the reported hydrolysis $t_{1/2}$ @ pH 7.4 and 70° using a 1:4 ethanol/water mixture is 10.2-10.4 h.

EXPOSURE LIMITS: NIOSH REL: IDLH 5,000 mg/m^3; OSHA PEL: TWA 15 mg/m^3; ACGIH TLV: TWA 10 mg/m^3.

SYMPTOMS OF EXPOSURE: In animals: chol inhibition; irrit eyes; liv and kid damage.

USE: Insecticide.

STRYCHNINE

SYNONYMS: Certox; Dolco mouse cereal; Kwik-kil; Mole death; Mole-nots; Mouse-rid; Mouse-tox; Pied piper mouse seed; RCRA waste number P108; Ro-dex; Sanaseed; **Strychnidin-10-one**; Strychnos; UN 1692.

CHEMICAL DESIGNATIONS: CAS: 57-24-9; DOT: 1692; mf: $C_{21}H_{22}N_2O_2$; fw: 334.42; RTECS: WL 2275000.

PROPERTIES: Colorless to white, odorless cryst. Mp: 286-288°, bp: 270° @ 5 mm, *d*: 1.36 @ 20/4°, pK_a: @ 20°: pK_1 = 6.0, pK_2 = 11.7, log K_{oc}: 2.45 (calcd), log K_{ow}: 1.93, S_o (g/L): sol in alcohol (6.67), amyl alcohol (4.55), benzene (5.56), chloroform (200), glycerol (3.13), methanol (3.85), toluene (5), S_w: 0.02 wt % @ 20°.

EXPOSURE LIMITS: NIOSH REL: IDLH 3 mg/m^3; OSHA PEL: TWA 0.15 mg/m^3.

SYMPTOMS OF EXPOSURE: Stiff neck, fasc musc, restle,ss, appre, acuity of perception, reflex excitability, cyan, tetanic convuls, opisthotonos.

USES: Destroying rodents and predatory animals; trapping fur-bearing animals; medicine.

STYRENE

SYNONYMS: Cinnamene; Cinnamenol; Cinnamol; Diarex HF77; **Ethyenyl benzene**; NCI-C02200; Phenethylene; Phenylethene; Phenylethylene; Styrene monomer; Styrol; Styrolene; Styron; Styropol; Styropor; UN 2055; Vinyl benzene; Vinyl benzol.

CHEMICAL DESIGNATIONS: CAS: 100-42-5; DOT: 2055; mf: C_8H_8; fw: 104.15; RTECS: WL 3675000.

PROPERTIES: Colorless to light yellow, oily liq with a sweet, penetrating odor. (Polymerizes readily in the presence of heat, light or a peroxide catalyst. Polymerization is exothermic and may become explosive). Mp: -30.6°, bp: 145.2°, *d*: 0.9060 @ 20/4°, fl p: 31°, lel: 1.1%, uel: 6.1%, K_H: 2.61 x 10^{-3} atm·m^3/mol (calcd), IP: 8.47 eV, 8.71 eV, log K_{oc}: 2.87 (calcd), log K_{ow}: 2.95, 3.16, S_o: sol in acetone, ethanol, benzene, ether, carbon disulfide and petroleum ether, S_w: 300 mg/L @ 20°, vap d: 4.26 g/L @ 25°, 3.60 (air = 1), vp: 5 mm @ 20°.

TRANSFORMATION PRODUCTS: Irradiation of styrene in soln forms polystyrene. In a benzene soln, irradiation of polystyrene will result in depolymerization to presumably styrene. In the dark, styrene reacted with O_3 forming benzaldehyde, formaldehyde, benzoic acid and trace amounts of formic acid.

EXPOSURE LIMITS: NIOSH REL: 10-h TWA 50 ppm, 15-min C 100 ppm, IDLH 5,000 ppm; OSHA PEL: TWA 100 ppm, C 200 ppm, 5-min/3-h peak 600 ppm; ACGIH TLV: TWA 50 ppm, STEL 100 ppm.

SYMPTOMS OF EXPOSURE: Irrit eyes, nose; drow, weak, unstaedy gait; narco; defat derm.

USES: Preparation of polystyrene, styrene oxide, ethylbenzene, ethylcyclohexane, benzoic acid, synthetic rubber, resins, protective coatings and insulators.

SULFOTEPP

SYNONYMS: ASP 47; Bay E-393; Bayer-E 393; Bis-*O,O*-diethylphosphorothionic anhydride; Bladafum; Bladafume; Bladafun; Dithio; Dithione; Dithiophos; Dithiophosphoric acid tetraethyl ester; Dithiotep; E 393; ENT 16,273; Ethyl thiopyrophosphate; Lethalaire G-57; Pirofos; Plant dithio aerosol; Plantfume 103 smoke generator; Pyrophosphorodithioic acid tetraethyl ester; Pyrophosphorodithioic acid *O,O,O,O*-tetraethyl dithionopyrophosphate; RCRA waste number P109; Sulfatep; Sulfotep; TEDP; TEDTP; Tetraethyl dithionopyrophosphate; Tetraethyl dithiopyrophosphate; *O,O,O,O*-Tetraethyl dithiopyrophosphate; Thiodiphosphoric acid tetraethyl ester; **Thiophosphoric acid tetraethyl ester**; Thiopyrophosphoric acid tetraethyl ester; Thiotepp; UN 1704.

CHEMICAL DESIGNATIONS: CAS: 3689-24-5; DOT: 1704; mf: $C_8H_{20}O_5P_2S_2$; fw: 322.30; RTECS: XN 4375000.

PROPERTIES: Pale yellow liq with a garlic-like odor. bp: 136-139° @ 2 mm, *d*: 1.196 @ 25/4°, K_H: 2.88 x 10^{-6} atm·m^3/mol @ 20° (calcd), log K_{oc}: 2.87 (calcd), log K_{ow}: 3.02 (calcd), S_o: misc with most organic solvents, S_w: 0.0025 wt % @ 20°, vap d: 13.17 g/L @ 25°, 11.13 (air = 1), vp: 1.7 x 10^{-4} mm @ 20°.

EXPOSURE LIMITS: NIOSH REL: IDLH 35 mg/m^3; OSHA PEL: TWA 0.2 mg/m^3.

USE: Insecticide.

2,4,5-T

SYNONYMS: Amine 2,4,5-T for rice; BCF-bushkiller; Brush-off 445 low volatile brush killer; Brush-rhap; Brushtox; Dacamine; Debroussaillant concentre; Debroussaillant super concentre; Decamine 4T; Ded-weed brush killer; Ded-weed LV-6 brush-kil and T-5 brush-kil; Dinoxol; Envert-T; Estercide T-2 and T-245; Esteron; Esterone 245; Esteron 245 BE; Esteron brush killer; Farmco fence rider; Fence rider; Forron; Forst U 46; Fortex; Fruitone A; Inverton 245; Line rider; NA 2,765; Phortox; RCRA waste number U232; Reddon; Reddox; Spontox; Super D weedone; Tippon; Tormona; Transamine; Tributon; **(2,4,5-Trichlorophenoxy)-acetic acid**; Trinoxol; Trioxon; Trioxone; U 46; Veon; Veon 245; Verton 2T; Visko rhap low volatile ester; Weddar; Weedone; Weedone 2,4,5-T.

CHEMICAL DESIGNATIONS: CAS: 93-76-5; DOT: 2765; mf: $C_8H_5Cl_3O_3$; fw: 255.48; RTECS: AJ 8400000.

PROPERTIES: Pale brown cryst. Mp: 157-158°, *d*: 1.80 @ 20/20°, pK_a: 2.80-2.88, K_H: 4.87 x 10^{-8} atm·m^3/mol @ 25° (calcd), log K_{oc}: 1.72, 2.27, log K_{ow}: 0.60-3.40, S_o: sol in alcohol and slightly sol in petroleum ether, S_w: 220 mg/L @ 20°, vp: 3.75 x 10^{-5} mm @ 20°.

TRANSFORMATION PRODUCTS: 2,4,5-T degraded in anaerobic sludge by reductive dechlorination to 2,4,5-trichlorophenol, 3,4-dichlorophenol and 4-chlorophenol. An anaerobic methanogenic consortium, growing on 3-chlorobenzoate, metabolized 2,4,5-T to (2,5-dichlorophenoxy)acetic acid. Under aerobic conditions, 2,4,5-T degraded to 2,4,5-trichlorophenol and 3,5-dichlorocatechol which may further degrade to 4-chlorocatechol or *cis,cis*-2,4-dichloromuconic acid, 2-chloro-4-carboxymethylenebut-2-enolide, chlorosuccinic acid and succinic acid. 2,4,5-Trichlorophenol and 2,4,5-trichloroanisole were formed when 2,4,5-T was incubated in soil @ 25° under aerobic conditions ($t_{1/2}$ = 14 days). CO_2, chloride, dichloromaleic, oxalic and glycolic acids, were reported as ozonation products of 2,4,5-T in water @ pH 8. When 2,4,5-T (10^{-4} M) in an oxygenated, titanium dioxide (2 g/L) suspension was irradiated by sunlight ($\lambda \geq 340$ nm), 2,4,5-trichlorophenol and 2,4,5-trichlorophenyl formate formed as major intermediates. Other compounds identified as intermediates included 9 chlorinated aromatic hydrocarbons. Complete mineralization yielded hydrochloric acid, CO_2 and water.

EXPOSURE LIMITS: NIOSH REL: IDLH 5,000 mg/m^3; OSHA PEL: TWA 10 mg/m^3; ACGIH TLV: TWA 10 mg/m^3.

SYMPTOMS OF EXPOSURE: Animal studies: atax, skin irrit, acne-like rash, blood in stool.

USES: Plant hormone; defoliant; herbicide.

TCDD

SYNONYMS: Dioxin; Dioxin (herbicide contaminant); Dioxine; NCI-C03714; TCDBD; 2,3,7,8-TCDD; 2,3,7,8-Tetrachlorodibenzodioxin; 2,3,7,8-Tetrachlorodibenzo[*b,e*][1,4]dioxan; **2,3,7,8-Tetrachlorodibenzo[*b,e*][1,4]dioxin;** 2,3,7,8-Tetrachlorodibenzo-1,4-dioxin; 2,3,7,8-Tetrachlorodibenzo-*p*-dioxin; Tetradioxin.

CHEMICAL DESIGNATIONS: CAS: 1746-01-6; DOT: none assigned; mf: $C_{12}H_4Cl_4O_2$; fw: 321.98; RTECS: HP 3500000.

PROPERTIES: Colorless needles. Mp: 295°, bp: 421.2° (est), *d*: 1.827 @ 25° (est), K_H: 5.40 x 10^{-23} atm·m^3/mol @ 18-22° (calcd), IP: 9.148 eV (calcd), log K_{oc}: 6.66, log K_{ow}: 5.38-7.02, S_o: sol in fats, oils, acetone (110 mg/L), benzene (570 mg/L), chloroform (370 mg/L), *o*-dichlorobenzene (1,400 mg/L), methanol (10 mg/L) and *n*-octanol (50 mg/L), S_w: 0.317 μg/L @ 25°, vp: 6.4 x 10^{-10} mm @ 20°.

TRANSFORMATION PRODUCTS: Pure TCDD did not photolyze under UV light. However, in aq solns containing cationic, (1-hexadecylpyridinium chloride), anionic (sodium dodecyl sulfate) and nonionic (methanol) surfactants, TCDD decomposed into the end product tentatively identified as 2-phenoxyphenol. The time required for total TCDD decomposition using the cationic, anionic and nonionic solns was 4, 8 and 16 h, respectively. TCDD photodegrades rapidly in alcoholic solns by reductive dechlorination. In water, however, the reaction was very slow. ^{14}C-labeled TCDD on a silica plate was exposed to summer sunlight at Beltsville, MD for 20 h. A polar product was formed which was not identified. An identical experiment using soil demonstrated photodegradation occurred but to a smaller extent. No photoproducts were identified. TCDD was dehalogenated by a soln composed of poly(ethylene glycol), potassium carbonate and sodium peroxide. After 2 h @ 85°, >99.9% of the applied TCDD decomposed. Chemical intermediates identified include tri-, di- and monochloro[*b,e*]dibenzo[1,4]dioxin, dibenzodioxin, hydrogen, CO, methane, ethylene and acetylene. When TCDD in isooctane (3.1 x 10^{-6} M) was irradiated by UV light ($\lambda \leq 310$ nm) @ 31-33°, ≈ 10% was converted to 2,3,7-trichlorodibenzo-*p*-dioxin. In addition, 4,4′,5,5′-tetrachloro-2,2′-dihydroxybiphenyl was identified as a new photoproduct formed by the reductive rearrangement of TCDD.

USES: Occurs as an impurity in the herbicide 2,4,5-T (2,4,5-trichlorophenoxyacetic acid).

1,2,4,5-TETRABROMOBENZENE

SYNONYM: *sym*-Tetrabromobenzene.

CHEMICAL DESIGNATIONS: CAS: 636-28-2; DOT: none assigned; mf: $C_6H_2Br_4$; fw: 393.70; RTECS: none assigned.

PROPERTIES: Solid. Mp: 189°, bp: 329°, log K_{oc}: 4.82 (calcd), log K_{ow}: 5.13, S_o: sol in alcohol, benzene and ether, S_w: 0.040 mg/L.

TRANSFORMATION PRODUCTS: 1,2,4,5-Tetrabromobenzene (600 mg/L in distilled water) was stirred in the dark for 1 day. GC analysis of the soln showed

1,2,4-tribromobenzene as a major transformation product.

USES: Organic synthesis.

1,1,2,2-TETRABROMOETHANE

SYNONYMS: Acetylene tetrabromide; Muthmann's liquid; TBE; Tetrabromoacetylene; Tetrabromoethane; *sym*-Tetrabromoethane; UN 2504.

CHEMICAL DESIGNATIONS: CAS: 79-27-6; DOT: 2504; mf: $C_2H_2Br_4$; fw: 345.65; RTECS: KI 8225000.

PROPERTIES: Colorless to pale yellow liq with a pungent odor resembling camphor and iodoform. Mp: 0°, bp: 243.5°, *d*: 2.8748 @ 20/4°, K_H: 6.40 x 10^{-5} atm·m^3/mol @ 20° (calcd), log K_{oc}: 2.45 (calcd), log K_{ow}: 2.91 (calcd), S_o: sol in acetone and benzene; misc with acetic acid, alcohol, aniline, chloroform, ether, S_w: 651 mg/L @ 30°, vap d: 14.13 g/L @ 25°, 11.93 (air = 1), vp: 0.1 mm @ 20°.

SYMPTOMS OF EXPOSURE: Irrit eyes, nose; anor, nau; severe head; abdom pain; jaun; monocy.

USES: Solvent; in microscopy; separating minerals by density.

1,2,3,4-TETRACHLOROBENZENE

SYNONYM: 1,2,3,4-TCB.

CHEMICAL DESIGNATIONS: CAS: 634-66-2; DOT: none assigned; mf: $C_6H_2Cl_4$; fw: 215.89; RTECS: DB 9440000.

PROPERTIES: White cryst or needles. Mp: 46.6°, bp: 254°, K_H: 6.9 x 10^{-3} atm·m^3/mol @ 20°, log K_{oc}: 5.4, log K_{ow}: 4.37-4.83, S_o: sol in acetic acid, ether and lignoin, S_w: 5.92 mg/L @ 25°, vp: 2.6 x 10^{-2} mm @ 25°.

TRANSFORMATION PRODUCTS: Irradiation ($\lambda \geq 285$ nm) of 1,2,3,4-tetrachlorobenzene (1.1-1.2 mM/L) in an acetonitrile-water mixture containing acetone (concn = 0.553 mM/L) as a sensitizer gave the following products (% yield): 1,2,3-trichlorobenzene (9.2), 1,2,4-trichlorobenzene (32.6), 1,3-dichlorobenzene (5.2), 1,4-dichlorobenzene (1.5), 2,2′,3,3′,4,4′,5-heptachlorobiphenyl (2.52), 2,2′,3,3′,4,5,6′-heptachlorobiphenyl (1.22), 10 hexachlorobiphenyls (3.50), 5 pentachlorobiphenyls (0.87), dichlorophenyl cyanide, 2 trichloroacetophenones, trichlorocyanophenol, (trichlorophenyl)acetonitriles and 1-(trichlorophenyl)-2-propanone. Without acetone, the identified photolysis products (% yield) included 1,2,3-trichlorobenzene (7.8), 1,2,4-trichlorobenzene (26.8), 1,2-dichlorobenzene (0.5), 1,3-dichlorobenzene (0.7), 1,4-dichlorobenzene (30.4), 1,2,3,5-tetrachlorobenzene (2.26), 1,2,4,5-tetrachlorobenzene (0.72), 2,2′,3,3′,4,4′,5-heptachlorobiphenyl (< 0.01) and 2,2′,3,3′,4,5,6′-heptachlorobiphenyl (< 0.01). The sunlight irradiation of 1,2,3,4-tetrachlorobenzene (20 g) in a 100 ml borosilicate glass-stoppered Erlenmeyer flask for 56 days yielded 4,280 ppm heptachlorobiphenyl. A mixed culture of soil bacteria or a *Pseudomonas* sp. transformed 1,2,3,4-tetrachlorobenzene to 2,3,4,5-tetrachlorophenol.

USES: Dielectric fluids; organic synthesis.

1,2,3,5-TETRACHLOROBENZENE

SYNONYM: 1,2,3,5-TCB.

CHEMICAL DESIGNATIONS: CAS: 634-90-2; DOT: none assigned; mf: $C_6H_2Cl_4$; fw: 215.89; RTECS: DB 9445000.

PROPERTIES: Solid. Mp: 54.5°, bp: 246°, K_H: 1.58 x 10^{-3} atm·m^3/mol @ 25°, log K_{oc}: 6.0, log K_{ow}: 4.46-4.658, S_o: sol in alcohol, benzene, ether and lignoin, S_w: 5.19 mg/L @ 25°, vp: 0.07 mm @ 25° (est).

TRANSFORMATION PRODUCTS: Irradiation ($\lambda \geq$ 285 nm) of 1,2,3,5-tetrachlorobenzene (1.1-1.2 mM/L) in an acetonitrile-water mixture containing acetone (concn = 0.553 mM/L) as a sensitizer gave the following products (% yield): 1,2,3-trichlorobenzene (5.3), 1,2,4-trichlorobenzene (4.9), 1,3,5-trichlorobenzene (49.3), 1,3-dichlorobenzene (1.8), 2,3,4,4′,5,5′,6-heptachlorobiphenyl (1.41), 2,2′,3,4,4′,6,6′-heptachlorobiphenyl (1.10), 2,2′,3,3′,4,5′,6-heptachlorobiphenyl (4.50), 4 hexachlorobiphenyls (4.69), 1 pentachlorobiphenyl (0.64), trichloroacetophenone, 1-(trichlorophenyl)-2-propanone and (trichlorophenyl)acetonitrile. Without acetone, the identified photolysis products (% yield) included 1,2,3-trichlorobenzene (trace), 1,2,4-trichlorobenzene (24.3), 1,3,5-trichlorobenzene (11.7), 1,3-dichlorobenzene (0.5), 1,4-dichlorobenzene (3.3), 1,2,3,4,5-pentachlorobenzene (1.43), 1,2,3,4-tetrachlorobenzene (5.99), 2 heptachlorobiphenyls (1.40), 2 hexachlorobiphenyls (< 0.01) and 1 pentachlorobiphenyl (0.75). A mixed culture of soil bacteria or a *Pseudomonas* sp. transformed 1,2,3,5-tetrachlorobenzene to 2,3,4,6-tetrachlorophenol.

USES: Organic synthesis.

1,2,4,5-TETRACHLOROBENZENE

SYNONYMS: *sym*-Tetrachlorobenzene; RCRA waste number U207.

CHEMICAL DESIGNATIONS: CAS: 95-94-3; DOT: none assigned; mf: $C_6H_2Cl_4$; fw: 215.89; RTECS: DB 9450000.

PROPERTIES: White flakes or needles. Mp: 139-140°, bp: 243-246°, *d*: 1.858 @ 21/4°, fl p: 155°, K_H: 1.0 x 10^{-2} atm·m^3/mol @ 20°, log K_{oc}: 3.72-3.91, log K_{ow}: 4.56-4.67, S_o: sol in benzene, chloroform and ether, S_w: 0.465 mg/L @ 25°, vp: 0.005 mm @ 25° (est).

TRANSFORMATION PRODUCTS: Irradiation ($\lambda \geq$ 285 nm) of 1,2,4,5-tetrachlorobenzene (1.1-1.2 mM/L) in an acetonitrile-water mixture containing acetone (concn = 0.553 mmol/L) as a sensitizer gave the following products (% yield): 1,2,4-trichlorobenzene (25.3), 1,3-dichlorobenzene (8.1), 1,4-dichlorobenzene (3.6), 2,2′,3,4′,5,5′,6-heptachlorobiphenyl (4.19), 4 hexachlorobiphenyls (6.78), 4 pentachlorobiphenyls (2.33), 1 tetrachlorobiphenyl (0.32), 2,4,5-trichloroacetophenone and (2,4,5-trichlorophenyl)acetonitrile. Without acetone, the identified photolysis products (% yield) included 1,2,4-trichlorobenzene (27.7), 1,3-dichlorobenzene (0.3), 1,4-dichlorobenzene (8.5), 1,2,3,4,5-pentachlorobenzene (trace), 1,2,3,4-tetrachlorobenzene (0.45), 1,2,3,5-tetrachlorobenzene (1.11), 2,2′,3,4′,5,5′,6-heptachlorobiphenyl (1.24), 3 hexachlorobiphenyls (1.19) and 4 pentachlorobiphenyls (0.56). The sunlight irradiation of 1,2,4,5-tetrachlorobenzene (20 g) in a 100 ml borosilicate glass-stoppered Erlenmeyer flask for 28 days yielded 26 ppm heptachlorobiphenyl. A mixed culture of soil bacteria or a *Pseudomonas* sp. transformed 1,2,4,5-tetrachlorobenzene to 2,3,5,6-tetrachlorophenol.

USES: Insecticides; intermediate for herbicides and defoliants; electrical insulation; impregnant for moisture resistance.

1,1,2,2-TETRACHLOROETHANE

SYNONYMS: Acetosol; Acetylene tetrachloride; Bonoform; Cellon; 1,1-Dichloro-2,2-dichloroethane; Ethane tetrachloride; NCI-C03554; RCRA waste number U208; TCE; Tetrachlorethane; Tetrachloroethane; *sym*-Tetrachloroethane; UN 1702; Westron.

CHEMICAL DESIGNATIONS: CAS: 79-34-5; DOT: 1702; mf: $C_2H_2Cl_4$; fw: 167.85; RTECS: KI 8575000.

PROPERTIES: Colorless to pale yellow liq with a sweet chloroform-like odor. Mp: -42.5°, bp: 146.2°, *d*: 1.5953 @ 20/4°, K_H: 4.56 x 10^{-4} atm·m^3/mol @ 25°, IP: 11.1 eV, log K_{oc}: 1.663, 2.07, log K_{ow}: 2.39-2.56, S_o: sol in acetone, ethanol, ether, benzene, carbon tetrachloride, petroleum ether, carbon disulfide, dimethylformamide and oils; misc in alcohol and chloroform, S_w: 2,900 mg/L @ 20°, vap d: 6.86 g/L @ 25°, 5.79 (air = 1), vp: 5 mm @ 20°.

TRANSFORMATION PRODUCTS: Monodechlorination by microbes under laboratory conditions produced 1,1,2-trichloroethane. In an aq soln containing 0.100 M phosphate-buffered distilled water, 1,1,2,2-tetrachloroethane was abiotically transformed to 1,1,2-trichloroethane. This reaction was investigated over a temperature range of 30-95° at various pHs (5-9). Abiotic dehydrohalogenation of 1,1,2,2-tetrachloroethane was reported to yield 1,1,1-trichloroethylene ($t_{1/2}$ = 0.8 yr @ 20°). Under alkaline conditions, 1,1,2,2-tetrachloroethane dehydrohalogenated to trichloroethylene ($t_{1/2}$ = 146 days @ 25° and pH 7).

EXPOSURE LIMITS: NIOSH REL: lowest detectable limit; OSHA PEL: TWA 5 ppm; ACGIH TLV: TWA 1 ppm.

SYMPTOMS OF EXPOSURE: Nau, vomit, abdom pain, tremor fingers, jaun, enl tend liv, derm, monocy, kid damage, pares.

USES: Solvent for chlorinated rubber; insecticide and bleach manufacturing; paint, varnish and rust remover manufacturing; degreasing, cleansing and drying of metals; denaturant for ethyl alcohol; preparation of 1,1-dichloro-ethylene; extractant and solvent for oils and fats; insecticides; weed killer; fumigant; intermediate in the manufacture of other chlorinated hydrocarbons; herbicide.

TETRACHLOROETHYLENE

SYNONYMS: Ankilostin; Antisol 1; Carbon bichloride; Carbon dichloride; Dee-Solv; Didakene; Dow-per; ENT 1,860; Ethylene tetrachloride; Fedal-UN;

NCI-C04580; Nema; PCE; PER; Perawin; PERC; Perchlor; Perchlorethylene; Perchloroethylene; Perclene; Perclene D; Percosolv; Perk; Perklone; Persec; RCRA waste number U210; Tetlen; Tetracap; Tetrachlorethylene; **Tetrachloroethene**; 1,1,2,2-Tetrachloroethylene; Tetraleno; Tetralex; Tetravec; Tetroguer; Tetropil; UN 1897.

CHEMICAL DESIGNATIONS: CAS: 127-18-4; DOT: 1897; mf: C_2Cl_4; fw: 165.83; RTECS: KX 3850000.

PROPERTIES: Colorless liq with a chloroform or sweet ethereal odor. Mp: -19°, bp: 121.2°, *d*: 1.6227 @ 20/4°, K_H: 1.46 x 10^{-2} atm·m^3/mol @ 20°, IP: 9.32 eV, 9.71 eV, log K_{oc}: 2.322-2.63, log K_{ow}: 2.10-2.88, S_o: sol in ethanol, benzene, ether and oils, S_w: 150 mg/L @ 20°, vap d: 6.78 g/L @ 25°, 5.72 (air = 1), vp: 14 mm @ 20°.

TRANSFORMATION PRODUCTS: Sequential dehalogenation by microbes under laboratory conditions produced trichloroethylene, *cis*-1,2-dichloroethylene, *trans*-1,2-dichloroethylene and vinyl chloride. A microcosm composed of aquifer water and sediment collected from uncontaminated sites in the Everglades biotransformed tetrachloroethylene to *cis*- and *trans*-1,2-dichloroethylene. Microbial degradation to trichloroethylene under anaerobic conditions or using mixed cultures was also reported. In a continuous-flow mixed-film methanogenic column study, tetrachloroethylene degraded to trichloroethylene with traces of vinyl chloride, dichloroethylene isomers and CO_2. Photolysis in the presence of NO_x yielded phosgene (carbonyl chloride) with minor amounts of carbon tetrachloride, dichloroacetyl chloride and trichloroacetyl chloride. In sunlight, photolysis products reported include chlorine, hydrogen chloride and trichloroacetic acid. Tetrachloroethylene reacts with O_3 forming a mixture of phosgene and trichloroacetyl chloride ($t_{1/2}$ = 8 days). Reported photooxidation products in the troposphere include trichloroacetyl chloride and phosgene. Phosgene is hydrolyzed readily to hydrogen chloride and CO_2. The reported hydrolysis $t_{1/2}$ @ 25° and pH of 9.9 is 9.9 x 10^8 yr.

EXPOSURE LIMITS: NIOSH REL: lowest feasible limit; OSHA PEL: TWA 100 ppm, C 200 ppm, 5-min/3-h peak 300 ppm; ACGIH TLV: TWA 50 ppm, STEL 200 ppm.

SYMPTOMS OF EXPOSURE: Irrit eyes, nose, throat; nau; flush face, neck; verti, dizz, inco; head; som; eryt.

USES: Dry cleaning fluid; degreasing and drying metals and other solids; solvent for waxes, greases, fats, oils, gums; manufacturing printing inks and paint removers; preparation of fluorocarbons and trichloroacetic acid; vermifuge; heat-transfer medium; organic synthesis.

TETRAETHYL PYROPHOSPHATE

SYNONYMS: Bis-*O,O*-diethylphosphoric anhydride; Bladan; **Diphosphoric acid tetraethyl ester**; ENT 18,771; Ethyl pyrophosphate; Fosvex; Grisol; Hept; Hexamite; Killax; Kilmite 40; Lethalaire G-52; Lirohex; Mortopal; NA 2,783; Nifos; Nifos T; Nifost; Pyrophosphoric acid tetraethyl ester; RCRA waste number P111; TEP; TEPP; Tetraethyl diphosphate; Tetraethyl pyrofosfaat; Tetrastigmine;

Tetron; Tetron-100; Vapotone.

CHEMICAL DESIGNATIONS: CAS: 107-49-3; DOT: 2784; mf: $C_8H_{10}O_7P_2$; fw: 290.20; RTECS: UX 6825000.

PROPERTIES: Colorless to amber liq with a fruity odor. Mp: 0°, bp: 135-138° @ 1 mm, *d*: 1.185 @ 20/4°, S_o: misc in acetone, benzene, carbon tetrachloride, chloroform, ethanol, ethylene glycol; glycerol, methanol, propylene glycol, toluene and xylene, S_w: misc, vap d: 11.86 g/L @ 25°, 10.02 (air = 1), vp: 1.55 x 10^{-4} mm @ 20°.

TRANSFORMATION PRODUCTS: Though no products were identified, tetraethyl pyrophosphate is quickly hydrolyzed by water ($t_{1/2}$ = 7 h @ 25°).

EXPOSURE LIMITS: NIOSH REL: IDLH 10 mg/m^3; OSHA PEL: TWA 0.05 mg/m^3.

SYMPTOMS OF EXPOSURE: Eye pain, vision, tears, head, chest, cyan, anor, nau, vomit, diarr, local sweat, weak, twitch, para, Cheyne-Stokes resp, convul, low BP, card.

USES: Insecticide for mites and aphids; rodenticide.

TETRAHYDROFURAN

SYNONYMS: Butylene oxide; Cyclotetramethylene oxide; Diethylene oxide; Furanidine; Hydrofuran; NCI-C60560; Oxacyclopentane; Oxolane; RCRA waste number U213; Tetramethylene oxide; THF; UN 2506.

CHEMICAL DESIGNATIONS: CAS: 109-99-9; DOT: 2056; mf: C_4H_8O; fw: 72.11; RTECS: LU 5950000.

PROPERTIES: Colorless liq with an ether-like odor. Mp: -108°, -65°, bp: 67°, *d*: 0.8892 @ 20/4°, fl p: -17.2°, lel: 1.8%, uel: 11.8%, K_H: 7.06 x 10^{-5} atm·m^3/mol @ 25°, IP: 9.54 eV, log K_{ow}: 0.46, S_o: sol in alcohols, ketones, esters, ethers and hydrocarbons, S_w: 4.2 mol/L @ 25°, vap d: 2.95 g/L @ 25°, 2.49 (air = 1), vp: 145 mm @ 20°.

EXPOSURE LIMITS: NIOSH REL: IDLH 20,000 ppm; OSHA PEL: TWA 200 ppm; ACGIH TLV: TWA 200 ppm, STEL 250 ppm.

SYMPTOMS OF EXPOSURE: Irrit eyes, upper resp; nau; dizz; head.

USES: Solvent for uncured rubber and polyvinylchlorides, vinyl chloride copolymers, vinylidene chloride copolymers, natural resins; topcoating solns; cellophane; polymer coating; magnetic tapes; adhesives; printing inks; organic synthesis.

1,2,4,5-TETRAMETHYLBENZENE

SYNONYMS: Durene; Durol; *sym*-1,2,4,5-Tetramethylbenzene.

CHEMICAL DESIGNATIONS: CAS: 95-93-2; DOT: none assigned; mf: $C_{10}H_{14}$; fw: 134.22; RTECS: DC 0500000.

PROPERTIES: Colorless cryst or scales with a camphor-like odor. Mp: 79.2°, bp: 196.8°, *d*: 0.8380 @ 81/4°, fl p: 54°, K_H: 2.49 x 10^{-2} atm·m^3/mol @ 25° (calcd), log K_{oc}: 3.79 (calcd), log K_{ow}: 4.00, S_o: sol in acetone, alcohol, benzene and ether, S_w: 3.48 ppm @ 25°, vp: 0.49 mm @ 25°.

USES: Plasticizers; polymers; fibers; organic synthesis.

TETRANITROMETHANE

SYNONYMS: NCI-C55947; RCRA waste number P112; Tetan; TNM; UN 1510.

CHEMICAL DESIGNATIONS: CAS: 509-14-8; DOT: 1510; mf: CN_4O_8; fw: 196.03; RTECS: PB 4025000.

PROPERTIES: Colorless to pale yellow liq or solid with a pungent odor. Mp: 14.2°, bp: 126°, *d*: 1.6380 @ 20/4°, S_o: misc with alcohol and ether, vap d: 8.01 g/L @ 25°, 6.77 (air = 1), vp: 8.4 mm @ 20°.

EXPOSURE LIMITS: NIOSH REL: IDLH 5 ppm; OSHA PEL: TWA 1 ppm.

SYMPTOMS OF EXPOSURE: Irrit eyes, nose, throat; dizz, head; chest pain, dysp, methhemoglobinuria, cyan; skin burns.

USES: Laboratory reagent for detecting double bonds in organic compounds; oxidizer in rocket propellants; diesel fuel booster.

TETRYL

SYNONYMS: *N*-Methyl-*N*,2,4,6-tetranitroaniline; ***N*-Methyl-*N*,2,4,6-tetranitrobenzenamine**; Nitramine; Picrylmethylnitramine; Picrylnitromethylamine; Tetralit; Tetralite; Tetril; 2,4,6-Tetryl; Trinitrophenylmethylnitramine; 2,4,6-Trinitrophenylmethylnitramine; 2,4,6-Trinitrophenyl-*N*-methylnitramine; UN 0208.

CHEMICAL DESIGNATIONS: CAS: 479-45-8; DOT: none assigned; mf: $C_7H_5N_5O_8$; fw: 287.15; RTECS: BY 6300000.

PROPERTIES: Colorless to pale yellow cryst. Mp: 131-132°, bp: explodes @ 187°, *d*: 1.57 @ 19/4°, K_H: < 1.89 x 10^{-3} atm·m^3/mol @ 20° (calcd), log K_{oc}: 2.37 (calcd), log K_{ow}: 2.04 (calcd), S_o: sol in acetone, benzene, glacial acetic acid and ether, S_w: 0.02 wt % @ 20°, vp: < 1 mm @ 20°.

EXPOSURE LIMITS: OSHA PEL: TWA 1.5 mg/m^3.

SYMPTOMS OF EXPOSURE: Sens derm, itch, eryt, edema on nasal folds, cheeks, neck; kera; epis; irido-cyclitis; sneez; cough, coryza; irrit, ftg, mal, head, lass, insom, nau, vomit; anem.

USES: As an indicator in analytical chemistry; explosives.

THIOPHENE

SYNONYMS: CP 34; Divinylene sulfide; Huile H50; Huile HSO; Thiacyclopentadiene; Thiaphene; Thiofuran; Thiofuram; Thiofurfuran; Thiole; Thiophen; Thiotetrole; UN 2414; USAF EK-1860.

CHEMICAL DESIGNATIONS: CAS: 110-02-1; DOT: 2414; mf: C_4H_4S; fw: 84.14; RTECS: XM 7350000.

PROPERTIES: Colorless liq with an aromatic odor. Mp: -38.2°, bp: 84.2°, *d*: 1.0649 @ 20/4°, fl p: -1.1°, K_H: 2.93 x 10^{-3} atm·m^3/mol @ 25° (calcd), IP: 8.860 eV, log K_{oc}: 1.73 (calcd), log K_{ow}: 1.81, S_o: sol in acetone, alcohol, benzene, ether and pyrimidine, S_w: 3,015 ppm @ 25°, vap d: 3.44 g/L @ 25°, 2.90 (air = 1), vp: 60 mm @ 20°.

USES: Solvent; manufacturing resins, dyes and pharmaceuticals.

THIRAM

SYNONYMS: Aatack; Accelerator thiuram; Aceto TETD; Arasan; Arasan 70; Arasan 75; Arasan-M; Arasan 42-S; Arasan-SF; Arasan-SF-X; Aules; Bis(dimethylamino)carbonothioyl disulfide; Bis(dimethylthiocarbamoyl) disulfide; Bis(dimethylthiocarbamyl) disulfide; Chipco thiram 75; Cyuram DS; α,α′-Dithiobis(dimethylthio) formamide; *N,N′*-(Dithiodicarbonothioyl)bis(*N*-methylmethanamine); Ekagom TB; ENT 987; Falitram; Fermide; Fernacol; Fernasan; Fernasan A; Fernide; Flo pro T seed protectant; Hermal; Hermat TMT; Heryl; Hexathir; Kregasan; Mercuram; Methyl thiram; Methyl thiuramdisulfide; Methyl tuads; NA 2,771; Nobecutan; Nomersan; Normersan; NSC 1,771; Panoram 75; Polyram ultra; Pomarsol; Pomersol forte; Pomasol; Puralin; RCRA waste number U244; Rezifilm; Royal TMTD; Sadoplon; Spotrete; Spotrete-F; SQ 1489; Tersan; Tersan 75; Tetramethyldiurane sulphite; Tetramethylthiuram bisulfide; Tetramethylthiuram bisulphide; Tetramethylenethiuram disulfide; **Tetramethylthioperoxydicarbonic diamide**; Tetramethylthiocarbamoyl disulfide; Tetramethylthioperoxydicarbonic diamide; Tetramethylthiuram disulfide; Tetramethylthiuram disulphide; *N,N*-Tetramethylthiuram disulfide; *N,N,N′,N′*-Tetramethylthiuram disulfide; Tetramethylthiuran disulfide; Tetramethylthiurane disulphide; Tetramethylthiurum disulfide; Tetramethylthiurum disulphide; Tetrapom; Tetrasipton; Tetrathiuram disulfide; Tetrathiuram disulphide; Thillate; Thimer; Thiosan; Thiotex; Thiotox; Thiram 75; Thiramad; Thiram B; Thirasan; Thiulix; Thiurad; Thiuram; Thiuram D; Thiuramin; Thiuram M; Thiuram M rubber accelerator; Thiuramyl; Thylate; Tirampa; Tiuramyl; TMTD; TMTDS; Trametan; Tridipam; Tripomol; TTD; Tuads; Tuex; Tulisan; USAF B-30; USAF EK-2089; USAF P-5; Vancida TM-95; Vancide TM; Vulcafor TMTD; Vulkacit MTIC.

CHEMICAL DESIGNATIONS: CAS: 137-26-8; DOT: 2771mf: $C_6H_{12}N_2S_4$; fw: 269.35; RTECS: JO 1400000.

PROPERTIES: Colorless to white to cream colored cryst. Mp: 155.6°, bp: 310-315° @ 15 mm, *d*: 1.29 @ 20/4°, fl p: 88.9°, S_o: sol in chloroform, acetone (1.2 wt %), alcohol (< 0.2 wt %), benzene (2.5 wt %) and ether (< 0.2 wt %); slightly sol in carbon disulfide, S_w: 30 ppm.

TRANSFORMATION PRODUCTS: In soils. microbes converted thiram to carbon disulfide and dimethylamine.

EXPOSURE LIMITS: NIOSH REL: IDLH 1,500 mg/m^3; OSHA PEL: TWA 5 mg/m^3.

SYMPTOMS OF EXPOSURE: Irrit muc memb, derm; with ethanol consumption: flush, eryt, pruritis, urticaria, head, nau, vomit, diarr, weak, dizz, diff breath.

USES: Vulcanizer; seed disinfectant; rubber accelerator; animal repellant; fungicide; bacteriostat in soap.

TOLUENE

SYNONYMS: Antisal 1a; Methacide; **Methylbenzene**; Methylbenzol; NCI-C07272; Phenylmethane; RCRA waste number U220; Toluol; Tolu-sol; UN 1294.

CHEMICAL DESIGNATIONS: CAS: 108-88-3; DOT: 1294; mf: C_7H_8; fw: 92.14; RTECS: XS 5250000.

PROPERTIES: Colorless, water-white liq with a pleasant odor similar to benzene. Mp: -95°, bp: 110.6°, *d*: 0.8669 @ 20/4°, fl p: 4.4°, lel: 1.3%, uel: 7.1%, pK_a: ≈ 35, K_H: 6.74 x 10^{-3} atm·m^3/mol @ 25°, IP: 8.82 eV, log K_{oc}: 1.57-2.25, log K_{ow}: 2.21-2.80, S_o: sol in acetone, carbon disulfide and ligroin; misc with acetic acid, ethanol, benzene, ether, chloroform and other solvents, S_w: 515 mg/L @ 20°, vap d: 3.77 g/L @ 25°, 3.18 (air = 1), vp: 22 mm @ 20°.

TRANSFORMATION PRODUCTS: A mutant of *Pseudomonas putida* oxidized toluene to (+)-*cis*-2,3-dihydroxy-1-methylcyclohexa-1,4-diene. Other metabolites identified in the microbial degradation of toluene include *cis*-2,3-dihydroxy-2,3-dihydrotoluene, 3-methylcatechol, benzyl alcohol, benzaldehyde, benzoic acid and catechol. In a methanogenic aquifer material, toluene degraded completely to CO_2. In anoxic groundwater near Bemidji, MI, toluene anaerobically biodegraded to the intermediate benzoic acid. Methylmuconic acid was reported to be the biooxidation product of toluene by *Nocardia corallina* V-49 using *n*-hexadecane as the substrate. With methane as the substrate and *Methylosinus trichosporium* OB3b as the microorganism, *p*-hydroxytoluene and benzoic acid are the products of biooxidation. In addition, *Methyloccus capsulatus* was reported to bioxidize toluene to benzyl alcohol and cresol. Pure microbial cultures isolated from soil hydroxylated toluene to 2- and 4-hydroxytoluene. Synthetic air containing gaseous nitrous acid and exposed to artificial sunlight (λ = 300-450 nm) photooxidized toluene into methyl nitrate, peroxyacetal nitrate and a nitro aromatic compound tentatively identified as a nitrophenol or nitrocresol. A *n*-hexane soln containing toluene and spread as a thin film (4 mm) on cold water (10°) was irradiated by a mercury medium pressure lamp. In 3 h, 26% of the toluene photooxidized into benzaldehyde, benzyl alcohol, benzoic acid and *m*-cresol. Methane and ethane were reported as products of the gas-phase photolysis of toluene (λ = 2537 Å). Products identified from the reaction of toluene with NO and hydroxyl radicals include benzaldehyde, benzyl alcohol, *m*-nitrotoluene, *p*-methylbenzoquinone, *o*-, *m*- and *p*-cresol. Irradiation of toluene (80 ppm) on titanium dioxide in the presence of oxygen (20%) and moisture by UV light (λ = 200-300 nm) resulted in the formation of benzaldehyde and CO_2. CO_2 concentrations increased linearly with the increase in relative humidity. However, the concentration of benzaldehyde decreased with an increase in relative humidity. An identical experiment but without moisture resulted in the formation of benzaldehyde, CO_2, hydrogen cyanide and nitrotoluenes. In an atmosphere containing moisture and

NO_2 (80 ppm), cresols, benzaldehyde, CO_2 and nitrotoluenes were the photoirradiation products. Under atmospheric conditions, the gas-phase reaction with hydroxyl radicals and NO_x resulted in the formation of benzaldehyde, benzyl nitrate, *m*-nitrotoluene, *o*-, *m*- and *p*-cresol. Irradiation of toluene in the presence of chlorine yielded benzyl hydroperoxide, benzaldehyde, peroxybenzoic acid, CO, CO_2 and other unidentified products. The photooxidation of toluene in the presence of NO_x (NO and NO_2) yielded small amounts of formaldehyde and traces of acetaldehyde or other low molecular weight carbonyls. Other photooxidation products not previously mentioned include phenol, phthalaldehydes and benzoyl alcohol. A CO_2 yield of 8.4% was achieved when toluene adsorbed on silica gel was irradiated with light (λ > 290 nm) for 17 h. In activated sludge, 26.3% of the applied toluene mineralized to CO_2 after 5 days.

EXPOSURE LIMITS: NIOSH REL: 10-h TWA 100 ppm, 10-min C 200 ppm, IDLH 2,000 ppm; OSHA PEL: TWA 200 ppm, C 300 ppm, 10-min peak 500 ppm; ACGIH TLV: TWA 100 ppm, STEL 150 ppm.

SYMPTOMS OF EXPOSURE: Ftg, weak; conf, euph, dizz; head; dil pup, lac; ner; musc ftg; insom; pares; derm; photo.

USES: Manufacture of caprolactum, saccharin, medicines, dyes, perfumes, benzoic acid, trinitrotoluene (TNT), *o*-nitrotoluene, *p*-nitrotoluene, *o*-toluenesulfonic acid, *p*-toluenesulfonic acid, *o*-xylene, *p*-xylene, benzyl chloride, benzal chloride, benzotrichloride, halogenated toluenes and many other organic compounds; solvent for paints and coatings, gums, resins, rubber, oils and vinyl compounds; adhesive solvent in plastic toys and model airplanes; diluent and thinner for nitrocellulose lacquers; detergent manufacturing; aviation gasoline and high-octane blending stock; preparation of toluene diisocyanate for polyurethane resins.

2,4-TOLUENE DIISOCYANATE

SYNONYMS: Desmodur T80; **2,4-Diisocyanato-1-methylbenzene**; Diisocyanatoluene; 2,4-Diisocyanotoluene; Isocyanic acid methylphenylene ester; Isocyanic acid 4-methyl-*m*-phenylene ester; Hylene T; Hylene TCPA; Hylene TLC; Hylene TM; Hylene TM-65; Hylene TRF; 4-Methylphenylene diisocyanate; 4-Methylphenylene isocyanate; Mondur TD; Mondur TD-80; Mondur TDS; Nacconate 100; NCI-C50533; Niax TDI; Niax TDI-P; RCRA waste number U223; Rubinate TDI 80/20; TDI; 2,4-TDI; TDI-80; Toluene 2,4-diisocyanate; Toluylene-2,4-diisocyanate; Tolyene-2,4-diisocyanate; Tolylene-2,4-diisocyanate; 2,4-Tolylene diisocyanate; *m*-Tolylene diisocyanate.

CHEMICAL DESIGNATIONS: CAS: 584-84-9; DOT: 2078; mf: $C_9H_6N_2O_2$; fw: 174.15; RTECS: CZ 6300000.

PROPERTIES: Colorless to yellow liq with a pungent odor. Mp: 19.5-21.5°, bp: 251°, *d*: 1.2244 @ 20/4°, fl p: 127°, lel: 0.9%, uel: 9.5%, S_o: misc with acetone, alcohol (decomp), benzene, carbon tetrachloride, diglycol monomethyl ether, ether and kerosene and olive oil, vap d: 7.12 g/L @ 25°, 6.01 (air = 1), vp: 0.01 mm @ 20°.

TRANSFORMATION PRODUCTS: Slowly reacts with water releasing CO_2.

EXPOSURE LIMITS: NIOSH REL: 10-h TWA 35 μg/m^3, 10-min C 140 μg/m^3, IDLH 10 ppm; OSHA PEL: C 0.02 ppm; ACGIH TLV: TWA 0.005 ppm.

SYMPTOMS OF EXPOSURE: Irrit nose, throat; choke, parox cough; retster pain; nau, abdom pain; bronspas; pulm edema; asth; conj; derm.

USES: Manufacturing polyurethane foams and other plastics; cross-linking agent for nylon 6.

o-TOLUIDINE

SYNONYMS: 1-Amino-2-methylbenzene; 2-Amino-1-methylbenzene; 2-Aminotoluene; *o*-Aminotoluene; C.I. 37,077; 1-Methyl-2-aminobenzene; 2-Methyl-1-aminobenzene; 2-Methylaniline; *o*-Methylaniline; **2-Methylbenzenamine**; *o*-Methylbenzenamine; 2-Toluidine; *o*-Tolylamine; UN 1708.

CHEMICAL DESIGNATIONS: CAS: 95-53-4; DOT: 1708; mf: C_7H_9N; fw: 107.16; RTECS: XU 2975000.

PROPERTIES: Colorless liq becoming reddish-brown on exposure to air and light. Mp: -14.7°, bp: 200.2°, *d*: 0.9984 @ 20/4°, fl p: 85°, lel: 1.5%, pK_a: 4.45 @ 25°, K_H: 3.01 x 10^{-3} atm·m^3/mol @ 25° (calcd), IP: 7.6 eV, log K_{oc}: 2.61 (calcd), log K_{ow}: 1.29, 1.32, S_o: sol in alcohol and ether, S_w: 16.33 g/L @ 25°, vap d: 4.38 g/L @ 25°, 3.70 (air = 1), vp: 0.1 mm @ 20°.

EXPOSURE LIMITS: NIOSH REL: IDLH 100 ppm; OSHA PEL: TWA 5 ppm; ACGIH TLV: TWA 2 ppm.

SYMPTOMS OF EXPOSURE: Anoxia, head, cyan, weak, dizz, drow, micro hema, eye burns, derm.

USES: Manufacture of dyes; vulcanization accelerator; organic synthesis.

TOXAPHENE

SYNONYMS: Agricide maggot killer (F); Alltex; Alltox; Attac 4-2; Attac 4-4; Attac 6; Attac 6-3; Attac 8; Camphechlor; Camphochlor; Camphoclor; Chem-phene M5055; Chlorinated camphene; Chloro-camphene; Clor chem T-590; Compound 3,956; Crestoxo; Crestoxo 90; ENT 9,735; Estonox; Fasco-terpene; Geniphene; Gy-phene; Hercules 3,956; Hercules toxaphene; Huilex; Kamfochlor; M 5,055; Melipax; Motox; NA 2,761; NCI-C00259; Octachlorocamphene; PCC; Penphene; Phenacide; Phenatox; Phenphane; Polychlorcamphene; Polychlorinated camphenes; Polychlorocamphene; RCRA waste number P123; Strobane-T; Strobane T-90; Synthetic 3956; Texadust; Toxakil; Toxon 63; Toxyphen; Vertac 90%; Vertac toxaphene 90.

CHEMICAL DESIGNATIONS: CAS: 8001-35-2; DOT: 2761; mf: $C_{10}H_{10}Cl_8$; fw: 413.82; RTECS: XW 5250000.

PROPERTIES: Yellow, waxy solid with a chlorine-like odor. Mp: 65-90°, bp: > 120° (decomp), *d*: 1.6 @ 20/4°, fl p: 28.9° in soln, lel: 1.1% (in solvent), uel: 6.4% (in solvent), K_H: 6.3 x 10^{-2} atm·m^3/mol, log K_{oc}: 3.18 (calcd), log K_{ow}: 3.23-5.50,

S_o: very sol in most solvents, S_w: 550 μg/L @ 20°, vp: 0.2-0.4 mm @ 25°.

TRANSFORMATION PRODUCTS: Dehydrochlorination will occur after prolonged exposure to sunlight releasing hydrochloric acid. Two compounds isolated from toxaphene, 2-*exo*, 3-*exo*, 5,5,6-*endo*, 8,9,10,10-nonachloroborane and 2-*exo*, 3-*exo*, 5,5,6-*endo*, 8,10,10-octachloroborane were irradiated with UV light ($\lambda > 290$ nm) in a neutral aq soln and on a silica gel surface. Both compounds underwent reductive dechlorination, dehydrochlorination and/or oxidation to yield numerous products including bicyclo[2.1.1]hexane derivatives.

EXPOSURE LIMITS: NIOSH REL: IDLH 200 mg/m^3; OSHA PEL: TWA 0.5 mg/m^3; ACGIH TLV: STEL 1 mg/m^3 STEL.

SYMPTOMS OF EXPOSURE: Nau, conf, agitation, tremors, convuls, unconscious, dry red skin.

USES: Pesticide used primarily on cotton, lettuce, tomatoes, corn, peanuts, wheat and soybean.

1,3,5-TRIBROMOBENZENE

SYNONYM: *sym*-Tribromobenzene.

CHEMICAL DESIGNATIONS: CAS: 626-39-1; DOT: none assigned; mf: $C_6H_3Br_3$; fw: 314.80; RTECS: none assigned.

PROPERTIES: Solid. Mp: 121-122°, bp: 271° @ 765 mm, log K_{oc}: 4.05 (calcd), log K_{ow}: 4.51, S_o: sol in benzene, chloroform and ether, S_w: 2.51 x 10^{-6} M @ 25°.

USES: Organic synthesis.

TRIBUTYL PHOSPHATE

SYNONYMS: Celluphos 4; **Phosphoric acid tributyl ester**; TBP; Tri-*n*-butyl phosphate.

CHEMICAL DESIGNATIONS: CAS: 126-73-8; DOT: none assigned; mf: $C_{12}H_{27}O_4P$; fw: 266.32; RTECS: TC 7700000.

PROPERTIES: Colorless to pale yellow, odorless liq. Mp: < -80°, bp: 292°, *d*: 0.982 @ 20/4°, fl p: 146° (OC), log K_{oc}: 2.29 (calcd), log K_{ow}: 4.00, S_o: sol in benzene; misc with alcohol and ether, S_w: 280 ppm @ RT (commercial mixture), vap d: 10.89 g/L @ 25°, 9.19 (air = 1).

TRANSFORMATION PRODUCTS: Indigenous microbes in Mississippi River water degraded tributyl phosphate to CO_2. After 4 wk, 90.8% of the theoretical CO_2 had evolved.

EXPOSURE LIMITS: NIOSH REL: IDLH 130 mg/m^3; OSHA PEL: TWA 5 mg/m^3; ACGIH TLV: TWA 0.2 ppm.

SYMPTOMS OF EXPOSURE: Eyes, resp, skin irrithead, nau.

USES: Plasticizer for lacquers, plastics, cellulose esters and vinyl resins; heat-exchange liquid; solvent extraction of metal ions from soln of reactor products; pigment grinding assistant; antifoaming agent; solvent for nitrocellulose and cellulose acetate.

1,2,3-TRICHLOROBENZENE

SYNONYMS: 1,2,3-TCB; UN 2321.

CHEMICAL DESIGNATIONS: CAS: 87-61-6; DOT: 2321 (liq); mf: $C_6H_3Cl_3$; fw: 181.45; RTECS: DC 2095000.

PROPERTIES: White cryst or platelets. Mp: 53-54°C, bp: 218-219°, *d*: 1.69, fl p: 113°, K_H: 8.9 x 10^{-3} atm·m^3/mol @ 20°, log K_{oc}: 3.87 (calcd), log K_{ow}: 4.04-4.11, S_o: sol in benzene and ether, S_w: 12 ppm @ 22°, vp: 1 mm @ 40.0°.

TRANSFORMATION PRODUCTS: Under aerobic conditions, soil microbes are capable of degrading 1,2,3-trichlorobenzene to 1,3- and 2,3-dichlorobenzene and CO_2. A mixed culture of soil bacteria or a *Pseudomonas* sp. transformed 1,2,3-trichlorobenzene to 2,3,4-, 3,4,5- and 2,3,6-trichlorophenol. The sunlight irradiation of 1,2,3-trichlorobenzene (20 g) in a 100 ml borosilicate glass-stoppered Erlenmeyer flask for 56 days yielded 32 ppm pentachlorobiphenyl.

USES: The isomeric mixture is used to control termites; organic synthesis.

1,2,4-TRICHLOROBENZENE

SYNONYMS: 1,2,4-TCB; *unsym*-Trichlorobenzene; UN 2321.

CHEMICAL DESIGNATIONS: CAS: 120-82-1; DOT: 2321; mf: $C_6H_3Cl_3$; fw: 181.45; RTECS: DC 2100000.

PROPERTIES: Colorless liq with an odor similar to *o*-dichlorobenzene. Mp: 17°, bp: 213.5°, *d*: 1.4542 @ 20/4°, fl p: 105°, lel: 2.5% @ 150°, uel: 6.6% @ 150°, K_H: 1.42 x 10^{-3} atm·m^3/mol @ 25°, log K_{oc}: 2.94-4.61, log K_{ow}: 3.93-4.23, S_o: sol in ether and oils, S_w: 19 ppm @ 22°, vap d: 7.42 g/L @ 25°, 6.26 (air = 1), vp: 0.4 mm @ 25°.

TRANSFORMATION PRODUCTS: Under aerobic conditions, biodegradation products may include 2,3-dichlorobenzene, 2,4-dichlorobenzene, 2,5-dichlorobenzene, 2,6-dichlorobenzene and CO_2. A CO_2 yield of 9.8% was achieved when 1,2,4-trichlorobenzene adsorbed on silica gel was irradiated with light ($\lambda > 290$ nm) for 17 h. In activated sludge, < 0.1% mineralized to CO_2 after 5 days. A mixed culture of soil bacteria or a *Pseudomonas* sp. transformed 1,2,4-trichlorobenzene to 2,4,5- and 2,4,6-trichlorophenol. The sunlight irradiation of 1,2,4-trichlorobenzene (20 g) in a 100 ml borosilicate glass-stoppered Erlenmeyer flask for 56 days yielded 9,770 ppm 2,4,5,2′,5′-pentachlorobiphenyl.

EXPOSURE LIMITS: ACGIH TLV: TWA 5 ppm.

USES: Solvent in chemical manufacturing; dyes and intermediates; dielectric fluid; synthetic transformer oils; lubricants; heat-transfer medium; insecticides.

1,3,5-TRICHLOROBENZENE

SYNONYMS: 1,3,5-TCB; *sym*-Trichlorobenzene; UN 2321.

CHEMICAL DESIGNATIONS: CAS: 108-70-3; DOT: 2321 (liq); mf: $C_6H_3Cl_3$; fw: 181.45; RTECS: DC 2100100.

PROPERTIES: Cryst. Mp: 63-64°, bp: 208° @ 763 mm, fl p: 107°, K_H: 1.9 x 10^{-3} atm·m^3/mol @ 20°, log K_{oc}: 2.85, log K_{ow}: 3.93-4.40, S_o: sol in benzene, ether, lignoin, glacial acetic acid, carbon disulfide and petroleum ether, S_w: 5.8 ppm @ 20°, vp: 0.58 mm @ 25°.

TRANSFORMATION PRODUCTS: Under aerobic conditions, soil microbes are capable of degrading 1,3,5-trichlorobenzene to 1,4- and 2,4-dichlorobenzene and CO_2. A mixed culture of soil bacteria or a *Pseudomonas* sp. transformed 1,3,5-trichlorobenzene to 2,4,6-trichlorophenol. The sunlight irradiation of 1,3,5-trichlorobenzene (20 g) in a 100 ml borosilicate glass-stoppered Erlenmeyer flask for 56 days yielded 160 ppm pentachlorobiphenyl.

USES: Organic synthesis.

1,1,1-TRICHLOROETHANE

SYNONYMS: Aerothene; Aerothene TT; Baltana; Chloroethene; Chloroethene NU; Chlorothane NU; Chlorothene; Chlorothene NU; Chlorothene VG; Chlorten; Genklene; Inhibisol; Methyl chloroform; Methyltrichloromethane; NCI-C04626; RCRA waste number U226; Solvent III; α-T; 1,1,1-TCA; 1,1,1-TCE; α-Trichloroethane; Tri-ethane; UN 2831.

CHEMICAL DESIGNATIONS: CAS: 71-55-6; DOT: 2831; mf: $C_2H_3Cl_3$; fw: 133.40; RTECS: KJ 2975000.

PROPERTIES: Colorless, watery liq with an odor similar to chloroform. Mp: -30.4°, bp: 74.1°, *d*: 1.3390 @ 20/4°, fl p: ≤ 25°, lel: 7.5%, uel: 12.5%, K_H: 1.5 x 10^{-2} atm·m^3/mol @ 20°, log K_{oc}: 2.017, 2.18, log K_{ow}: 2.18-2.49, S_o: sparingly sol in ethanol; freely sol in carbon disulfide, benzene, ether, methanol and carbon tetrachloride; misc in alcohol and chloroform, S_w: 480-1,360 mg/L @ 20°, vap d: 5.45 g/L @ 25°, 4.60 (air = 1), vp: 90 mm @ 20°.

TRANSFORMATION PRODUCTS: Microbial degradation by sequential dehalogenation under laboratory conditions produced 1,1-dichloroethane, *cis*-1,2-dichloroethylene, *trans*-1,2-dichloroethylene, chloroethane and vinyl chloride. Hydrolysis products via dehydrohalogenation included acetic acid, 1,1-dichloroethylene and hydrochloric acid. The reported half-lives for this reaction @ 20 and 25° are 0.5-2.5 yr and 1.1 yr, respectively. In an anoxic aquifer beneath a landfill in Ottawa, Ontario, Canada, there was evidence that 1,1,1-trichloroethane was biotransformed to 1,1-dichloroethane, 1,1-dichloroethylene and vinyl chloride. An anaerobic species of *Clostridium* biotransformed 1,1,1-trichloroethane to 1,1-dichloroethane, acetic acid and unidentified products. Reported photooxidation products include phosgene, chlorine, hydrochloric acid and CO_2. Acetyl chloride and trichloroacetaldehyde have also been reported as photooxidation products. 1,1,1-Trichloroethane may react with hydroxyl radicals in the atomosphere producing chlorine atoms and chlorine oxides. A microcosm composed of aquifer water and sediment collected from uncontaminated sites in the Everglades biotransformed 1,1,1-trichloroethane to 1,1-dichloroethylene.

EXPOSURE LIMITS: NIOSH REL: 15-min C 350 ppm; OSHA PEL: TWA 350 ppm; ACGIH TLV: TWA 350 ppm, STEL 450 ppm.

SYMPTOMS OF EXPOSURE: Head, lass, CNS depres, poor equi, irrit eyes, derm, card arrhy.

USES: Organic synthesis; solvent for metal cleaning of precision instruments; textile processing; aerosol propellants; pesticide.

1,1,2-TRICHLOROETHANE

SYNONYMS: Ethane trichloride; NCI-C04579; RCRA waste number U227; β-T; 1,1,2-TCA; 1,2,2-Trichloroethane; β-Trichloroethane; Vinyl trichloride.

CHEMICAL DESIGNATIONS: CAS: 79-00-5; DOT: 2831; mf: $C_2H_3Cl_3$; fw: 133.40; RTECS: KJ 3150000.

PROPERTIES: Colorless liq with a pleasant odor. Mp: -36.5°, bp: 113.8°, *d*: 1.4397 @ 20/4°, lel: 6%, uel: 15.5%, K_H: 9.09 x 10^{-4} atm·m^3/mol @ 25°, log K_{oc}: 1.75, log K_{ow}: 2.18, S_o: sol in ethanol and chloroform, S_w: 4,500 mg/L @ 20°, vap d: 5.45 g/L @ 25°, 4.60 (air = 1), vp: 19 mm @ 20°.

TRANSFORMATION PRODUCTS: If abiotic dehydrohalogenation occurs, the products would be 1,1-dichloroethylene and hydrochloric acid ($t_{1/2}$ = 170 yr @ 20°). Vinyl chloride was reported to be a biodegradation product from an anaerobic digester at a wastewater treatment facility. Under aerobic conditions, *Pseudomonas putida* oxidized 1,1,2-trichloroethane to chloroacetic and glyoxylic acids. Simultaneously, 1,1,2-trichloroethane is reduced to vinyl chloride exclusively. Under alkaline conditions, 1,1,2-trichloroethane hydrolyzed to 1,2-dichloroethylene ($t_{1/2}$ = 139.2 yr @ 25° and pH 7).

EXPOSURE LIMITS: OSHA PEL: TWA 10 ppm.

SYMPTOMS OF EXPOSURE: Irrit eyes, nose; CNS depres; liv and kid damage.

USES: Solvent for fats, oils, resins, waxes, resins and other products; organic synthesis.

TRICHLOROETHYLENE

SYNONYMS: Acetylene trichloride; Algylen; Anamenth; Benzinol; Blacosolv; Blancosolv; Cecolene; Chlorilen; 1-Chloro-2,2-dichloroethylene; Chlorylea; Chlorylen; Circosolv; Crawhaspol; Densinfluat; 1,1-Dichloro-2-chloroethylene; Dow-tri; Dukeron; Ethinyl trichloride; Ethylene trichloride; Fleck-flip; Flock-flip; Fluate; Gemalgene; Germalgene; Lanadin; Lethurin; Narcogen; Narkogen; Narkosoid; NCI-C04546; Nialk; Perm-a-chlor; Perm-a-clor; Petzinol; Philex; RCRA waste number U228; TCE; Threthylen; Threthylene; Trethylene; Tri; Triad; Trial; Triasol; Trichloran; Trichloren; **Trichloroethene**; 1,1,2-Trichloroethene; 1,2,2-Trichloroethene; 1,1,2-Trichloroethylene; 1,2,2-Trichloroethylene; Tri-clene; Trielene; Trieline; Triklone; Trilen; Trilene; Triline; Trimar; Triol; Tri-plus; Tri-plus M; UN 1710; Vestrol; Vitran; Westrosol.

CHEMICAL DESIGNATIONS: CAS: 79-01-6; DOT: 1710; mf: C_2HCl_3; fw: 131.39; RTECS: KX 4550000.

PROPERTIES: Clear, colorless, liq with a chloroform-like odor. Mp: -84.8°, bp: 87.2°, *d*: 1.4642 @ 20/4°, fl p: 32.2°, lel: 8% @ 25°, 7.8% @ 100°, uel: 10.5% @ 25°,

52% @ 100°, K_H: 9.9 x 10^{-3} atm·m^3/mol @ 20°, IP: 9.45 eV, 9.47 eV, 9.94 eV, log K_{oc}: 1.81-2.15, log K_{ow}: 2.29-3.30, S_o: sol in acetone, ethanol, chloroform and ether, S_w: 1,100 mg/L @ 20°, vap d: 5.37 g/L @ 25°, 4.54 (air = 1), vp: 57.8 mm @ 20°.

TRANSFORMATION PRODUCTS: Microbial degradation of trichloroethylene via sequential dehalogenation produced *cis*- and *trans*-1,2-dichloroethylene and vinyl chloride. Anoxic microcosms in sediment and water degraded trichloroethylene to 1,2-dichloroethylene and then to vinyl chloride. Trichloroethylene in soil samples collected from Des Moines, IA anaerobically degraded to 1,2-dichloroethylene. In a methanogenic aquifer, trichloroethylene biodegraded to 1,2-dichloroethylene and vinyl chloride. Dichloroethylene was reported as a biotransformation product under anaerobic conditions and in experimental systems using mixed or pure cultures. Under aerobic conditions, CO_2 was the principal degradation product in experiments containing pure and mixed cultures. A microcosm composed of aquifer water and sediment collected from uncontaminated sites in the Everglades biotransformed trichloroethylene to *cis*- and *trans*-1,2-dichloroethylene. Trichloroethylene biodegraded to dichloroethylene in both anaerobic and aerobic soils of different soil types. Under anaerobic conditions, nonmethanogenic fermenters and methanogens degraded trichloroethylene to chloroethane, methane, 1,1-dichloroethylene, 1,2-dichloroethylene and vinyl chloride. Titanium dioxide suspended in an aq soln and irradiated with UV light (λ = 365 nm) converted trichloroethylene to CO_2 at a significant rate. Under smog conditions, indirect photolysis via hydroxyl radicals will yield phosgene, dichloroacetyl chloride and formyl chloride. These compounds are readily hydrolyzed to hydrochloric acid, CO, CO_2 and dichloroacetic acid. Though not identified, dichloroacetic acid and hydrogen chloride were reported to be aq photodecomposition products. An aq soln containing 300 ng/μL trichloroethylene and a colloidal platinum catalyst was irradiated with UV light. After 12 h, 7.4 ng/μL trichloroethylene and 223.9 ng/μL ethane was detected. A duplicate experiment was performed but 1 g zinc was added to the system. After 5 h, 259.9 ng/μL ethane was formed and trichloroeth-ylene was non-detectable. The reported hydrolysis $t_{1/2}$ @ 25° and pH 7 is 1.3 x 10^{-6} yr.

EXPOSURE LIMITS: NIOSH REL: 10-h TWA 25 ppm; OSHA PEL: TWA 100 ppm, C 200 ppm, 5-min/2-h peak 300 ppm; ACGIH TLV: TWA 50 ppm, STEL 200 ppm.

SYMPTOMS OF EXPOSURE: Head, verti, vis dist, tremors, somnolence, nau, vomit, irrit eyes, derm, card arrhy; pares.

USES: Dry cleaning fluid; degreasing and drying metals and electronic parts; extraction solvent for oils, waxes and fats; solvent for cellulose esters and ethers; removing caffeine from coffee; refrigerant and heat exchange liquid; fumigant; diluent in paints and adhesives; textile processing; aerospace operations (flushing liquid oxygen); anesthetic; medicine; organic synthesis.

TRICHLOROFLUOROMETHANE

SYNONYMS: Algofrene type 1; Arcton 9; Electro-CF 11; Eskimon 11; F 11; FC

11; Fluorocarbon 11; Fluorotrichloromethane; Freon 11; Freon 11A; Freon 11B; Freon HE; Freon MF; Frigen 11; Genetron 11; Halocarbon 11; Isceon 11; Isotron 11; Ledon 11; Monofluorotrichloromethane; NCI-C04637; RCRA waste number U121; Refrigerant 11; Trichloromonofluoromethane; Ucon 11; Ucon fluorocarbon 11; Ucon refrigerant 11.

CHEMICAL DESIGNATIONS: CAS: 75-69-4; DOT: 1078; mf: CCl_3F; fw: 137.37; RTECS: PB 6125000.

PROPERTIES: Colorless, odorless liq. Mp: -111°, bp: 23.63°, *d*: 1.487 @ 20/4°, K_H: 5.83 x 10^{-3} atm·m^3/mol @ 25°, IP: 11.77 eV, log K_{oc}: 2.20, 2.13, log K_{ow}: 2.53, S_o: sol in ethanol, ether and other solvents, S_w: 1,100 mg/L @ 20°, vap d: 5.61 g/L @ 25°, 4.74 (air = 1), vp: 687 mm @ 20°.

EXPOSURE LIMITS: NIOSH REL: IDLH 10,000 ppm; OSHA PEL: TWA 1,000 ppm; ACGIH TLV: C 1,000 ppm.

SYMPTOMS OF EXPOSURE: Inco, tremors, derm, frost, card arrhy, card arrest.

USES: Aerosol propellant; refrigerant; solvent; blowing agent for polyurethane foams; fire extinguishing; chemical intermediate; organic synthesis.

2,4,5-TRICHLOROPHENOL

SYNONYMS: Collunosol; Dowicide 2; Dowicide B; NCI-C61187; Nurelle; Phenachlor; Preventol I; RCRA waste number U230; 2,4,5-TCP; 2,4,5-TCP-Dowicide 2.

CHEMICAL DESIGNATIONS: CAS: 95-95-4; DOT: 2020; mf: $C_6H_3Cl_3O$; fw: 197.45; RTECS: SN 1400000.

PROPERTIES: Colorless cryst or yellow to gray flakes with a strong disinfectant or phenolic odor. Mp: 57°, bp: 252°, *d*: 1.678 @ 25/4°, pK_a: 7.37 @ 25°, K_H: 1.76 x 10^{-7} atm·m^3/mol @ 25° (est), log K_{oc}: 2.56, log K_{ow}: 3.72-4.19, S_o: sol in ethanol and ligroin, S_w: 1,190 mg/kg @ 25°, vp: 2.2 x 10^{-2} mm @ 25°.

TRANSFORMATION PRODUCTS: Chloroperoxidase, a fungal enzyme isolated from *Caldariomyces fumago*, chlorinated 2,4,5-trichlorophenol to give 2,3,4,6-tetrachlorophenol. When 2,4,5-trichlorophenol (10^{-4} M) in an oxygenated, titanium dioxide (2 g/L) suspension was irradiated by sunlight ($\lambda \geq 340$ nm), complete mineralization to CO_2 and water and chloride ions was observed.

USES: Fungicide; bactericide; organic synthesis.

2,4,6-TRICHLOROPHENOL

SYNONYMS: Dowicide 2S; NCI-C02904; Omal; Phenachlor; RCRA waste number F027; 2,4,6-TCP; 2,4,6-TCP-Dowicide 25.

CHEMICAL DESIGNATIONS: CAS: 88-06-2; DOT: 2020; mf: $C_6H_3Cl_3O$; fw: 197.45; RTECS: SN 1575000.

PROPERTIES: Colorless needles or yellow solid with a strong phenolic odor. Mp: 69.5°, bp: 246°, *d*: 1.4901 @ 75/4°, pK_a: 7.42 @ 25°, K_H: 9.07 x 10^{-8}

atm·m^3/mol @ 25° (est), log K_{oc}: 3.03, log K_{ow}: 2.80-3.72, S_o: sol in ethanol and ether, S_w: 800 mg/L @ 25°, vp: 1.7 x 10^{-2} mm @ 25°.

TRANSFORMATION PRODUCTS: Titanium dioxide suspended in an aq soln and irradiated with UV light (λ = 365 nm) converted 2,4,6-trinitrophenol to CO_2 at a significant rate. An aq soln containing chloramine reacted with 2,4,6-trichlorophenol to yield the following intermediate products after 2 h @ 25°: 2,6-dichloro-1,4-benzoquinone-4-(*N*-chloro)imine and 4,6-dichloro-1,2-benzoquinone-2-(*N*-chloro)imine. In anaerobic sludge, 2,4,6-trichlorophenol degraded to *p*-chlorophenol. A CO_2 yield of 65.8% was achieved when 2,4,6-trichlorophenol adsorbed on silica gel was irradiated with light (λ > 290 nm) for 17 h. In activated sludge, only 0.3% mineralized to CO_2 after 5 days.

USES: Manufacture of fungicides, bactericides, antiseptics, germicides; wood and glue preservatives; in textiles to prevent mildew; defoliant; disinfectant; organic synthesis.

1,2,3-TRICHLOROPROPANE

SYNONYMS: Allyl trichloride; Glycerin trichlorohydrin; Glycerol trichlorohydrin; Glyceryl trichlorohydrin; NCI-C60220; Trichlorohydrin.

CHEMICAL DESIGNATIONS: CAS: 96-18-4; DOT: none assigned; mf: $C_3H_5Cl_3$; fw: 147.43; RTECS: TZ 9275000.

PROPERTIES: Colorless liq with a strong odor. Mp: -14.7°, bp: 156.8°, *d*: 1.3889 @ 20/4°, fl p: 73.3°, lel: 3.2%, uel: 12.6%, K_H: 3.18 x 10^{-4} atm·m^3/mol @ 25° (calcd), S_o: sol in alcohol, chloroform and ether, vap d: 6.03 g/L @ 25°, 5.09 (air = 1), vp: 3.4 mm @ 20°.

EXPOSURE LIMITS: NIOSH REL: IDLH 1,000 ppm; OSHA PEL: TWA 50 ppm; ACGIH TLV: 10 ppm.

SYMPTOMS OF EXPOSURE: Irrit eyes, throat; CNS depres; liv inj; skin irrit.

USES: Solvent; degreaser; paint and varnish removers.

1,1,2-TRICHLOROTRIFLUOROETHANE

SYNONYMS: Fluorocarbon 113; Freon 113; Frigen 113 TR-T; Halocarbon 113; Kaiser chemicals 11; R-113; Refrigerant 113; 1,1,2-Trichloro-1,2,2-trifluoroethane; Trichlorotrifluoroethane; 1,1,2-Trifluoro-1,2,2-trichloroethane; TTE; Ucon-113; Ucon fluorocarbon 113; Ucon 113/halocarbon 113.

CHEMICAL DESIGNATIONS: CAS: 76-13-1; DOT: 1078; mf: $C_2Cl_3F_3$; fw: 187.38; RTECS: KJ 4000000.

PROPERTIES: Colorless liq. Mp: -36.4°, bp: 47.7°, *d*: 1.5635 @ 20/4°, K_H: 3.33 x 10^{-1} atm·m^3/mol @ 20° (calcd), IP: 11.99 eV, log K_{oc}: 2.59 (calcd), log K_{ow}: 2.57, S_o: sol in alcohol, benzene and ether, S_w: 136 mg/L @ 10°, vap d: 7.66 g/L @ 25°, 6.47 (air = 1), vp: 270 mm @ 20°.

TRANSFORMATION PRODUCTS: In an anoxic aquifer beneath a landfill in Ottawa, Ontario, Canada, there was evidence to suggest that 1,1,2-trichlorotrifluoroethane underwent reductive dehalogenation to give 1,2-difluoro-1,1,2-trichloroethylene and 1,2-dichloro-1,1,2-trifluoroethane. It was proposed that the latter compound was further degraded via dehydrodehalogenation to give 1-chloro-1,1,2-trifluoroethylene.

EXPOSURE LIMITS: NIOSH REL: IDLH 4,500 ppm; OSHA PEL: TWA 1,000 ppm.

SYMPTOMS OF EXPOSURE: Irrit throat, drow, derm.

USES: Fire extinguishers; dry-cleaning solvent; manufacture of chlorotrifluoroethylene; polymer intermediate; blowing agent; drying electronic parts and precision equipment; solvent drying.

TRI-*o*-CRESYL PHOSPHATE

SYNONYMS: *o*-Cresyl phosphate; Phosflex 179-C; **Phosphoric acid tris(2-methylphenyl) ester**; Phosphoric acid tri-2-tolyl ester; Phosphoric acid tri-*o*-cresyl ester; TCP; TOCP; TOFK; *o*-Tolyl phosphate; TOTP; Tricresyl phosphate; Triorthocresyl phosphate; Tri-2-methylphenyl phosphate; Tris(*o*-cresyl)phosphate; Tris(*o*-methylphenyl)phosphate; Tris(*o*-tolyl)phosphate; Tri-2-tolyl phosphate; Tri-*o*-tolyl phosphate.

CHEMICAL DESIGNATIONS: CAS: 78-30-8; DOT: 2574; mf: $C_{21}H_{21}O_4P$; fw: 368.37; RTECS: TD 0350000.

PROPERTIES: Colorless liq. Mp: -30°, bp: 410°, *d*: 1.955 @ 20/4°, fl p: 225°, log K_{oc}: 3.37 (calcd), log K_{ow}: 5.11 (commercial mixture containing tricresyl phosphates), S_o: sol in acetic acid, alcohol, ether and toluene, S_w: 3.1 ppm @ 25°, vap d: 15.06 g/L @ 25°, 12.72 (air = 1).

TRANSFORMATION PRODUCTS: Tri-*o*-cresyl phosphate hydrolyzed rapidly in Lake Ontario water presumably to di-*o*-cresyl phosphate. A commercial mixture containing tricresyl phosphates was completely degraded by indigenous microbes in Mississippi River to CO_2. After 4 wk, 82.1% of the theoretical CO_2 had evolved. When an aq soln containing a mixture of isomers (0.1 mg/L) and chlorine (3-1,000 mg/L) was stirred in the dark @ 20° for 24 h, the benzene ring was substituted with 1-3 chlorine atoms.

EXPOSURE LIMITS: NIOSH REL: IDLH 40 mg/m^3; OSHA PEL: TWA 0.1 mg/m^3.

SYMPTOMS OF EXPOSURE: GI; peripheral neuropathy; cramps in calves; pores in feet or hands; weak, bilat, foot or wrist drop, para.

USES: Plasticizer in lacquers, varnishes, polyvinylchloride, polystyrene, nitrocellulose; waterproofing; hydraulic fluid and heat exchange medium; fire retardant for plastics; solvent mixtures; additive to extreme pressure lubricants.

TRIETHYLAMINE

SYNONYMS: Diethylaminoethane; ***N,N*-Diethylethanamine**; TEN; UN 1296.

CHEMICAL DESIGNATIONS: CAS: 121-44-8; DOT: 1296; mf: $C_6H_{15}N$; fw:

101.19; RTECS: YE 0175000.

PROPERTIES: Colorless liq with a strong ammoniacal odor. Mp: -114.7°, bp: 89.3°, *d*: 0.7275 @ 20/4°, fl p: -6.7° (OC), lel: 1.2%, uel: 8.0%, pK_a: 10.72 @ 25°, K_H: 4.79 x 10^{-4} atm·m^3/mol @ 20° (calcd), IP: 7.50 eV, log K_{ow}: 1.44, 1.45, S_o: sol in acetone and benzene; misc with alcohol and ether, S_w: 15 g/L @ 20°, vap d: 4.14 g/L @ 25°, 3.49 (air = 1), vp: 54 mm @ 20°.

TRANSFORMATION PRODUCTS: Triethylamine reacted with NO_x in the dark forming diethylnitrosamine. In an outdoor chamber, photooxidation by natural sunlight yielded the following products: diethylnitramine, diethylformamide, diethylacetamide, ethylacetamide, diethylhydroxylamine, O_3, acetaldehyde and peroxyacetylnitrate.

EXPOSURE LIMITS: NIOSH REL: IDLH 1,000 ppm; OSHA PEL: TWA 25 ppm; ACGIH TLV: TWA 10 ppm, STEL 15 ppm.

SYMPTOMS OF EXPOSURE: Irrit eyes, resp sys, skin.

USES: Curing and hardening polymers; catalytic solvent in chemical synthesis; accelerator activators for rubber; wetting, penetrating and waterproofing agents of quaternary ammonium types; corrosion inhibitor; propellant.

TRIFLURALIN

SYNONYMS: Agreflan; Agriflan 24; Crisalin; Digermin; 2,6-Dinitro-*N,N*-dipropyl-4-trifluoromethylaniline; **2,6-Dinitro-*N,N*-dipropyl-4-(trifluoromethyl)benzenamine**; 2,6-Dinitro-*N,N*-di-*n*-propyl-α,α,α-trifluoro-*p*-toluidine; 4-(Di-*n*-propylamino)-3,5-dinitro-1-trifluoromethylbenzene; *N,N*-Di-*n*-propyl-2,6-dinitro-4-trifluoromethylaniline; *N,N*-Dipropyl-4-trifluoromethyl-2,6-dinitroaniline; Elancolan; L-36,352; Lilly 36,352; NCI-C00442; Nitran; Olitref; Trefanocide; Treficon; Treflam; Treflan; Treflanocide elancolan; Trifluoralin; α,α,α-Trifluoro-2,6-dinitro-*N,N*-dipropyl-*p*-toluidine; Trifluraline; Triflurex; Trikepin; Trim.

CHEMICAL DESIGNATIONS: CAS: 1582-09-8; DOT: none assigned; mf: $C_{13}H_{16}F_3N_3O_4$; fw: 335.29; RTECS: XU 9275000.

PROPERTIES: Yellow cryst. Mp: 46-47°, bp: 139-140° @ 4.2 mm, K_H: 4.84 x 10^{-5} atm·m^3/mol @ 23°, log K_{oc}: 2.94-4.49, log K_{ow}: 5.28, 5.34, S_o: freely sol in Stoddard solvent, acetone, chloroform, methanol, ether and ethanol, S_w: 4 ppm @ 27°, vp: 1.1 x 10^{-4} mm @ 25°.

TRANSFORMATION PRODUCTS: Irradiation of trifluralin in *n*-hexane by laboratory light produced α,α,α-trifluoro-2,6-dinitro-*N*-propyl-*p*-toluidine and α,α,α-trifluoro-2,6-dinitro-*p*-toluidine. Anaerobic degradation in a Crowley silt loam yielded α,α,α-trifluoro-N^4,N^4-dipropyl-5-nitrotoluene-3,4-diamine and α,α,α-trifluoro-N^4,N^4-dipropyltoluene-3,4,5-triamine. In addition to the 2 photoproducts mentioned above, the sunlight irradiation of trifluralin in water yielded 2-ethyl-7-nitro-5-trifluoromethylbenzimidazole, 2,3-dihydroxy-2-ethyl-7-nitro-1-propyl-5-trifluoromethylbenzimidazoline and 2-ethyl-7-nitro-5-trifluoromethylbenzimidazole. 2-Amino-6-nitro-α,α,α-trifluoro-*p*-toluidine and 2-ethyl-5-nitro-7-trifluoromethylbenzimidazole also were reported as major products under acidic and basic conditions, respectively. When trifluralin was released in the atmosphere on a sunny day, it was rapidly converted to the photochemical 2,6-dinitro-*N*-propyl-α,α,α-trifluoro-*p*-toluidine (est $t_{1/2}$ = 20 min). ^{14}C-labeled trifluralin on a silica plate was exposed to summer sunlight 7.5 h. Although 52% of trifluralin was recovered, no photodegradation products were identified. In soil, no significant photodegradation of trifluralin was observed after 9 h of irradiation. Reported degradation products in aerobic soils include α,α,α-trifluoro-2,6-dinitro-*N*-propyl-*p*-toluidine, α,α,α-trifluoro-2,6-dinitro-*p*-toluidine, α,α,α-trifluoro-5-nitrotoluene-3,4-diamine and α,α,α-trifluoro-*N,N*-dipropyl-5-nitrotoluene-3,4-diamine. Anaerobic degradation products identified include α,α,α-trifluoro-*N,N*-dipropyl-5-nitrotoluene-3,4-diamine, α,α,α-trifluoro-*N,N*-dipropyltoluene-3,4,5-triamine, α,α,α-trifluorotoluene-3,4,5-triamine and α,α,α-trifluoro-*N*-propyltoluene-3,4,5-triamine. α,α,α-Trifluoro-5-nitro-*N*-propyl-toluene-3,4-diamine was identified in both aerobic and anaerobic soils. The following compounds were reported as major soil metabolites: α,α,α-trifluoro-2,6-dinitro-*N*-propyl-*p*-toluidine, α,α,α-trifluoro-2,6-dinitro-*p*-toluidine, α,α,α-trifluoro-5-nitrotoluene-3,4-diamine, α,α,α-trifluorotoluene-3,4,5-triamine, 2-ethyl-7-nitro-1-propyl-5-(trifluoromethyl)benzimidazole, 2-ethyl-7-nitro-5-(trifluoromethyl)benzimidazole, 7-nitro-1-propyl-5-(trifluoromethyl)benzimidazole, 4-(dipropylamino)-3,5-dinitrobenzoic acid, 2,2′-azoxybis(α,α,α-trifluoro-6-nitro-*N*-propyl-*p*-toluidine), 2,2′-azobis(α,α,α-trifluoro-6-nitro-*N*-propyl-*p*-toluidine), 2,6-dinitro-*N,N*-dipropyl-4-(trifluoromethyl)-*m*-anisidine and α,α,α-trifluoro-2′,6′-dinitro-*N*-propyl-*p*-propionotoluidine.

USE: Herbicide.

1,2,3-TRIMETHYLBENZENE

SYNONYM: Hemimellitene.

CHEMICAL DESIGNATIONS: CAS: 526-73-8; DOT: none assigned; mf: C_9H_{12}; fw: 120.19; RTECS: DC 3300000.

PROPERTIES: Colorless liq. Mp: -25.4°, bp: 176.1°, *d*: 0.8944 @ 20/4°, fl p: 53° (90.5% soln), K_H: 3.18 x 10^{-3} atm·m^3/mol @ 25° (calcd), log K_{oc}: 3.34 (calcd), log K_{ow}: 3.55, 3.66, S_o: sol in acetone, alcohol, benzene and ether, S_w: 75.2 ppm @ 25°, vap d: 4.91 g/L @ 25°, 4.15 (air = 1), vp: 1.51 mm @ 25°.

TRANSFORMATION PRODUCTS: Glyoxal, methylglyoxal and biacetyl were produced from the photooxidation of 1,2,3-trimethylbenzene by hydroxyl radicals in air @ 25°.

USES: Organic synthesis.

1,2,4-TRIMETHYLBENZENE

SYNONYMS: *asym*-Trimethylbenzene; Pseudocumene; Pseudocumol.

CHEMICAL DESIGNATIONS: CAS: 95-63-6; DOT: none assigned; mf: C_9H_{12}; fw: 120.19; RTECS: DC 3325000.

PROPERTIES: Colorless liq. Mp: -43.8°, bp: 169.3°, *d*: 0.8758 @ 20/4°, fl p: 54.4°, lel: 0.9%, uel: 6.4%, K_H: 5.7 x 10^{-3} atm·m^3/mol @ 25°, log K_{oc}: 3.57 (calcd), log K_{ow}: 3.78, S_o: sol in acetone, alcohol, benzene and ether, S_w: 57 ppm @ 20°, vap d: 4.91 g/L @ 25°, 4.15 (air = 1), vp: 2.03 mm @ 25°.

TRANSFORMATION PRODUCTS: Glyoxal, methylglyoxal and biacetyl were produced from the photooxidation of 1,2,4-trimethylbenzene by hydroxyl radicals in air @ 25°. In anoxic groundwater near Bemidji, MI, 1,2,4-trimethylbenzene anaerobically biodegraded to the intermediate 3,4-dimethylbenzoic acid and the tentatively identified compounds 2,4- and/or 2,5-dimethylbenzoic acid.

USES: Manufacture of dyes, resins, perfumes, trimellitic anhydride, pseudocumidine.

1,3,5-TRIMETHYLBENZENE

SYNONYMS: Mesitylene; Fleet-X; TMB; *sym*-Trimethylbenzene; Trimethylbenzol; UN 2325.

CHEMICAL DESIGNATIONS: CAS: 108-67-8; DOT: 2325; mf: C_9H_{12}; fw: 120.19; RTECS: OX 6825000.

PROPERTIES: Colorless liq with a peculiar odor. Mp: -44.7°, bp: 164.7°, *d*: 0.8652 @ 20/4°, fl p: 50°, K_H: 3.93 x 10^{-3} atm·m^3/mol @ 25° (calcd), log K_{oc}: 3.21 (calcd), log K_{ow}: 3.42, S_o: sol in acetone; misc with alcohol, benzene and ether, S_w: 97.0 mg/L @ 25°, vap d: 4.91 g/L @ 25°, 4.15 (air = 1), vp: 2.42 mm @ 25°.

TRANSFORMATION PRODUCTS: Glyoxal, methylglyoxal and biacetyl were produced from the photooxidation of 1,3,5-trimethylbenzene by hydroxyl radicals in air @ 25°. In anoxic groundwater near Bemidji, MI, 1,3,5-trimethylbenzene anaerobically biodegraded to the intermediate tentatively identified as 3,5-dimethylbenzoic acid.

USES: UV oxidation stabilizer for plastics; manufacturing anthraquinone dyes.

1,1,3-TRIMETHYLCYCLOHEXANE

SYNONYM: 1,3,3-Trimethylcyclohexane.

CHEMICAL DESIGNATIONS: CAS: 3073-66-3; DOT: none assigned; mf: C_9H_{18}; fw: 126.24; RTECS: none assigned.

PROPERTIES: Liq. Bp: 138-139°, *d*: 0.7664 @ 20/4°, S_w: 1.77 ppm @ 25°, vap d: 4.36 g/L @ 25°, 3.68 (air = 1).

USES: Organic synthesis; gasoline component.

1,1,3-TRIMETHYLCYCLOPENTANE

SYNONYM: 1,3,3-Trimethylcyclopentane.

CHEMICAL DESIGNATIONS: CAS: 4516-69-2; DOT: none assigned; mf: C_8H_{16}; fw: 112.22; RTECS: none assigned.

PROPERTIES: Colorless liq. Mp: -14.4°, bp: 115-116°, *d*: 0.7703 @ 20/4°, K_H: 1.57 atm·m^3/mol @ 25° (calcd), S_w: 3.73 ppm @ 25°, vap d: 4.59 g/L @ 25°, 3.87 (air = 1), vp: 39.7 mm @ 25°.

USES: Organic synthesis; gasoline component.

2,2,5-TRIMETHYLHEXANE

SYNONYM: 2,5,5-Trimethylhexane.

CHEMICAL DESIGNATIONS: CAS: 3522-94-9; DOT: none assigned; mf: C_9H_{20}; fw: 128.26; RTECS: none assigned.

PROPERTIES: Colorless liq. Mp: -105.8°, bp: 124°, *d*: 0.7072 @ 20/4°, fl p: 12.7°, K_H: 2.42 atm·m^3/mol @ 25° (calcd), log K_{ow}: 3.88 (calcd), S_o: sol in acetone, alcohol, benzene, ether and lignoin, S_w: 0.54 ppm @ 25°, vap d: 5.24 g/L @ 25°, 4.43 (air = 1), vp: 16.5 mm @ 25°.

USES: Motor fuel additive; organic synthesis.

2,2,4-TRIMETHYLPENTANE

SYNONYMS: Isobutyltrimethylmethane; Isooctane; UN 1262.

CHEMICAL DESIGNATIONS: CAS: 540-84-1; DOT: 1262mf: C_8H_{18}; fw: 114.23; RTECS: SA 3320000.

PROPERTIES: Colorless liq with a gasoline-like odor. Mp: -107.4°, bp: 99.2°, *d*: 0.6919 @ 20/4°, fl p: -12°, lel: 1.1%, uel: 6.0%, K_H: 3.01 atm·m^3/mol @ 25°, IP: 9.86 eV, log K_{ow}: 5.83, S_o: sol in acetone, alcohol, benzene, chloroform, ether, carbon disulfide, carbon tetrachloride, dimethylformamide, toluene, xylene and oils except castor oil, S_w: 0.56-2.05 ppm @ 25°, vap d: 4.67 g/L @ 25°, 3.94 (air = 1), vp: 49.3 mm @ 25°.

USES: Determining octane numbers of fuels; solvent and thinner; in spectrophotometric analysis.

2,3,4-TRIMETHYLPENTANE

SYNONYMS: None.

CHEMICAL DESIGNATIONS: CAS: 565-75-3; DOT: none assigned; mf: C_8H_{18}; fw: 114.23; RTECS: none assigned.

PROPERTIES: Colorless liq. Mp: -109.2°, bp: 113.4°, K_H: 2.98 atm·m^3/mol @ 25° (calcd), log K_{ow}: 3.78 (calcd), S_o: sol in acetone, alcohol, benzene, chloroform and ether, S_w: 1.36 ppm @ 25°, vap d: 4.67 g/L @ 25°, 3.94 (air = 1), vp: 27.0 mm @ 25°.

USES: Organic synthesis.

2,4,6-TRINITROTOLUENE

SYNONYMS: Entsufon; 1-Methyl-2,4,6-trinitrobenzene; **2-Methyl-1,3,5-tri-**

nitrobenzene; NCI-C56155; TNT; α-TNT; TNT-tolite; Tolit; Tolite; Trilit; Trinitrotoluene; *sym*-Trinitrotoluene; Trinitrotoluol; α-Trinitrotoluol; *sym*-Trinitrotoluol; Tritol; Triton; Trotyl; Trotyl oil; UN 0209.

CHEMICAL DESIGNATIONS: CAS: 118-96-7; DOT: 1356; mf: $C_7H_5N_3O_6$; fw: 227.13; RTECS: XU 0175000.

PROPERTIES: Colorless to light yellow monoclinic cryst. Mp: 82°, bp: explodes @ 240°, *d*: 1.654 @ 20/4°, fl p: explodes, log K_{oc}: 2.48 (calcd), log K_{ow}: 2.25 (calcd), S_o: sol in acetone, benzene, ether and pyrimidine, S_w: 200 mg/L @ 15°, vp: 0.05 mm @ 85°.

TRANSFORMATION PRODUCTS: 4-Amino-2,6-dinitrotoluene and 2-amino-4,6-dinitrotoluene, detected in contaminated groundwater beneath the Hawthorne Naval Ammunition Depot, NV, were reported to have formed from the microbial degradation of 2,4,6-trinitrotoluene.

EXPOSURE LIMITS: OSHA PEL: TWA 1.5 mg/m^3; ACGIH TLV: TWA 0.5 mg/m^3.

SYMPTOMS OF EXPOSURE: Jaun; cyan; sneez; cough, sore throat; peri neur, musc pain; liv and kid damage; card; cataract; derm; leucyt, anem.

USES: High explosive; intermediate in dyestuffs and photographic chemicals.

TRIPHENYL PHOSPHATE

SYNONYMS: Celluflex TPP; **Phosphoric acid triphenyl ester**; Phenyl phosphate; TPP.

CHEMICAL DESIGNATIONS: CAS: 115-86-6; DOT: none assigned; mf: $C_{18}H_{15}O_4P$; fw: 326.29; RTECS: TC 8400000.

PROPERTIES: Colorless, odorless solid. Mp: 50-51°, bp: 245° @ 11 mm, *d*: 1.2055 @ 50/4°, fl p: 220°, K_H: 5.88 x 10^{-2} atm·m^3/mol @ 20-25° (calcd), log K_{oc}: 3.72 (calcd), log K_{ow}: 5.27 (calcd), S_o: sol in alcohol, benzene, chloroform and ether, S_w: 0.73 mg/L @ 24°, vp: $<$ 0.1 mm @ 20°.

TRANSFORMATION PRODUCTS: When an aq soln containing triphenyl phosphate (0.1 mg/L) and chlorine (3-1,000 mg/L) was stirred in the dark @ 20° for 24 h, the benzene ring was substituted with 1-3 chlorine atoms.

EXPOSURE LIMITS: OSHA PEL: TWA 3 mg/m^3.

SYMPTOMS OF EXPOSURE: Minor changes in blood enzymes; in animals: musc weakness, para.

USES: Camphor substitute in celluloid; impregnating roofing paper; plasticizer in lacquers and varnishes; renders acetylcellulose, airplane "dope", nitrocellulose, stable and fireproof; gasoline additives; insecticides; flotation agents; antioxidants, stabilizers and surfactants.

VINYL ACETATE

SYNONYMS: **Acetic acid ethenyl ester**; Acetic acid ethylene ester; Acetic acid vinyl ester; 1-Acetoxyethylene; Ethenyl acetate; Ethenylethanoate; UN 1301; VAC;

VAM; Vinyl acetate H.Q.

CHEMICAL DESIGNATIONS: CAS: 108-05-4; DOT: 1301; mf: $C_4H_6O_2$; fw: 86.09; RTECS: AK 0875000.

PROPERTIES: Colorless, watery liq with a pleasant fruity odor. Mp: -93.2°, bp: 72.2°, *d*: 0.9317 @ 20/4°, fl p: -8°, lel: 2.6%, uel: 13.4%, K_H: 4.81 x 10^{-4} atm·m^3/mol (calcd), IP: 9.19 eV, log K_{oc}: 0.45 (est), log K_{ow}: 0.73, S_o: sol in acetone, ethanol, benzene, chloroform and ether, S_w: 20 g/L @ 20°, vap d: 3.52 g/L @ 25°, 2.97 (air = 1), vp: 83 mm @ 20°.

TRANSFORMATION PRODUCTS: Anticipated hydrolysis products would include acetic acid and vinyl alcohol. Slowly polymerizes in light to a colorless, transparent mass.

EXPOSURE LIMITS: ACGIH TLV: TWA 10 ppm, STEL 20 ppm.

USES: Manufacture of polyvinyl acetate, polyvinyl alcohol, polyvinyl chloride-acetate resins; used particularly in latex paint; paper coatings; adhesives; textile finishing; safety glass interlayers.

VINYL CHLORIDE

SYNONYMS: **Chlorethene**; Chlorethylene; Chloroethene; 1-Chloroethene; Chloroethylene; 1-Chloroethylene; Ethylene monochloride; Monochloroethene; Monochloroethylene; MVC; RCRA waste number U043; Trovidur; UN 1086; VC; VCM; Vinyl C monomer; Vinyl chloride monomer.

CHEMICAL DESIGNATIONS: CAS: 75-01-4; DOT: 1086; mf: C_2H_3Cl; fw: 62.50; RTECS: KU 9625000.

PROPERTIES: Colorless liquified compressed gas with a faint, sweetish odor. Mp: -153.8°, bp: -13.4°, *d*: 0.9106 @ 20/4°, fl p: -78°, lel: 3.6%, uel: 33%, K_H: 5.6 x 10^{-2} atm·m^3/mol @ 25°, IP: 9.995 eV, log K_{oc}: 0.39 (calcd), log K_{ow}: 0.60, S_o: sol in ethanol, carbon tetrachloride and ether, S_w: 1,100 mg/L @ 25°, vap d: 2.55 g/L @ 25°, 2.16 (air = 1), vp: 2,560 mm @ 20°.

TRANSFORMATION PRODUCTS: Irradiation of vinyl chloride in the presence of NO_2 for 160 min produced formic acid, hydrochloric acid, CO, formaldehyde, O_3 with trace amounts of formyl chloride and nitric acid. In the presence of O_3, however, vinyl chloride photooxidized to CO, formaldehyde, formic acid and small amounts of hydrochloric acid. Reported photooxidation products in the troposphere include hydrogen chloride and and/or formyl chloride. In the presence of moisture, formyl chloride will decompose to CO and hydrochloric acid. In natural surface waters, vinyl chloride was resistant to biological and chemical degradation. Under anaerobic or aerobic conditions, degradation to CO_2 was reported in experimental systems containing mixed or pure cultures. The anaerobic degradation of vinyl chloride dissolved in groundwater by static microcosms was enhanced by the presence of nutrients (methane, methanol, ammonium phosphate, phenol). Methane and ethylene were reported as the biodegradation end products. In a laboratory experiment, it was observed that the leaching of vinyl chloride monomer from polyvinyl chloride pipe into water reacted with chlorine to form chloroacetaldehyde, chloroacetic acid and other unidentified compounds.

EXPOSURE LIMITS: NIOSH REL: lowest detectable level; OSHA PEL: TWA 1 ppm, 15-min C 5 ppm; ACGIH TLV: TWA 5 ppm.

SYMPTOMS OF EXPOSURE: Weak abdom pain, GI bleeding, hematomegaly, pal or cyan of extrem.

USES: Manufacture of polyvinyl chloride and copolymers; adhesives for plastics; refrigerant; extraction solvent; organic synthesis.

WARFARIN

SYNONYMS: 3-(Acetonylbenzyl)-4-hydroxycoumarin; 3-(α-Acetonylbenzyl)-4-hydroxycoumarin; Athrombine-K; Athrombin-K; Brumolin; Compound 42; Co-rax; Coumadin; Coumafen; Coumafene; Cov-r-tox; D-con; Dethmor; Dethnel; Eastern states duocide; Fasco fascrat powder; 1-(4′-Hydroxy-3′-coumarinyl)-1-phenyl-3-butanone; **4-Hydroxy-3-(3-oxo-1-phenylbutyl)-2*H*-1-benzopyran-2-one**; 4-Hydroxy-3-(1-phenyl-3-oxobutyl)coumarin; Kumader; Kumadu; Kypfarin; Liqua-tox; Mar-frin; Martin's mar-frin; Maveran; Mouse pak; 3-(1′-Phenyl-2′-acetylethyl)-4-hydroxycoumarin; 3-α-Phenyl-β-acetylethyl-4-hydroxycoumarin; Prothromadin; Rat-a-way; Rat-b-gon; Rat-gard; Rat-kill; Rat & mice bait; Rat-mix; Rat-o-cide #2; Rat-ola; Ratorex; Ratox; Ratoxin; Ratron; Ratron G; Rats-no-more; Rat-trol; Rattunal; Rax; RCRA waste number P001; Rodafarin; Ro-deth; Rodex; Rodex blox; Rosex; Rough & ready mouse mix; Solfarin; Spraytrol brand roden-trol; Temus W; Tox-hid; Twin light rat away; Vampirinip II; Vampirin III; Waran; W.A.R.F. 42; Warfarat; Warfarin plus; Warfarin Q; Warf compound 42; Warficide.

CHEMICAL DESIGNATIONS: CAS: 81-81-2; DOT: 3027; mf: $C_{19}H_{16}O_4$; fw: 308.33; RTECS: GN 4550000.

PROPERTIES: Colorless and odorless cryst. Mp: 161°, bp: decomp, log K_{oc}: 2.96 (calcd), log K_{ow}: 3.20 (calcd), S_o: sol in alcohol, benzene, dioxane and acetone; moderately sol in methanol, ethanol, isopropanol and some oils, S_w: 17 mg/L @ 20°.

EXPOSURE LIMITS: NIOSH REL: IDLH 200 mg/m^3; OSHA PEL: TWA 0.1 mg/m^3.

SYMPTOMS OF EXPOSURE: Hema, back pain, hemat arms, legs; epis, bleeding lips, muc memb hemorr, abdom pain, vomit, fecal blood; petechial rash; abnormal hematology.

USES: Rodenticide.

o-XYLENE

SYNONYMS: 1,2-Dimethylbenzene; *o*-Dimethylbenzene; *o*-Methyltoluene; UN 1307; 1,2-Xylene; ortho-Xylene; *o*-Xylol.

CHEMICAL DESIGNATIONS: CAS: 95-47-6; DOT: 1307; mf: C_8H_{10}; fw: 106.17; RTECS: ZE 2450000.

PROPERTIES: Clear, colorless liq. Mp: -25.2°, bp: 144.4°, *d*: 0.88011 @ 20/4°, fl p: 17°, lel: 1%, uel: 6%, pK_a: > 15, K_H: 5.35 x 10^{-3} atm·m^3/mol @ 25°, IP: 8.56 eV, 9.04 eV, log K_{oc}: 2.11-2.41, log K_{ow}: 2.77-3.16, S_o: sol in acetone, ethanol, benzene and ether, S_w: 152 mg/L @ 20°, vap d: 4.34 g/L @ 25°, 3.66 (air = 1), vp: 6.6 mm @ 25°.

TRANSFORMATION PRODUCTS: Synthetic air containing gaseous nitrous acid and exposed to artificial sunlight (λ = 300-450 nm) photooxidized *o*-xylene into biacetyl, peroxyacetal nitrate and methyl nitrate. A *n*-hexane soln containing *o*-xylene and spread as a thin film (4 mm) on cold water (10°) was irradiated by a mercury medium pressure lamp. In 3 h, 13.6% of the *o*-xylene photooxidized into *o*-methylbenzaldehyde, *o*-benzyl alcohol, *o*-benzoic acid and *o*-methylacetophenone. Irradiation of *o*-xylene ($\lambda \approx$ 2537 Å) @ 35° and 6 mm isomerizes to *m*-xylene. Glyoxal, methylglyoxal and biacetyl were produced from the photooxidation of *o*-xylene by hydroxyl radicals in air @ 25°. Under atmospheric conditions, the gas-phase reaction of *o*-xylene with hydroxyl radicals and NO_x resulted in the formation of *o*-tolualdehyde, *o*-methylbenzyl nitrate, nitro-*o*-xylenes, 2,3- and 3,4-dimethylphenol. Major products reported from the photooxidation of *o*-xylene with NO_x not previously mentioned include formaldehyde, acetaldehyde, peroxyacetyl nitrate, glyoxal and methylglyoxal. Biodegradation products of the commercial product containing mixed xylenes include α-hydroxy-*p*-toluic acid, *p*-methylbenzyl alcohol, benzyl alcohol, 4-methylcatechol, *m*- and *p*-toluic acids. In anoxic groundwater near Bemidji, MI, *o*-xylene anaerobically biodegraded to the intermediate *o*-toluic acid.

EXPOSURE LIMITS: For an isomeric mixture: NIOSH REL: IDLH 1,000 ppm; OSHA PEL: TWA 100 ppm; ACGIH TLV: TWA 100 ppm, STEL 150 ppm.

SYMPTOMS OF EXPOSURE: Dizz, excitement, drow, inco, staggering gait; irrit eyes, nose, throat; corneal vacuolization; anor, nau, vomit, abdom pain; derm.

USES: Preparation of phthalic acid, phthalic anhydride, terephthalic acid, isophthalic acid; solvent for alkyd resins, lacquers, enamels, rubber cements; manufacture of dyes, pharmaceuticals and insecticides; motor fuels.

m-XYLENE

SYNONYMS: 1,3-Dimethylbenzene; *m*-Dimethylbenzene; *m*-Methyltoluene; UN 1307; 1,3-Xylene; *m*-Xylol.

CHEMICAL DESIGNATIONS: CAS: 108-38-3; DOT: 1307; mf: C_8H_{10}; fw: 106.17; RTECS: ZE 2275000.

PROPERTIES: Clear, colorless, watery liq with a sweet odor. Mp: -47.9°, bp: 139.1°, *d*: 0.86407 @ 20/4°, fl p: 25°, lel: 1.1%, uel: 7%, pK_a: > 15, K_H: 6.3 x 10^{-3} atm·m^3/mol @ 25°, IP: 8.58 eV, 9.05 eV, log K_{oc}: 3.20, log K_{ow}: 3.13, 3.20, S_o: sol in acetone, ethanol, benzene and ether, S_w: 146-173 mg/L @ 25°, vap d: 4.34 g/L @ 25°, 3.66 (air = 1), vp: 8.3 mm @ 25°.

TRANSFORMATION PRODUCTS: Microbial degradation produced 3-methylbenzyl alcohol, 3-methylbenzaldehyde, *m*-toluic acid and 3-methylcatechol. *m*-Toluic acid acid was reported to be the biooxidation product of *m*-xylene by

Nocardia corallina V-49 using *n*-hexadecane as the substrate. Reported biodegradation products of the commercial product containing mixed xylenes include α-hydroxy-*p*-toluic acid, *p*-methylbenzyl alcohol, benzyl alcohol, 4-methylcatechol, *m*- and *p*-toluic acids. In anoxic groundwater near Bemidji, MI, *m*-xylene anaerobically biodegraded to the intermediate *m*-toluic acid. Synthetic air containing gaseous nitrous acid and exposed to artificial sunlight (λ = 300-450 nm) photooxidized *o*-xylene into biacetyl, peroxyacetal nitrate and methyl nitrate. A *n*-hexane soln containing *m*-xylene and spread as a thin film (4 mm) on cold water (10°) was irradiated by a mercury medium pressure lamp. In 3 h, 25% of the *m*-xylene photooxidized into *m*-methylbenzaldehyde, *m*-benzyl alcohol, *m*-benzoic acid and *m*-methylacetophenone. Irradiation of *m*-xylene yields the isomer *p*-xylene. Glyoxal, methylglyoxal and biacetyl were produced from the photooxidation of *m*-xylene by hydroxyl radicals in air @ 25°. Under atmospheric conditions, the gas-phase reaction with hydroxyl radicals and NO_x resulted in the formation of *m*-tolualdehyde, *m*-methylbenzyl nitrate, nitro-*m*-xylenes, 2,4- and 2,6-dimethylphenol. The photooxidation of *m*-xylene in the presence of NO_x (NO and NO_2) yielded small amounts of formaldehyde and a trace of acetaldehyde. *m*-Tolualdehyde and nitric acid also were identified as photooxidation products of *m*-xylene with NO_x.

EXPOSURE LIMITS: For an isomeric mixture: NIOSH REL: IDLH 1,000 ppm; OSHA PEL: TWA 100 ppm; ACGIH TLV: TWA 100 ppm, STEL 150 ppm.

SYMPTOMS OF EXPOSURE: Dizz, excitement, drow, inco, staggering gait; irrit eyes, nose, throat; corneal vacuolization; anor, nau, vomit, abdom pain; derm.

USES: Solvent; preparation of isophthalic acid, intermediate for dyes; insecticides; aviation fuel.

p-XYLENE

SYNONYMS: Chromar; 1,4-Dimethylbenzene; *p*-Dimethylbenzene; *p*-Methyltoluene; Scintillar; UN 1307; 1,4-Xylene; *p*-Xylol.

CHEMICAL DESIGNATIONS: CAS: 106-42-3; DOT: 1307; mf: C_8H_{10}; fw: 106.17; RTECS: ZE 2625000.

PROPERTIES: Clear, colorless, liq with a sweet odor. Mp: 13.3°, bp: 138.3°, *d*: 0.86100 @ 20/4°, fl p: 27.2° (OC), lel: 1.1%, uel: 7%, pK_a: > 15, K_H: 6.3 x 10^{-3} atm·m^3/mol @ 25°, IP: 8.44 eV, 8.99 eV, log K_{oc}: 2.31, 2.42, log K_{ow}: 3.15, 3.18, S_o: sol in acetone, ethanol and benzene, S_w: 180-200 mg/L @ 25°, vap d: 4.34 g/L @ 25°, 3.66 (air = 1), vp: 8.8 mm @ 25°.

TRANSFORMATION PRODUCTS: Microbial degradation of *p*-xylene produced 4-methylbenzyl alcohol, 4-methylbenzaldehyde, *p*-toluic acid and 4-methylcatechol. Dimethyl-*cis,cis*-muconic acid and 2,3-dihydroxy-*p*-toluic acid were reported to be biooxidation products of *p*-xylene by *Nocardia corallina* V-49 using *n*-hexadecane as the substrate. Reported biodegradation products of the commercial product containing mixed xylenes include α-hydroxy-*p*-toluic acid, *p*-methylbenzyl alcohol, benzyl alcohol, 4-methylcatechol, *m*- and *p*-toluic acids. In anoxic groundwater near Bemidji, MI, *p*-xylene anaerobically biodegraded to the intermediate *p*-toluic acid. A *n*-hexane soln containing *m*-xylene and spread as a thin film (4 mm) on cold water (10°) was irradiated by a mercury medium pressure lamp. In 3 h, 18.5% of the *p*-xylene photooxidized into *p*-methylbenzaldehyde, *p*-benzyl alcohol, *p*-benzoic acid and *p*-methylacetophenone. Glyoxal and methylglyoxal were produced from the photooxidation of *p*-xylene by hydroxyl radicals in air @ 25°. Under atmospheric conditions, the gas-phase reaction with hydroxyl radicals and NO_x resulted in the formation of *p*-tolualdehyde.

EXPOSURE LIMITS: For an isomeric mixture: NIOSH REL: IDLH 1,000 ppm; OSHA PEL: TWA 100 ppm; ACGIH TLV: TWA 100 ppm, STEL 150 ppm.

SYMPTOMS OF EXPOSURE: Dizz, excitement, drow, inco, staggering gait; irrit eyes, nose, throat; corneal vacuolization; anor, nau, vomit, abdom pain; derm.

USES: Preparation of terephthalic acid for polyester resins and fibers (Dacron, Mylar and Terylene), vitamins, pharmaceuticals and insecticides.

CAS Index

74-96-4 Ethyl bromide
74-97-5 Bromochloromethane
74-98-6 Propane
74-99-7 Propyne
75-00-3 Chloroethane
75-01-4 Vinyl chloride
75-04-7 Ethylamine
75-05-5 Acetonitrile
75-07-0 Acetaldehyde
75-08-1 Ethyl mercaptan
75-09-2 Methylene chloride
75-15-0 Carbon disulfide
75-25-2 Bromoform
75-27-4 Bromodichloromethane
75-28-5 2-Methylpropane
75-31-0 Isopropylamine
75-34-3 1,1-Dichloroethane
75-35-4 1,1-Dichloroethylene
75-43-4 Dichlorofluoromethane
75-52-5 Nitromethane
75-56-9 Propylene oxide
75-61-6 Dibromodifluoromethane
75-63-8 Bromotrifluoromethane
75-65-0 *tert*-Butyl alcohol
75-69-4 Trichlorofluoromethane
75-71-8 Dichlorodifluoromethane
75-83-2 2,2-Dimethylbutane
76-01-7 Pentachloroethane
76-06-2 Chloropicrin
76-11-9 1,1-Difluorotetrachloroethane
76-12-0 1,2-Difluorotetrachloroethane
76-13-1 1,1,2-Trichlorotrifluoroethane
76-22-2 Camphor
76-44-8 Heptachlor
77-47-4 Hexachlorocyclopentadiene
77-78-1 Dimethyl sulfate
78-30-8 Tri-*o*-cresyl phosphate
78-59-1 Isophorone
78-78-4 2-Methylbutane
78-79-5 2-Methyl-1,3-butadiene
78-83-1 Isobutyl alcohol
78-87-5 1,2-Dichloropropane
78-92-2 *sec*-Butyl alcohol
78-93-3 2-Butanone
79-00-5 1,1,2-Trichloroethane
79-01-6 Trichloroethylene

79-06-1 Acrylamide
79-20-9 Methyl acetate
79-24-3 Nitroethane
79-27-6 1,1,2,2-Tetrabromoethane
79-29-8 2,3-Dimethylbutane
79-34-5 1,1,2,2-Tetrachloroethane
79-46-9 2-Nitropropane
80-62-6 Methyl methacrylate
81-81-2 Warfarin
83-26-1 Pindone
83-32-9 Acenaphthene
84-66-2 Diethyl phthalate
84-74-2 Di-*n*-butyl phthalate
85-01-8 Phenanthrene
85-44-9 Phthalic anhydride
85-68-7 Benzyl butyl phthalate
86-30-6 *n*-Nitrosodiphenylamine
86-73-7 Fluorene
86-88-4 ANTU
87-61-6 1,2,3-Trichlorobenzene
87-68-3 Hexachlorobutadiene
87-86-5 Pentachlorophenol
88-06-2 2,4,6-Trichlorophenol
88-72-2 2-Nitrotoluene
88-74-4 2-Nitroaniline
88-75-5 2-Nitrophenol
88-89-1 Picric acid
90-04-0 *o*-Anisidine
91-17-8 Decahydronaphthalene
91-20-3 Naphthalene
91-57-6 2-Methylnaphthalene
91-58-7 2-Chloronaphthalene
91-59-8 2-Naphthylamine
91-94-1 3,3′-Dichlorobenzidine
92-52-4 Biphenyl
92-67-1 4-Aminobiphenyl
92-87-5 Benzidine
92-93-3 4-Nitrobiphenyl
93-37-8 2,7-Dimethylquinoline
93-76-5 2,4,5-T
94-75-7 2,4-D
95-47-6 *o*-Xylene
95-48-7 2-Methylphenol
95-50-1 1,2-Dichlorobenzene
95-53-4 *o*-Toluidine
95-57-8 2-Chlorophenol

95-63-6 1,2,4-Trimethylbenzene
95-93-2 1,2,4,5-Tetramethylbenzene
95-94-3 1,2,4,5-Tetrachlorobenzene
95-95-4 2,4,5-Trichlorophenol
96-12-8 1,2-Dibromo-3-chloropropane
96-14-0 3-Methylpentane
96-18-4 1,2,3-Trichloropropane
96-33-3 Methyl acrylate
96-37-7 Methylcyclopentane
98-00-0 Furfuryl alcohol
98-01-1 Furfural
98-06-6 *tert*-Butylbenzene
98-82-8 Isopropylbenzene
98-83-9 α-Methylstyrene
98-95-3 Nitrobenzene
99-08-1 3-Nitrotoluene
99-09-2 3-Nitroaniline
99-65-0 1,3-Dinitrobenzene
99-99-0 4-Nitrotoluene
100-00-5 *p*-Chloronitrobenzene
100-01-6 4-Nitroaniline
100-02-7 4-Nitrophenol
100-25-4 1,4-Dinitrobenzene
100-37-8 2-Diethylaminoethanol
100-41-4 Ethylbenzene
100-42-5 Styrene
100-44-7 Benzyl chloride
100-51-6 Benzyl alcohol
100-61-8 Methylaniline
100-63-0 Phenylhydrazine
100-74-3 4-Ethylmorpholine
101-55-3 4-Bromophenyl phenyl ether
101-84-8 Phenyl ether
103-65-1 *n*-Propylbenzene
104-51-8 *n*-Butylbenzene
104-94-9 *p*-Anisidine
105-46-4 *sec*-Butyl acetate
105-67-9 2,4-Dimethylphenol
106-35-4 3-Heptanone
106-37-6 1,4-Dibromobenzene
106-42-3 *p*-Xylene
106-44-5 4-Methylphenol
106-46-7 1,4-Dichlorobenzene
106-47-8 4-Chloroaniline
106-50-3 *p*-Phenylenediamine
106-51-4 *p*-Quinone

106-89-8 Epichlorohydrin
106-92-3 Allyl glycidyl ether
106-93-4 Ethylene dibromide
106-97-8 *n*-Butane
106-98-9 1-Butene
106-99-0 1,3-Butadiene
107-02-8 Acrolein
107-05-1 Allyl chloride
107-06-2 1,2-Dichloroethane
107-07-3 Ethylene chlorohydrin
107-08-4 1-Iodopropane
107-13-1 Acrylonitrile
107-15-3 Ethylenediamine
107-18-6 Allyl alcohol
107-20-0 Chloroacetaldehyde
107-31-3 Methyl formate
107-49-3 Tetraethyl pyrophosphate
107-83-5 2-Methylpentane
107-87-9 2-Pentanone
108-03-2 1-Nitropropane
108-05-4 Vinyl acetate
108-08-7 2,4-Dimethylpentane
108-10-1 4-Methyl-2-pentanone
108-18-9 Diisopropylamine
108-20-3 Isopropyl ether
108-21-4 Isopropyl acetate
108-24-7 Acetic anhydride
108-31-6 Maleic anhydride
108-38-3 *m*-Xylene
108-60-1 Bis(2-chloroisopropyl)ether
108-67-8 1,3,5-Trimethylbenzene
108-70-3 1,3,5-Trichlorobenzene
108-83-8 Diisobutyl ketone
108-84-9 *sec*-Hexyl acetate
108-86-1 Bromobenzene
108-87-2 Methylcyclohexane
108-88-3 Toluene
108-90-7 Chlorobenzene
108-93-0 Cyclohexanol
108-94-1 Cyclohexanone
108-95-2 Phenol
109-60-4 *n*-Propyl acetate
109-66-0 *n*-Pentane
109-67-1 1-Pentene
109-73-9 *n*-Butylamine
109-79-5 *n*-Butyl mercaptan

109-86-4 Methyl cellosolve
109-87-5 Methylal
109-89-7 Diethylamine
109-94-4 Ethyl formate
109-99-9 Tetrahydrofuran
110-02-1 Thiophene
110-19-0 Isobutyl acetate
110-43-0 2-Heptanone
110-49-6 Methyl cellosolve acetate
110-54-3 *n*-Hexane
110-75-8 2-Chloroethyl vinyl ether
110-80-5 2-Ethoxyethanol
110-82-7 Cyclohexane
110-83-8 Cyclohexene
110-86-1 Pyridine
110-91-8 Morpholine
111-15-9 2-Ethoxyethyl acetate
111-44-4 Bis(2-chloroethyl)ether
111-65-9 *n*-Octane
111-66-0 1-Octene
111-76-2 2-Butoxyethanol
111-84-2 *n*-Nonane
111-91-1 Bis(2-chloroethoxy)methane
112-40-3 *n*-Dodecane
115-11-7 2-Methylpropene
115-86-6 Triphenyl phosphate
117-81-7 Bis(2-ethylhexyl)phthalate
117-84-0 Di-*n*-octyl phthalate
118-52-5 1,3-Dichloro-5,5-dimethylhydantoin
118-74-1 Hexachlorobenzene
118-96-7 2,4,6-Trinitrotoluene
120-12-7 Anthracene
120-72-9 Indole
120-82-1 1,2,4-Trichlorobenzene
120-83-2 2,4-Dichlorophenol
121-14-2 2,4-Dinitrotoluene
121-44-8 Triethylamine
121-69-7 Dimethylaniline
121-75-5 Malathion
122-66-7 1,2-Diphenylhydrazine
123-31-9 Hydroquinone
123-42-2 Diacetone alcohol
123-51-3 Isoamyl alcohol
123-73-9 Crotonaldehyde
123-86-4 *n*-Butyl acetate
123-91-1 Dioxane

123-92-2 Isoamyl acetate
124-18-5 *n*-Decane
124-40-3 Dimethylamine
124-48-1 Dibromochloromethane
126-73-8 Tributyl phosphate
126-99-8 Chloroprene
127-18-4 Tetrachloroethylene
127-19-5 *N,N*-Dimethylacetamide
129-00-0 Pyrene
131-11-3 Dimethyl phthalate
132-64-9 Dibenzofuran
134-32-7 1-Naphthylamine
135-98-8 *sec*-Butylbenzene
137-26-8 Thiram
140-88-5 Ethyl acrylate
141-43-5 Ethanolamine
141-78-6 Ethyl acetate
141-79-7 Mesityl oxide
142-29-0 Cyclopentene
142-82-5 *n*-Heptane
143-50-0 Kepone
144-62-7 Oxalic acid
151-56-4 Ethylenimine
156-60-5 *trans*-1,2-Dichloroethylene
191-24-2 Benzo[*ghi*]perylene
192-97-2 Benzo[*e*]pyrene
193-39-5 Indeno[1,2,3-*cd*]pyrene
205-99-2 Benzo[*b*]fluoranthene
206-44-0 Fluoranthene
207-08-9 Benzo[*k*]fluoranthene
208-96-8 Acenaphthylene
218-01-9 Chrysene
287-92-3 Cyclopentane
291-64-5 Cycloheptane
299-84-3 Ronnel
300-76-5 Naled
309-00-2 Aldrin
319-84-6 α-BHC
319-85-7 β-BHC
319-86-8 δ-BHC
330-54-1 Diuron
463-82-1 2,2-Dimethylpropane
479-45-8 Tetryl
496-11-7 Indan
496-15-1 Indoline
504-29-0 2-Aminopyridine

509-14-8 Tetranitromethane
526-73-8 1,2,3-Trimethylbenzene
528-29-0 1,2-Dinitrobenzene
532-27-4 α-Chloroacetophenone
534-52-1 4,6-Dinitro-*o*-cresol
538-93-2 Isobutylbenzene
540-84-1 2,2,4-Trimethylpentane
540-88-5 *tert*-Butyl acetate
541-73-1 1,3-Dichlorobenzene
541-85-5 5-Methyl-3-heptanone
542-88-1 *sym*-Dichloromethyl ether
542-92-7 Cyclopentadiene
556-52-5 Glycidol
562-49-2 3,3-Dimethylpentane
563-45-1 3-Methyl-1-butene
565-59-3 2,3-Dimethylpentane
565-75-3 2,3,4-Trimethylpentane
583-60-8 *o*-Methylcyclohexanone
584-84-9 2,4-Toluene diisocyanate
589-34-4 3-Methylhexane
589-81-1 3-Methylheptane
591-49-1 1-Methylcyclohexene
591-76-4 2-Methylhexane
591-78-6 2-Hexanone
591-93-5 1,4-Pentadiene
592-41-6 1-Hexene
600-25-9 1-Chloro-1-nitropropane
606-20-2 2,6-Dinitrotoluene
608-93-5 Pentachlorobenzene
613-12-7 2-Methylanthracene
621-64-7 *n*-Nitrosodi-*n*-propylamine
624-83-9 Methyl isocyanate
626-38-0 *sec*-Amyl acetate
626-39-1 1,3,5-Tribromobenzene
627-13-4 *n*-Propyl nitrate
627-20-3 *cis*-2-Pentene
628-63-7 *n*-Amyl acetate
634-66-2 1,2,3,4-Tetrachlorobenzene
634-90-2 1,2,3,5-Tetrachlorobenzene
636-28-2 1,2,4,5-Tetrabromobenzene
646-04-8 *trans*-2-Pentene
691-37-2 4-Methyl-1-pentene
763-29-1 2-Methyl-1-pentene
832-69-9 1-Methylphenanthrene
872-55-9 2-Ethylthiophene
959-98-8 α-Endosulfan

1024-57-3 Heptachlor epoxide
1031-07-8 Endosulfan sulfate
1563-66-2 Carbofuran
1582-09-8 Trifluralin
1640-89-7 Ethylcyclopentane
1746-01-6 TCDD
1929-82-4 Nitrapyrin
2040-96-2 Propylcyclopentane
2104-64-5 EPN
2207-01-4 *cis*-1,2-Dimethylcyclohexane
2216-34-4 4-Methyloctane
2234-13-1 Octachloronaphthalene
2698-41-1 *o*-Chlorobenzylidenemalononitrile
2921-88-2 Chlorpyrifos
3073-66-3 1,1,3-Trimethylcyclohexane
3522-94-9 2,2,5-Trimethylhexane
3689-24-5 Sulfotepp
3741-00-2 Pentylcyclopentane
4516-69-2 1,1,3-Trimethylcyclopentane
5103-71-9 *trans*-Chlordanc
5103-74-2 *cis*-Chlordane
6443-92-1 *cis*-2-Heptene
6876-23-9 *trans*-1,2-Dimethylcyclohexane
7005-72-3 4-Chlorophenyl phenyl ether
7421-93-4 Endrin aldehyde
7664-41-7 Ammonia
7786-34-7 Mevinphos
8001-35-2 Toxaphene
10061-01-5 *cis*-1,3-Dichloropropylene
10061-02-6 *trans*-1,3-Dichloropropylene
11096-82-5 PCB-1260
11097-69-1 PCB-1254
11104-28-2 PCB-1221
11141-16-5 PCB-1232
12672-29-6 PCB-1248
12674-11-2 PCB-1016
14686-13-6 *trans*-2-Heptene
33213-65-9 *β*-Endosulfan
53469-21-9 PCB-1242

Bibliography

Abd-El-Bary, M.F., Hamoda, M.F., Tanisho, S., and N. Wakao. "Henry's Constants for Phenol over its Diluted Aqueous Solution," *J. Chem. Eng. Data*, 31(2):229-230 (1986).

Abdelmagid, H.M., and M.A. Tabatabai. "Decomposition of Acrylamide in Soils," *J. Environ. Qual.*, 11(4):701-704 (1982).

Abdul, S.A., Gibson, T.L., and D.N. Rai. "Statistical Correlations for Predicting the Partition Coefficient for Nonpolar Organic Contaminants between Aquifer Organic Carbon and Water," *Haz. Waste Haz. Mater.*, 4(3):211-222 (1987).

Abraham, M.H. "Thermodynamics of Solution of Homologous Series of Solutes in Water," *J. Chem. Soc. Faraday Trans. I*, 80:153-181 (1984).

Abraham, T., Bery, V., and A.P. Kudchadker. "Densities of Some Organic Substances," *J. Chem. Eng. Data*, 16(3):355-356 (1971).

"A Compendium of Technologies Used in the Treatment of Hazardous Wastes," Office of Research and Development, U.S. EPA Report-625/8-87-014, 49 p.

Adewuyi, Y.G., and G.R. Carmichael. "Kinetics of Hydrolysis and Oxidation of Carbon Disulfide by Hydrogen Peroxide in Alkaline Medium and Application to Carbonyl Sulfide," *Environ. Sci. Technol.*, 21(2):170-177 (1987).

Adrian, P., Lahaniatis, E.S., Andreux, F., Mansour, M., Scheunert, I., and F. Korte. "Reaction of the Soil Pollutant 4-Chloroaniline with the Humic Acid Monomer Catechol," *Chemosphere*, 18(7/8):1599-1609 (1989).

Affens, W.A., and G.W. McLaren. "Flammability Properties of Hydrocarbon Solutions in Air," *J. Chem. Eng. Data*, 17(4):482-488 (1972).

Akimoto, H., and H. Takagi. "Formation of Methyl Nitrite in the Surface Reaction of Nitrogen Dioxide and Methanol. 2. Photoenhancement," *Environ. Sci. Technol.*, 20(4):387-393 (1986).

Aksnes, G., and K. Sandberg. "On the Oxidation of Benzidine and *o*-Dianisidine with Hydrogen Peroxide and Acetylcholine in Alkaline Solution," *Acta Chem. Scand.*, 11:876-880 (1957).

Alexander, D.M. "The Solubility of Benzene in Water," *J. Phys. Chem.*, 63(6):1021-1022 (1959).

Ali, S. "Degradation and Environmental Fate of Endosulfan Isomers and Endosulfan Sulfate in Mouse, Insect, and Laboratory Ecosystem," PhD Thesis, University of Illinois, Ann Arbor, MI (1978).

"A Literature Survey Oriented Towards Adverse Environmental Effects Resultant from the Use of Azo Compounds, Brominated Hydrocarbons, EDTA, Formaldehyde Resins, and *o*-Nitrochloro-benzene," Office of Toxic Substances, U.S. EPA Report-560/2-76-005 (1976), 480 p.

Allison, L.E. "Organic Carbon" in *Methods of Soil Analysis, Part 2.*, Black, C., Evans, D., White, J., Ensminger, L., and F. Clark, eds. (Madison, WI: American Society of Agronomy, 1965), pp 1367-1378.

Almgren, M., Grieser, F., Powell, J.R., and J.K. Thomas. "A Correlation between the Solubility of Aromatic Hydrocarbons in Water and Micellar Solutions, with Their Normal Boiling Points," *J. Chem. Eng. Data*, 24(4):285-287 (1979).

Altshuller, A.P. "Measurements of the Products of Atmospheric Photochemical

Reactions in Laboratory Studies and in Ambient Air-Relationships between Ozone and other Products," *Atmos. Environ.*, 17(12):2383-2427 (1983).

Altshuller, A.P., and H.E. Everson. "The Solubility of Ethyl Acetate in Water," *J. Am. Chem. Soc.*, 75(7):1727 (1953).

Altshuller, A.P., Kopczynski, S.L., Lonneman, W.A., Sutterfield, F.D., and D.L. Wilson. "Photochemical Reactivities of Aromatic Hydrocarbon-Nitrogen Oxide and Related Systems," *Environ. Sci. Technol.*, 4(1):44-49 (1970).

Ambrose, D., Ellender, J.H., Lees, E.B., Sprake, C.H.S., and R. Townsend. "Thermodynamic Properties of Organic Compounds. XXXVIII. Vapor Pressures of Some Aliphatic Ketones," *J. Chem. Thermodyn.*, 7(5):453-472 (1975).

Amidon, G.L., Yalkowsky, S.H., Anik, S.T., and S.C. Valvani. "Solubility of Nonelectrolytes in Polar Solvents. V. Estimation of the Solubility of Aliphatic Monofunctional Compounds in Water using a Molecular Surface Area Approach," *J. Phys. Chem.*, 79(21):2239-2246 (1975).

Amoore, J.E., and R.G. Buttery. "Partition Coefficients and Comparative Olfactometry," *Chem. Sens. Flavour*, 3(1):57-70 (1978).

Amoore, J.E., and E. Hautala. "Odor as an Aide to Chemical Safety: Odor Thresholds Compared with Threshold Limit Values and Volatilities for 214 Industrial Chemicals in Air and Water Dilution," *J. Appl. Toxicol.*, 3(6):272-290 (1983).

Andersson, H.F., Dahlberg, J.A., and R. Wettstrom. "On the Formation of Phosgene and Trichloroacetylchloride in the Non-Sensitized Photooxidation of Perchloroethylene in Air," *Acta Chem. Scand. A*, 29:473-474 (1975).

Andersson, J.T., Häussler, R., and K. Ballschmiter. "Chemical Degradation of Xenobiotics. III. Simulation of the Biotic Transformation of 2-Nitrophenol and 3-Nitrophenol," *Chemosphere*, 15(2):149-152 (1986).

Ando, M., and Y. Sayato. "Studies on Vinyl Chloride Migrating into Drinking Water from Polyvinyl Chloride Pipe and Reaction between Vinyl Chloride and Chlorine," *Water Res.*, 18(3):315-318 (1984).

Andon, R.J.L., Biddiscombe, D.P., Cox, F.D., Handley, R., Harrop, D., Herington, E.F.G., and J.F. Martin. "Thermodynamic Properties of Organic Oxygen Compounds. Part 1. Preparation of Physical Properties of Pure Phenol, Cresols and Xylenols," *J. Chem. Soc. (London)*, (1960), pp 5246-5254.

Andrews, L.J., and R.M. Keefer. "Cation Complexes of Compounds Containing Carbon-Carbon Double Bonds. IV. The Argentation of Aromatic Hydrocarbons," *J. Am. Chem. Soc.*, 71(11):3644-3647 (1949).

Andrews, L.J., and R.M. Keefer. "Cation Complexes of Compounds Containing Carbon-Carbon Double Bonds. VI. The Argentation of Substituted Benzenes," *J. Am. Chem. Soc.*, 72(7):3113-3116 (1950).

Andrews, L.J., and R.M. Keefer. "Cation Complexes of Compounds Containing Carbon-Carbon Double Bonds. VII. Further Studies on the Argentation of Substituted Benzenes," *J. Am. Chem. Soc.*, 72(11):5034-5037 (1950).

Anliker, R., and P. Moser. "The Limits of Bioaccumulation of Organic Pigments in Fish: Their Relation to the Partition Coefficient and the Solubility in Water and Octanol," *Ecotoxicol. Environ. Safety*, 13(1):43-52 (1987).

Antelo, J.M., Arce, F., Barbadillo, F., Casado, J., and A. Varela. "Kinetics and

Mechanism of Ethanolamine Chlorination," *Environ. Sci. Technol.*, 15(8):912-917 (19881).

Appleton, H.T., and H.C. Sikka. "Accumulation, Elimination, and Metabolism of Dichlorobenzidine in the Bluegill Sunfish," *Environ. Sci. Technol.*, 14(1):50-54 (1980).

Aquan-Yuen, M., Mackay, D., and W.-Y. Shiu. "Solubility of Hexane, Phenanthrene, Chlorobenzene, and *p*-Dichlorobenzene in Aqueous Electrolyte Solutions," *J. Chem. Eng. Data*, 24(1):30-34 (1979).

"Aquatic Fate Process Data for Organic Priority Pollutants," Office of Water Regulations and Standards, U.S. EPA Report-440/4-81-014 (1982), 407 p.

Archer, T.E., Nazer, I.K., and D.G. Crosby. "Photodecomposition of Endosulfan and Related Products in Thin Films by Ultraviolet Light Irradiation," *J. Agric. Food Chem.*, 20(5):954-956 (1972).

Arey, J., Zielinska, B., Atkinson, R., Winer, A.M., Randahl, T., and J.N. Pitts, Jr. "The Formation of Nitro-PAH from the Gas-Phase Reactions of Fluoranthene and Pyrene with the OH Radical in the Presence of NO_x," *Atmos. Environ.*, 20(12):2339-2345 (1986).

Arthur, M.F., and J.I. Frea. "2,3,7,8-Tetrachloro-*p*-dioxin: Aspects of Its Important Properties and Its Potential Biodegradation in Soils," *J. Environ. Qual.*, 18(1):1-11 (1989).

Arunachalam, K.D., and M. Lakshmanan. "Microbial Uptake and Accumulation of (^{14}C Carbofuran) 1,3-Dihydro-2,2-Dimethyl-7 Benzofuranylmethyl Carbamate in Twenty Fungal Strains Isolated by Miniecosystem Studies," *Bull. Environ. Contam. Toxicol.*, 41(1):127-134 (1988).

Atkinson, R. "Gas-Phase Tropospheric Chemistry of Organic Compounds: A Review," *Atmos. Environ.*, 24A(1):1-41 (1990).

Atkinson, R., Perry, R.A., and J.N. Pitts, Jr. "Rate Constants for the Reactions of the OH Radical with $(CH_3)_2NH$, $(CH_3)_3N$, and $C_2H_5NH_2$ Over the Temperature Range 298-426°K," *J. Chem. Phys.*, 68(4):1850-1853 (1978).

Atkinson, R., Tuazon, E.C., Wallington, T.J., Aschmann, S.M., Arey, J., Winer, A.M., and J.N. Pitts, Jr. "Atmospheric Chemistry of Aniline, *N,N*-Dimethylaniline, Pyridine, 1,3,5-Triazine, and Nitrobenzene," *Environ. Sci. Technol.*, 21(1):64-72 (1987).

Atlas, E., Foster, R., and C.S. Giam. "Air-Sea Exchange of High Molecular Weight Organic Pollutants: Laboratory Studies," *Environ. Sci. Technol.*, 16(5):283-286 (1982).

Babers, F.H. "The Solubility of DDT in Water Determined Radiometrically," *J. Am. Chem. Soc.*, 77(17):4666 (1955).

Bachmann, A., Wijnen, W.P., de Bruin, W., Huntjens, J.L.M., Roelofsen, W., and A.J.B. Zehnder. "Biodegradation of *Alpha-* and *Beta*-Hexachlorocyclohexane in a Soil Slurry under Different Redox Conditions," *Appl. Environ. Microbiol.*, 54(1):143-149 (1988).

Baek, N.H., and P.R. Jaffe. "The Degradation of Trichloroethylene in Mixed Methanogenic Cultures," *J. Environ. Qual.*, 18(4):515-518 (1989).

Bailey, G.W., and J.L. White. "Herbicides: A Compilation of Their Physical, Chemical, and Biological Properties," *Res. Rev.*, 10:97-122 (1965).

Baird, R., Carmona, L., and R.L. Jenkins. "Behavior of Benzidine and Other Aromatic Amines in Aerobic Wastewater Treatment," *J. Water Poll. Control Fed.*, 49(7):1609-1615 (1977).

Baker, E.G. "Origin and Migration of Oil," *Science*, 129(3353):871-874 (1959).

Ballschiter, K., and C. Scholz. "Mikrobieller Abbau von Chlorier Aromaten: VI. Bildung von Dichlorphenolen in Mikromolarer Lösung durch *Pseudomonas* sp.," *Chemosphere*, 9(7/8):457-467 (1980).

Balson, E.W. "Studies in Vapour Pressure Measurement, Part III. - An Effusion Manometer Sensitive to 5 x 10^{-6} Millimetres of Mercury: Vapour Pressure of DDT and Other Slightly Volatile Substances," *Trans. Faraday Soc.*, 43:54-60 (1947).

Banerjee, S., Howard, P.H., and S.S. Lande. "General Structure-Vapor Pressure Relationships for Organics," *Chemosphere*, 21(10/11):1173-1180 (1990).

Banerjee, P., Piwoni, M.D., and K. Ebeid. "Sorption of Organic Contaminants to a Low Carbon Subsurface Core," *Chemosphere*, 14(8):1057-1067 (1985).

Banerjee, S. "Solubility of Organic Mixtures in Water," *Environ. Sci. Technol.*, 18(8):587-591 (1984).

Banerjee, S., Howard, P.H., Rosenburg, A.M., Dombrowski, A.E., Sikka, H., and D.L. Tullis. "Development of a General Kinetic Model for Biodegradation and its Application to Chlorophenols and Related Compounds," *Environ. Sci. Technol.*, 18(6):416-422 (1984).

Banerjee, S., Sikka, H.C., Gray, R., and C.M. Kelly. "Photodegradation of 3,3′-Dichlorobenzidine," *Environ. Sci. Technol.*, 12(13):1425-1427 (1978).

Banerjee, S., Yalkowsky, S.H., and S.C. Valvani. "Water Solubility and Octanol/Water Partition Coefficients of Organics. Limitations of the Solubility-Partition Coefficient Correlation," *Environ. Sci. Technol.*, 14(10):1227-1229 (1980).

Barbeni, M., Pramauro, E., Pelizzetti, E., Vincenti, M., Borgarello, E., and N. Serpone. "Sunlight Photodegradation of 2,4,5-Trichloro-phenoxyacetic acid and 2,4,5-Trichlorophenol on TiO_2. Identification of Intermediates and Degradation Pathway," *Chemosphere*, 16(6):1165-1179 (1987).

Barlas, H., and H. Parlar. "Reactions of Naphthaline with Nitrogen Dioxide in UV-Light," *Chemosphere*, 16(2/3):519-520 (1987).

Barrio-Lage, G.A., Parsons, F.Z., Narbaitz, R.M., Lorenzo, P.A., and H.E. Archer. "Enhanced Anaerobic Biodegradation of Vinyl Chloride in Ground Water," *Environ. Toxicol. Chem.*, 9(1):403-415 (1990).

Barrio-Lage, G.A., Parsons, F.Z., Nassar, R.S., and P.A. Lorenzo. "Biotransformation of Trichloroethylene in a Variety of Subsurface Materials," *Environ. Toxicol. Chem.*, 6(8):571-578 (1987).

Barrio-Lage, G., Parsons, F.Z., Nassar, R.S., and P.A. Lorenzo. "Sequential Dehalogenation of Chlorinated Ethenes," *Environ. Sci. Technol.*, 20(1):96-99 (1986).

Baxter, R.M. "Reductive Dechlorination of Certain Chlorinated Organic Compounds by Reduced Hematin Compared with Their Behaviour in the Environment," *Chemosphere*, 21(4/5):451-458 (1990).

Baxter, R.M., and D.A. Sutherland. "Biochemical and Photochemical Processes in the Degradation of Chlorinated Biphenyls," *Environ. Sci. Technol.*, 18(8):608-610 (1984).

Bedard, D.L., Wagner, R.E., Brennan, M.J., Haberl, M.L., and J.F. Brown, Jr. "Extensive Degradation of Aroclors and Environmentally Transformed Polychlorinated Biphenyls by *Alcaligenes eutrophus* H850," *Appl. Environ. Microbiol.*, 53(5):1094-1102 (1987).

Belay, N., and L. Daniels. "Production of Ethane, Ethylene, and Acetylene from Halogenated Hydrocarbons by Methanogenic Bacteria," *Appl. Environ. Microbiol.*, 53(7):1604-1610 (1987).

Bell, G.H. "Solubilities of Normal Aliphatic Acids, Alcohols and Alkanes in Water," *Chem. Phys. Lipids*, 10:1-10 (1973).

Benezet, H.I., and F. Matsumura. "Isomerization of γ-BHC to a α-BHC in the Environment," *Nature*, 243:480-481 (1973).

Ben-Naim, A., and J. Wilf. "Solubilities and Hydrophobic Interactions in Aqueous Solutions of Monoalkylbenzene Molecules," *J. Phys. Chem.*, 84(6):583-586 (1980).

Bennett, D., and W.J. Canady. "Thermodynamics of Solution of Naphthalene in Various Water-Ethanol Mixtures," *J. Am. Chem. Soc.*, 106(4):910-915 (1984).

Bennett, G.M., and W.G. Philip. "The Influence of Structure on the Solubilities of Ethers. Part I. Aliphatic Ethers," *J. Chem. Soc.*, 131:1930-1937 (1928).

Benson, J.M., Brooks, A.L., Cheng, Y.S., Henderson, T.R., and J.E. White. "Environmental Transformation of 1-Nitropyrene on Glass Surfaces," *Atmos. Environ.*, 19(7):1169-1174 (1985).

Benson, W.R. "Photolysis of Solid and Dissolved Dieldrin," *J. Agric. Food Chem.*, 19(1):66-72 (1971).

Berg, G.L., Ed. *The Farm Book* (Willoughby, OH: Meister Publishing Co., 1983), 440 p.

Berry, D.F., Madsen, E.L., and J.-M. Bollag. "Conversion of Indole to Oxindole under Methanogenic Conditions," *Appl. Environ. Microbiol.*, 53(1):180-182 (1987).

Bevenue, A., and H. Beckman. "Pentachlorophenol: A Discussion of its Properties and its Occurrence as a Residue in Human and Animal Tissues," *Res. Rev.*, 19:83-134 (1967).

Bhavnagary, H.M., and M. Jayaram. "Determination of Water Solubilities of Lindane and Dieldrin at Different Temperatures," *Bull. Grain Technol.*, 12(2):95-99 (1974).

Bidleman, T.F. "Estimation of Vapor Pressures for Nonpolar Organic Compounds by Capillary Gas Chromatography," *Anal. Chem.*, 56(13):2490-2496 (1984).

Bielaszczyk, E., Czerwińska, E., Janko, Z., Kotarski, A., Kowalik, R., Kwiatkowski, M., and J. Zoledziowska. "Aerobic Reduction of Some Nitrochlorosubstituted Benzene Compounds by Microorganisms," *Acta Microbiol. Pol.*, 16:243-248 (1967).

Biggar, J.W., and I.R. Riggs. "Apparent Solubility of Organochlorine Insecticides in Water at Various Temperatures," *Hilgardia*, 42(10):383-391 (1974).

Bobra, A., Mackay, D., and W.-Y. Shiu. "Distribution of Hydrocarbons Among Oil, Water, and Vapor Phases During Oil Dispersant Toxicity Tests," *Bull.*

Environ. Contam. Toxicol., 23(4/5):558-565 (1979).

Boehm, P.D., and J.G. Quinn. "Solubilization of Hydrocarbons by the Dissolved Organic Matter in Sea Water," *Geochim. Cosmochim. Acta*, 37(11):2459-2477 (1973).

Bohon, R.L., and W.F. Claussen. "The Solubility of Aromatic Hydrocarbons in Water," *J. Am. Chem. Soc.*, 73(4):1571-1578 (1951).

Bollag, J.-M., and S.-Y. Liu. "Hydroxylations of Carbaryl by Soil Fungi," *Nature*, 236:177-178 (1972).

Borello, R., Minero, C., Pramauro, E., Pelizzetti, E., Serpone, N., and H. Hidaka. "Photocatalytic Degradation of DDT Mediated in Aqueous Semiconductor Slurries by Simulated Sunlight," *Environ. Toxicol. Chem.*, 8(11):997-1002 (1989).

Borsetti, A.P., and J.A. Roach. "Identification of Kepone Alteration Products in Soil and Mullet," *Bull. Environ. Contam. Toxicol.*, 20(2):241-247 (1978).

Botré, C., Memoli, A., and F. Alhaique. "TCDD Solubilization and Photodecomposition in Aqueous Solutions," *Environ. Sci. Technol.*, 12(3):335-336 (1978).

Boule, P., Guyon, C., and J. Lemaire. "Photochemistry and Environment. 4. Photochemical Behavior of Monochlorophenols in Dilute Aqueous Solution," *Chemosphere*, 11(12):1179-1188 (1982).

Bove, J.L., and J. Arrigo. "Formation of Polycyclic Aromatic Hydrocarbon Via Gas Phase Benzyne," *Chemosphere*, 14(1):99-101 (1985).

Bove, J.L., and P. Dalven. "Pyrolysis of Phthalic Acid Esters: Their Fate," *Sci. Tot. Environ.*, 36:313-318 (1984).

Bowman, M.C., Acree, F., Jr., and M.K. Corbett. "Solubility of Carbon-14 DDT in Water," *J. Agric. Food Chem.*, 8(5):406-408 (1960).

Bowman, B.T., and W.W. Sans. "Determination of Octanol-Water Partitioning Coefficients (K_{ow}) of 61 Organophosphorous and Carbamate Insecticides and Their Relationship to Respective Water Solubility (S) Values," *J. Environ. Sci. Health*, B18(6):667-683 (1983).

Bowman, B.T., and W.W. Sans. "Effect of Temperature on the Water Solubility of Insecticides," *J. Environ. Sci. Health*, B20(6):625-631 (1985).

Boyd, S.A. "Adsorption of Substituted Phenols by Soil," *Soil Sci.*, 134:337-343 (1982).

Boyd, S.A., Kao, C.W., and J.M. Suflita. "Fate of 3,3′-Dichlorobenzidine in Soil: Persistence and Binding," *Environ. Tox. Chem.*, 3:201-208 (1984).

Boyle, T.P., Robinson-Wilson, E.F., Petty, J.D., and W. Weber. "Degradation of Pentachlorophenol in Simulated Lenthic Environment," *Bull. Environ. Contam. Toxicol.*, 24(2):177-184 (1980).

Bradley, R.S., and T.G. Cleasby. "The Vapour Pressure and Lattice Energy of Some Aromatic Ring Compounds," *J. Chem. Soc. (London)*, (1953), pp 1690-1692.

Braker, W., and A.L. Mossman. *Matheson Gas Data Book* (East Rutherford, NJ: Matheson Gas Products, 1971), 574 p.

Branson, D.R. "Predicting the Fate of Chemicals in the Aquatic Environment from Laboratory Data," in *Estimating the Hazard of Chemical Substances to Aquatic Life*, Cairns, J., Jr., Dickson, K.L., and A.W. Maki, Eds. (Philadelphia, PA:

American Society for Testing and Materials, 1978), pp 55-70.

Briggs, G.G. "Theoretical and Experimental Relationships between Soil Adsorption, Octanol-Water Partition Coefficients, Water Solubilities, Bioconcentration Factors, and the Parachor," *J. Agric. Food Chem.*, 29(5):1050-1059 (1981).

Broadhurst, M.G. "Use and Replaceability of Polychlorinated Biphenyls," *Environ. Health Perspect.*, (October 1972), pp 81-102.

Brooke, D.N., A.J. Dobbs, and N. Williams. "Octanol:Water Partition Coefficients (P): Measurement, Estimation, and Interpretation, Particularly for Chemicals with $P > 10^5$," *Ecotoxicol. Environ. Safety*, 11(3):251-260 (1986).

Brooke, D., Nielsen, I., de Bruijn, J., and J. Hermens. "An Interlaboratory Evaluation of the Stir-Flask Method for the Determination of Octanol-Water Partition Coefficients (Log P_{ow})," *Chemosphere*, 21(1/2):119-133 (1990).

Brookman, G.T., Flanagan, M., and J.O. Kebe. "Literature Survey: Hydrocarbon Solubilities and Attenuation Mechanisms," API Publication 4414, (Washington, DC: American Petroleum Institute, 1985), 101 p.

Brooks, G.T. *Chlorinated Insecticides, Volume I, Technology and Applications* (Cleveland, OH: CRC Press, Inc., 1974), 249 p.

Brown, A.C., Canosa-Mas, C.E., and R.P. Wayne. "A Kinetic Study of the Reactions of OH with CH_3I and CF_3I," *Atmos. Environ.*, 24A(2):361-367 (1990).

Brown, D.S., and E.W. Flagg. "Empirical Prediction of Organic Pollutant Sorption in Natural Sediments," *J. Environ. Qual.*, 10(3):382-386 (1981).

Brown, L., and M. Rhead. "Liquid Chromatographic Determination of Acrylamide Monomer in Natural and Polluted Aqueous Environments," *Analyst*, 104(1238):391-399 (1979).

Brunelle, D.J., and D.A. Singleton. "Chemical Reaction of Polychlorinated Biphenyls on Soils with Poly(ethylene glycol)/KOH," *Chemosphere*, 14(2):173-181 (1985).

Bufalini, J.J., Gay, B.W., and S.L. Kopczynski. "Oxidation of *n*-Butane by the Photolysis of NO_2," *Environ. Sci. Technol.*, 5(4):333-336 (1971).

Bumpus, J.A., and S.D. Aust. "Biodegradation of DDT [1,1,1-Trichloro-2,2-bis(4-chlorophenyl)ethane] by the White Rot Fungus *Phanerochaete chrysosporium*," *Appl. Environ. Microbiol.*, 53(9):2001-2008 (1987).

Bunton, C.A., Fuller, N.A., Perry, S.G., and V.J. Shiner. "The Hydrolysis of Carboxylic Anhydrides. Part III. Reactions in Initially Neutral Solution," *J. Chem. Soc. (London)*, (May 1963) pp 2918-2926.

Burczyk, L., Walczyk, K., and R. Burczyk. "Kinetics of Reaction of Water Addition to the Acrolein Double-Bond in Dilute Aqueous Solution," *Przem. Chem.*, 47(10):625-627 (1968).

Burkhard, L.P., and D.W. Kuehl. "*n*-Octanol/Water Partition Coefficients by Reverse Phase Liquid Chromatography/Mass Spectrometry for Eight Tetrachlorinated Planar Molecules," *Chemosphere*, 15(2):163-167 (1986).

Burkhard, L.P., Kuehl, D.W., and G. D. Veith. "Evaluation of Reverse Phase Liquid Chromatography/Mass Spectrometry for Estimation of *n*-Octanol/Water Partition Coefficients for Organic Chemicals," *Chemosphere*, 14(10):1551-1560 (1985).

Burlinson, N.E., Lee, L.A., and D.H. Rosenblatt. "Kinetics and Products of

Hydrolysis of 1,2-Dibromo-3-chloropropane," *Environ. Sci. Technol.*, 16(9):627-632 (1982).

Burris, D.R., and W.G. MacIntyre. "A Thermodynamic Study of Solutions of Liquid Hydrocarbon Mixtures in Water," *Geochim. Cosmochim. Acta*, 50(7):1545-1549 (1986).

Burton, W.B., and G.E. Pollard. "Rate of Photochemical Isomerization of Endrin in Sunlight," *Bull. Environ. Contam. Toxicol.*, 12(1):113-116 (1974).

Butler, J.A.V., Thomson, D.W., and W.H. Maclennan. "The Free Energy of the Normal Aliphatic Alcohols in Aqueous Solution. Part I. The Partial Vapour Pressures of Aqueous Solutions of Methyl, *n*-Propyl, and *n*-Butyl Alcohols. Part II. The Solubilities of Some Normal Aliphatic Alcohols in Water. Part III. The Theory of Binary Solutions, and its Application to Aqueous-Alcoholic Solutions," *J. Chem. Soc.*, 136:674-686 (1933).

Butler, J.C., and W.P. Webb. "Upper Explosive Limits of Cumene," *J. Chem. Eng. Data*, 2(1):42-46 (1957).

Buttery, R.G., Ling, L.C., and D.G. Guadagni. "Volatilities of Aldehydes, Ketones, and Esters in Dilute Water Solution," *J. Agric. Food Chem.*, 17(2):385-389 (1969).

Butz, R.G., Yu, C.C., and Y.H. Atallah. "Photolysis of Hexachlorocyclopentadiene in Water," *Ecotoxicol. Environ. Safety*, 6(4):347-357 (1982).

Byast, T.H., and R.J. Hance. "Degradation of 2,4,5-T by South Vietnamese Soils Incubated in the Laboratory," *Bull. Environ. Contam. Toxicol.*, 14(1):71-76 (1975).

Callahan, M.A., Slimak, M.W., Gable, N.W., May, I.P., Fowler, C.F., Freed, J.R., Jennings, P., Durfee, R.L., Whitmore, F.C., Maestri, B., Mabey, W.R., Holt, B.R., and C. Gould. "Water-Related Environmental Fate of 129 Priority Pollutants Volumes I and II," U.S. EPA Report-440/4-79-029 (1979), 1160 p.

Calvert, J.G., and J.N. Pitts, Jr. *Photochemistry* (New York: John Wiley and Sons, Inc., 1966), 899 p.

Camilleri, P., Watts, S.A., and J.A. Boraston. "A Surface Area Approach to Determination of Partition Coefficients," *J. Chem. Soc. Perkin Trans. II*, (September 1988), pp 1699-1707.

Campbell, J.R., Luthy, R.G., and M.J.T. Carrondo. "Measurement and Prediction of Distribution Coefficients for Wastewater Aromatic Solutes," *Environ. Sci. Technol.*, 17(10):582-590 (1983).

Carlson, R.M., Carlson, R.E., and H.L. Kopperman. "Determination of Partition Coefficients by Liquid Chromatography," *J. Chromatogr.*, 107:219-223 (1975).

Caron, G., Suffet, I.H., and T. Belton. "Effect of Dissolved Organic Carbon on the Environmental Distribution of Nonpolar Organic Compounds," *Chemosphere*, 14(8):993-1000 (1985).

Carter, W.P.L., Winer, A.M., and J.N. Pitts, Jr. "Major Atmospheric Sink for Phenol and the Cresols. Reaction with the Nitrate Radical," *Environ. Sci. Technol.*, 15(7):829-831 (1981).

Casellato, F., Vecchi, C., Girelli, A., and B. Casu. "Differential Calorimetric Study of Polycyclic Aromatic Hydrocarbons," *Thermochim. Acta*, 6:361-368 (1973).

Casida, J.E., Holmstead, R.L., Khalifa, S., Knox, J.R., Ohsawa, T., Palmer, K.J., and R.Y. Wong. "Toxaphene Insecticide: A Complex Biodegradable Mixture," *Science*, 183:520-521 (1974).

Castro, C.E., and N.O. Belser. "Biodehalogenation: Oxidative and Reductive Metabolism of 1,1,2-Trichloroethane by *Pseudomonas putida* - Biogeneration of Vinyl Chloride," *Environ. Toxicol. Chem.*, 9(6):707-714 (1990).

Castro, C.E., and N.O. Belser. "Biodehalogenation. Reductive Dehalogenation of the Biocides Ethylene Dibromide, 1,2-Dibromo-3-chloropropane, and 2,3-Dibromobutane in Soil," *Environ. Sci. Technol.*, 2(10):779-783 (1968).

Castro, C.E., and N.O. Belser. "Hydrolysis of *cis*- and *trans*-1,3-Dichloropropene in Wet Soil," *J. Agric. Food Chem.*, 14(1):69-70 (1966).

Caswell, R.L., DeBold, K.J., and L.S. Gilbert, Eds. *Pesticide Handbook*, 29th ed. (College Park, MD: The Entomological Society of America, 1981), 286 p.

Catalog Handbook of Fine Chemicals (Milwaukee, WI: Aldrich Chemical Co., 1988), 2212 p.

Catalog Handbook of Fine Chemicals (Milwaukee, WI: Aldrich Chemical Co., 1990), 2105 p.

Caturla, F., Martin-Martinez, J.M., Molina-Sabio, M., Rodriguez-Reinoso, F., and R. Torregrosa. "Adsorption of Substituted Phenols on Activated Carbon," *J. Colloid Interface Sci.*, 124(2):528-534 (1988).

Cavalieri, E., and E. Rogan. "Role of Radical Cations in Aromatic Hydrocarbon Carcinogenesis," *Environ. Health Perspect.*, 64:69-84 (1985).

Chapman, P.J. "An Outline of Reaction Sequences Used for the Bacterial Degradation of Phenolic Compounds," in *Degradation of Synthetic Organic Molecules in the Biosphere: Natural, Pesticidal, and Various Other Man-Made Compounds* (Washington, DC: National Academy of Sciences, 1972), pp 17-55.

"Chemical, Physical, and Biological Properties of Compounds Present at Hazardous Waste Sites," U.S. EPA Report-530/SW-89-010 (1985), 619 p.

Chen, P.N., Junk, G.A., and H.J. Svec. "Reactions of Organic Pollutants. I. Ozonation of Acenaphthylene and Acenaphthene," *Environ. Sci. Technol.*, 13(4):451-454 (1979).

Chey, W., and G.V. Calder. "Method for Determining Solubility of Slightly Soluble Organic Compounds," *J. Chem. Eng. Data*, 17(2):199-200 (1972).

Chin, Y.-P., Peven, C.S., and W.J. Weber. "Estimating Soil/Sediment Partition Coefficients for Organic Compounds by High Performance Reverse Phase Liquid Chromatography," *Water Res.*, 22(7):873-881 (1988).

Chin, Y.-P., Weber, W.J., Jr., and T.C. Voice. "Determination of Partition Coefficients and Aqueous Solubilities by Reverse Phase Chromatography - II. Evaluation of Partitioning and Solubility Models," *Water Res.*, 20(11):1443-1451 (1986).

Chiou, C.T., Freed, V.H., Schmedding, D.W., and R.L. Kohnert. "Partition Coefficients and Bioaccumulation of Selected Organic Chemicals," *Environ. Sci. Technol.*, 11(5):475-478 (1977).

Chiou, C.T., Malcolm, R.L., Brinton, T.I., and D.E. Kile. "Water Solubility Enhancement of Some Organic Pollutants and Pesticides by Dissolved Humic and Fulvic Acids," *Environ. Sci. Technol.*, 20(5):502-508 (1986).

Chiou, C.T., Peters, L.J., and V.H. Freed. "A Physical Concept of Soil-Water

Equilibria for Nonionic Organic Compounds," *Science*, 206:831-832 (1979).

Chiou, C.T., Porter, P.E., and D.W. Schmedding. "Partition Equilibria of Nonionic Organic Compounds between Organic Matter and Water," *Environ. Sci. Technol.*, 17(4):227-231 (1983).

Chiou, C.T., Schmedding D.W., and M. Manes. "Partitioning of Organic Compounds in Octanol-Water Systems," *Environ. Sci. Technol.*, 16(1):4-10 (1982).

Chiou, C.T., Shoup, T.D., and P.E. Porter. "Mechanistic Roles of Soil Humus and Minerals in the Sorption of Nonionic Organic Compounds from Aqueous and Organic Solutions," *Org. Geochem.*, 8(1):9-14 (1985).

Chou, S.F.J., and R.A. Griffin. "Soil, Clay and Caustic Soda Effects on Solubility, Sorption and Mobility of Hexachlorocyclopentadiene," Environmental Geology Notes 104, Illinois Department of Energy and Natural Resources (1983), 54 p.

Chou, S.-F.J., Griffin, R.A., Chou, M.-I.M., and R.A. Larson. "Products of Hexchlorocyclopentadiene (C-56) in Aqueous Solution," *Environ. Toxicol. Chem.*, 6(5):371-376 (1987).

Chou, J.T., and P.C. Jurs. "Computer Assisted Computation of Partition Coefficients from Molecular Structures using Fragment Constants," *J. Chem. Info. Comp. Sci.*, 19:172-178 (1979).

Choudhry, G.G., and O. Hutzinger. "Acetone-Sensitized and Nonsensitized Photolyses of Tetra-, Penta-, and Hexachloro-benzenes in Acetonitrile-Water Mixtures: Photoisomerization and Formation of Several Products Including Polychlorobiphenyls," *Environ. Sci. Technol.*, 18(4):235-241 (1984).

Choudhry, G.G., Sundström, G., Ruzo, L.O., and O. Hutzinger. "Photochemistry of Chlorinated Diphenyl Ethers," *J. Agric. Food Chem.*, 25(6):1371-1376 (1977).

"CHRIS Hazardous Chemical Data" U.S. Department of Transportation, U.S. Coast Guard, U.S. Government Printing Office (October, 1978).

Christensen, J.J., Hansen, L.D., and R.M. Izatt. *Handbook of Proton Ionization Heats* (New York: John Wiley and Sons, Inc., 1975), 269 p.

Christiansen, V.O., Dahlberg, J.A., and H.F. Andersson. "On the Nonsensitized Photo-oxidation of 1,1,1-Trichloroethane Vapor in Air," *Acta. Chem. Scand. (Series A)*, 26:3319-3324 (1972).

Claussen, W.F., and M.F. Polglase. "Solubilities and Structures in Aqueous Aliphatic Hydrocarbon Solutions," *J. Am. Chem. Soc.*, 74(19):4817-4819 (1952).

Clayton, G.D., and F.E. Clayton, Eds. *Patty's Industrial Hygiene and Toxicology*, 3rd ed. (New York: John Wiley and Sons, Inc., 1981), 2878 p.

Cleland, J.G. "Project Summary - Environmental Hazard Rankings of Pollutants Generated in Coal Gasification Processes," Office of Research and Development, U.S. EPA Report-600/S7-81-101 (1981), 19 p.

Cleland, J.G., and G.L. Kingsbury. "Multimedia Environmental Goals for Environmental Assessment, Volume II. MEG Charts and Background Information," Office of Research and Development, U.S. EPA Report-600/7-77-136b (1977), 454 p.

Coates, M., Connell, D.W., and D.M. Barron. "Aqueous Solubility and Octan-1-ol to Water Partition Coefficients of Aliphatic Hydrocarbons," *Environ. Sci. Technol.*, 19(7):628-632 (1985).

Coffee, R.D., Vogel, P.C., Jr., and J.J. Wheeler. "Flammability Characteristics of

Methylene Chloride (Dichloromethane)," *J. Chem. Eng. Data*, 17(1):89-93 (1972).

Connors, T.F., Stuart, J.D., and J.B. Cope. "Chromatographic and Mutagenic Analyses of 1,2-Dichloropropane and 1,3-Dichloro-propylene and Their Degradation Products," *Bull. Environ. Contam. Toxicol.*, 44(2):288-293 (1990).

Cooney, R.V., Ross, P.D., Bartolini, G.L., and J. Ramseyer. "*N*-Nitrosamine and *N*-Nitroamine Formation: Factors Influencing the Aqueous Reactions of Nitrogen Dioxide with Morpholine," *Environ. Sci. Technol.*, 21(1):77-83 (1987).

Cooper, W.J., Mehran, M., Riusech, D.J., and J.A. Joens. "Abiotic Transformations of Halogenated Organics. Elimination Reaction of 1,1,2,2-Tetrachloroethane and Formation of 1,1,2-Trichloroethane," *Environ. Sci. Technol.*, 21(11):1112-1114 (1987).

Corbett, M.D., and B.R. Corbett. "Metabolism of 4-Chloronitrobenzene by the Yeast *Rhodosporidium* sp.," *Appl. Environ. Microbiol.*, 41(4):942-949 (1981).

Corless, C.E., Reynolds, G.L., Graham, N.J.D., and R. Perry. "Ozonation of Pyrene in Aqueous Solution," *Wat. Res.*, 24(9):1119-1123 (1990).

Cox, D.P., and C.D. Goldsmith. "Microbial Conversion of Ethylbenzene to 1-Phenethanol and Acetophenone by *Nocardia tartaricans* ATCC 31190," *Appl. Environ. Microbiol.*, 38(3):514-520 (1979).

Cox, R.A., Derwent, R.G., and M.R. Williams. "Atmospheric Photooxidation Reactions. Rates, Reactivity, and Mechanism for Reaction of Organic Compounds with Hydroxyl Radicals," *Environ. Sci. Technol.*, 14(1):57-61 (1980).

Cox, R.A., Patrick, K.F., and S.A. Chant. "Mechanism of Atmospheric Photooxidation of Organic Compounds. Reactions of Alkoxy Radicals in Oxidation of *n*-Butane and Simple Ketones," *Environ. Sci. Technol.*, 15(5):587-592 (1981).

Cozzarelli, I.M., Eganhouse, R.P., and M.J. Baedecker. "Transformation of Monoaromatic Hydrocarbons to Organic Acids in Anoxic Groundwater Environment," *Environ. Geol. Water Sci.*, 16(2):135-141 (1990).

Criteria for a Recommended Standard. . .Occupational Exposure to Phenol (Cincinnati, OH: National Institute for Occupational Safety and Health, 1976), 167 p.

Crosby, D.G., and N. Hamadmad. "The Photoreduction of Pentachlorobenzenes," *J. Agric. Food Chem.*, 19(6):1171-1174 (1971).

Crosby, D.G., and E. Leitis. "The Photodecomposition of Trifluralin in Water," *Bull. Environ. Contam. Toxicol.*, 10(4):237-241 (1973).

Crosby, D.G., and K.W. Moilanen. "Vapor-Phase Photodecomposition of Aldrin and Dieldrin," *Arch. Environ. Contam. Toxicol.*, 2(1):62-74 (1974).

Crosby, D.G., Moilanen, K.W., and A.S. Wong. "Environmental Generation and Degradation of Dibenzodioxins and Dibenzofurans," *Environ. Health Perspect.*, Experimental Issue No. 5, (September 1973), pp 259-266.

Crosby, D.G., and H.O. Tutass. "Photodecomposition of 2,4-Dichlorophenoxyacetic acid," *J. Agric. Food Chem.*, 14(6):596-599 (1966).

Crummett, W.B., and R.H. Stehl. "Determination of Chlorinated Dibenzo-*p*-dioxins and Dibenzofurans in Various Materials," *Environ. Health Perspect.*, (September 1973), pp 15-25.

Cupitt, L.T. "Fate of Toxic and Hazardous Materials in the Air Environment,"

Office of Research and Development, U.S. EPA Report-600/3-80-084 (1980), 28 p.

Curtice, S., Felton, E.G., and H.W. Prengle, Jr., "Thermodynamics of Solutions. Low-Temperature Densities and Excess Volumes of *cis*-Pentene-2 and Mixtures," *J. Chem. Eng. Data*, 17(2):192-194 (1972).

Dagley, S. "Microbial Degradation of Stable Chemical Structures: General Features of Metabolic Pathways," in *Degradation of Synthetic Organic Molecules in the Biosphere: Natural, Pesticidal, and Various Other Man-Made Compounds* (Washington, DC: National Academy of Sciences, 1972), pp 1-16.

Davies, R.P., and A.J. Dobbs. "The Prediction of Bioconcentration in Fish," *Water Res.*, 18(10):1253-1262 (1984).

Davis, J.B., and R.L. Raymond. "Oxidation of Alkyl-Substituted Cyclic Hydrocarbons by a *Nocardia* During Growth on *n*-Alkanes," *Appl. Microbiol.*, 9:383-388 (1961).

Davis, W.W., Krahl, M.E., and G.H.A. Clowes. "Solubility of Carcinogenic and Related Hydrocarbons in Water," *J. Am. Chem. Soc.*, 64(1):108-110 (1942).

Dean, J.A. *Handbook of Organic Chemistry* (New York: McGraw-Hill, Inc., 1987), 957 p.

Dean, J.A., Ed. *Lange's Handbook of Chemistry*, 11th ed. (New York: McGraw-Hill, Inc., 1973), 1570 p.

Dearden, J.C. "Partitioning and Lipophilicity in Quantitative Structure-Activity Relationships," *Environ. Health Perspect.*, (September 1985), pp 203-228.

de Bruijn, J., Busser, F., Seinen, W., and J. Hermens. "Determination of Octanol/Water Partition Coefficients for Hydrophobic Organic Chemicals with the "Slow-Stirring" Method," *Environ. Toxicol. Chem.*, 8:499-512 (1989).

DeFoe, D.L., Holcombe, G.W., Hammermeister, D.E., and K.E. Biesinger. "Solubility and Toxicity of Eight Phthalate Esters to Four Aquatic Organisms," *Environ. Toxicol. Chem.*, 9:623-636 (1990).

DeKock A.C., and D.A. Lord. "A Simple Procedure for Determining Octanol-Water Partition Coefficients using Reverse Phase High Performance Liquid Chromatography (RPHPLC)," *Chemosphere*, 16(1):133-142 (1987).

DeLassus, P.T., and D.D. Schmidt. "Solubilities of Vinyl Chloride and Vinylidene Chloride in Water," *J. Chem. Eng. Data*, 26(3):274-276 (1981).

Deneer, J.W., Sinnige, T.L., Seinen, W., and J.L.M. Hermens. "A Quantitative Structure-Activity Relationship for the Acute Toxicity of Some Epoxy Compounds to the Guppy," *Aquat. Toxicol.*, 13(3):195-204 (1988).

Dennis, W.H., Jr., Chang, Y.H., and W.J. Cooper. "Catalytic Dechlorination of Organochlorine Compounds-Aroclor 1254," *Bull. Environ. Contam. Toxicol.*, 22(6):750-753 (1979).

Dexter, R.N., and S.P. Pavlou. "Mass Solubility and Aqueous Activity Coefficients of Stable Organic Chemicals in the Marine Environment: Polychlorinated Biphenyls," *Mar. Sci.*, 6:41-53 (1978).

DiGeronimo, M.J., and A.D. Antoine. "Metabolism of Acetonitrile and Propionitrile by *Nocardia rhodochrous* LL 100-21," *Appl. Environ. Microbiol.*, 31(6):900-906 (1976).

Dilling, W.L. "Interphase Transfer Processes. II. Evaporation Rates of Chloro

Methanes, Ethanes, Ethylenes, Propanes, and Propylenes from Dilute Aqueous Solutions. Comparisons with Theoretical Predictions," *Environ. Sci. Technol.*, 11(4):405-409 (1977).

Dilling, W.L., Lickly, L.C., Lickly, T.D., Murphy, P.G., and R.L. McKellar. "Organic Photochemistry. 19. Quantum Yields for *O,O*-Diethyl *O*-(3,5,6-Trichloro-2-pyridinyl) Phosphorothioate (Chlorpyrifos) and 3,5,6-Trichloro-2-pyridinol in Dilute Aqueous Solutions and Their Environmental Phototransformation Rates," *Environ. Sci. Technol.* 18(7):540-543 (1984).

Dilling, W.L., Tefertiller, N.B., and G.J. Kallos. "Evaporation Rates and Reactivities of Methylene Chloride, Chloroform, 1,1,1-Trichloroethane, Trichloroethylene, Tetrachloroethylene, and Other Chlorinated Compounds in Dilute Aqueous Solutions," *Environ. Sci. Technol.*, 9(9):833-837 (1975).

Dobbs, A.J., and M.R. Cull. "Volatilisation of Chemicals - Relative Loss Rates and the Estimation of Vapour Pressures," *Environ. Pollut. (Series B)*, 3(4):289-298 (1982).

Documentation of the Threshold Limit Values and Biological Exposure Indices (Cincinnati, OH: American Conference of Governmental Industrial Hygienists, 1986), 744 p.

D'Oliveira, J.-C., Al-Sayyed, G., and P. Pichat. "Photodegradation of 2- and 3-Chlorophenol in TiO_2 Aqueous Suspensions," *Environ. Sci. Technol.*, 24(7):990-996 (1990).

Dong, S., and P.K. Dasgupta. "Solubility of Gaseous Formaldehyde in Liquid Water and Generation of Trace Standard Gaseous Formaldehyde," *Environ. Sci. Technol.*, 20(6):637-640 (1986).

Doucette, W.J., and A.W. Andren. "Estimation of Octanol/Water Partition Coefficients: Evaluation of Six Methods for Highly Hydrophobic Aromatic Hydrocarbons," *Chemosphere*, 17(2):345-359 (1988).

Draper, W.M., and D.G. Crosby. "Solar Photooxidation of Pesticides in Dilute Hydrogen Peroxide," *J. Agric. Food Chem.*, 32(2):231-237 (1984).

Dreisbach, R.R. *Pressure-Volume-Temperature Relationships of Organic Compounds* (Sandusky, OH: Handbook Publishers, 1952), 349 p.

Drinking Water and Health (Washington, DC: National Academy of Sciences, 1977), 939 p.

Drinking Water and Health (Washington, DC: National Academy of Sciences, 1980), 415 p.

Drinking Water Health Advisory. Pesticides. (Chelsea, MI: Lewis Publishers, Inc., 1989), 819 p.

Dugan, P.R. *Biochemical Ecology of Water Pollution* (New York: Plenum Press, 1972), 159 p.

Dulin, D., Drossman, H., and T. Mill. "Products and Quantum Yields for Photolysis of Chloroaromatics in Water," *Environ. Sci. Technol.*, 20(1):72-77 (1986).

Eadie, B.J., Morehead, N.R., and P.F. Landrum. "Three-Phase Partitioning of Hydrophobic Organic Compounds in Great Lakes Waters," *Chemosphere*, 20(1/2):161-178 (1990).

Eadsforth, C.V. "Application of Reverse-Phase H.P.L.C. for the Determination of Partition Coefficients," *Pestic. Sci.*, 17(3):311-325 (1986).

Eckert, J.W. "Fungistatic and Phytotoxic Properties of Some Derivatives of Nitrobenzene," *Phytopathology*, 52:642-649 (1962).

Edwards, D.A., Luthy, R.G., and Z. Liu. "Solubilization of Polycyclic Aromatic Hydrocarbons in Micellar Nonionic Surfactant Solutions," *Environ. Sci. Technol.*, 25(1):127-13 (1991).

Eganhouse, R.P., and J.A. Calder. "The Solubility of Medium Weight Aromatic Hydrocarbons and the Effect of Hydrocarbon Co-solutes and Salinity," *Geochim. Cosmochim. Acta*, 40(5):555-561 (1976).

Eiceman, G.A., and V.J. Vandiver. "Adsorption of Polycyclic Aromatic Hydrocarbons on Fly Ash from a Municipal Incinerator and a Coal-Fired Power Plant," *Atmos. Environ.*, 17(3):461-465 (1983).

Eichelberger, J.W., and J.J. Lichtenberg. "Persistence of Pesticides in River Water," *Environ. Sci. Technol.*, 5(6):541-544 (1971).

Eisenhauer, H.R. "The Ozonization of Phenolic Wastes," *J. Water Poll. Control Fed.*, 40(11):1887-1899 (1968).

Eisenreich, S.J., Looney, B.B., and J.D. Thornton. "Airborne Organic Contaminants in the Great Lakes Ecosystem," *Environ. Sci. Technol.*, 15(1):30-38 (1981).

Eklund, G., Pedersen, J.R., and B. Strömberg. "Formation of Chlorinated Organic Compounds During Combustion of Propane in the Presence of HCl," *Chemosphere*, 16(1):161-166 (1987).

Ellgehausen, H., D'Hondt, C., and R. Fuerer. "Reversed-phase Chromatography as a General Method for Determining Octan-1-ol/Water Partition Coefficients," *Pestic. Sci.*, 12:219-227 (1981).

Elliott, S. "Effect of Hydrogen Peroxide on the Alkaline Hydrolysis of Carbon Disulfide," *Environ. Sci. Technol.*, 24(2):264-267 (1990).

Elliott, S. "The Solubility of Carbon Disulfide Vapor in Natural Aqueous Systems," *Atmos. Environ.*, 23(9):1977-1980 (1989).

Engelhardt, G., Wallnofer, P.R., and O. Hutzinger. "The Microbial Metabolism of Di-*n*-butyl phthalate and Related Dialkyl Phthalates," *Bull. Environ. Contam. Toxicol.*, 13(3):342-347 (1975).

Environmental Health Criteria 10: Carbon disulfide (Geneva: World Health Organization, 1979), 100 p.

Environmental Health Criteria 26: Styrene (Geneva: World Health Organization, 1983), 123 p.

Environmental Health Criteria 28: Acrylonitrile (Geneva: World Health Organization, 1983), 125 p.

Environmental Health Criteria 38: Heptachlor (Geneva: World Health Organization, 1984), 81 p.

Environmental Health Criteria 45: Camphechlor (Geneva: World Health Organization, 1984), 66 p.

Environmental Health Criteria 71: Pentachlorophenol (Geneva: World Health Organization, 1987), 236 p.

Eon, C., Pommier, C., and G. Guiochon. "Vapor Pressures and Second Virial Coefficients of Some Five-membered Heterocyclic Derivatives," *J. Chem. Eng. Data*, 16(4):408-410 (1971).

Epling, G.A., McVicar, W.M., and A. Kumar. "Borohydride-Enhanced

Photodehalogenation of Aroclor 1232, 1242, 1254, and 1260," *Chemosphere*, 17(7):1355-1362 (1988).

"Extremely Hazardous Substances-Superfund Chemical Profiles. Volume 1," (Park Ridge, NJ: Noyes Data Corp., 1988), 932 p.

Fairbanks, B.C., and G.A. O'Connor. "Effect of Sewage Sludge on the Adsorption of Polychlorinated Biphenyls by Three New Mexico Soils," *J. Environ. Qual.*, 13(2):297-300 (1984).

Fairbanks, B.C., O'Connor, G.A., and S.E. Smith. "Fate of Di-2-(ethylhexyl)phthalate in Three Sludge-amended New Mexico Soils," *J. Environ. Qual.*, 14(4):479-483 (1985).

Fairbanks, B.C., Schmidt, N.E., and G.A. O'Connor. "Butanol Degradation and Volatilization in Soils Amended with Spent Acid or Sulfuric Acid," *J. Environ. Qual.*, 14(1):83-86 (1985).

Farm Chemicals Handbook (Willoughby, OH: Meister Publishing Co., 1988).

Fathepure, B.Z., Tiedje, J.M., and S.A. Boyd. "Reductive Dechlorination of Hexachlorobenzene to Tri- and Dichlorobenzenes in Anaerobic Sewage Sludge," *Appl. Environ. Microbiol.*, 54(2):327-330 (1988).

Fatiadi, A.J. "Effects of Temperature and of Ultraviolet Radiation on Pyrene Adsorbed on Garden Soil," *Environ. Sci. Technol.*, 1(7):570-572 (1967).

Felsot, A., and P.A. Dahm. "Sorption of Organophosphorus and Carbamate Insecticides by Soil," *J. Agric. Food Chem.*, 27(3):557-563 (1979).

Fendinger, N.J., and D.E. Glotfelty. "Henry's Law Constants for Selected Pesticides, PAHs and PCBs," *Environ. Toxicol. Chem.*, 9(6):731-735 (1990).

Fendinger, N.J., Glotfelty, D.W., and H.P. Freeman. "Comparison of Two Experimental Techniques for Determining Air/Water Henry's Law Constants," *Environ. Sci. Technol.*, 23(12):1528-1531 (1989).

Fire Protection Guide on Hazardous Materials (Quincy, MA: National Fire Protection Association, 1984), 443 p.

Fischer, I., and L. Ehrenberg. "Studies of the Hydrogen Bond. II. Influence of the Polarizability of the Heteroatom," *Acta Chem. Scand.*, 2:669-677 (1948).

Fishbein, L. "An Overview of Environmental and Toxicological Aspects of Aromatic Hydrocarbons. III. Xylene," *Sci. Tot. Environ.*, 43(1/2):165-183 (1985).

Fishbein, L. "Potential Halogenated Industrial Carcinogenic and Mutagenic Chemicals. III. Alkane Halides, Alkanols, and Ethers," *Sci. Tot. Environ.*, 11:223-257 (1979).

Fishbein, L., and P.W. Albro. "Chromatographic and Biological Aspects of the Phthalate Esters," *J. Chromatogr.*, 70(2):365-412 (1972).

Fluka Catalog 1988/89 - Chemika-Biochemika (Ronkonkoma, NY: Fluka Chemical Corp., 1988), 1536 p.

Fluka Catalog 1989/90 - Chemika-Biochemika (Ronkonkoma, NY: Fluka Chemical Corp., 1990), 1480 p.

Fordyce. R.G., and E.C. Chapin. "Copolymerization. I. The Mechanism of Emulsion Copolymerization of Styrene and Acrylonitrile," *J. Am. Chem Soc.*, 69(3):581-583 (1947).

Foreman, W.T., and T.F. Bidleman. "Vapor Pressure Estimates of Individual Polychlorinated Biphenyls and Commercial Fluids using Gas Chromatographic

Retention Data," *J. Chromatogr.*, 330(2):203-216 (1985).
Fowler, L., Trump, W.N., and C.E. Vogler. "Vapor Pressure of Naphthalene - New Measurements between 40° and 180 °C," *J. Chem. Eng. Data*, 13(2):209-210 (1968).
Frankel, L.S., McCallum, K.S., and L. Collier. "Formation of Bis(chloromethyl) Ether from Formaldehyde and Hydrogen Chloride," *Environ. Sci. Technol.*, 8(4):356-359 (1974).
Franklin, J.L., Dillard, J.G., Rosenstock, H.M., Herron, J.T., Draxl K., and F.H. Field. "Ionization Potentials, Appearance Potentials and Heats of Formation of Gaseous Positive Ions," National Bureau of Standards Report NSRDS-NBS 26, U.S. Government Printing Office (1969), 289 p.
Franks, F. "Solute-Water Interactions and the Solubility Behaviour of Long-chain Paraffin Hydrocarbons," *Nature*, 210(5031):87-88 (1966).
Freed, V.H., Chiou, C.T., and R. Haque. "Chemodynamics: Transport and Behavior of Chemicals in the Environment - A Problem in Environmental Health," *Environ. Health Perspect.*, 20:55-70 (1977).
Freed, V.H., Chiou, C.T., and D.W. Schmedding. "Degradation of Selected Organophosphate Pesticides in Water and Soil," *J. Agric. Food Chem.*, 27(4):706-708 (1979).
Freeze, R.A., and J.A. Cherry. *Groundwater* (Englewood Cliffs, NJ: Prentice-Hall, Inc., 1974), 604 p.
Freitag, D., Ballhorn, L., Geyer, H., and F. Korte. "Environmental Hazard Profile of Organic Chemicals," *Chemosphere*, 14(10):1589-1616 (1985).
Freitag, D., Scheunert, I., Klein, W., and F. Korte. "Long-Term Fate of 4-Chloroaniline-^{14}C in Soil and Plants under Outdoor Conditions. A Contribution to Terrestrial Ecotoxicology of Chemicals," *J. Agric. Food Chem.*, 32(2):203-207 (1984).
Fries, G.F. "Degradation of Chlorinated Hydrocarbons under Anaerobic Conditions," in *Fate of Organic Pesticides in the Aquatic Environment, Advances in Chemistry Series*, R.F. Gould, Ed. (Washington, DC: American Chemical Society, 1972), pp 256-270.
Friesen, K.J., Sarna, L.P., and G.R.B. Webster. "Aqueous Solubility of Polychlorinated Dibenzo-*p*-dioxins Determined by High Pressure Liquid Chromatography," *Chemosphere*, 14(9):1267-1274 (1985).
Fujita, T., Iwasa, J., and C. Hansch. "A New Substituent Constant, π, Derived from Partition Coefficients," *J. Am. Chem. Soc.*, 86(23):5175-5180 (1964).
Fukano, I., and Y. Obata. "Solubility of Phthalates in Water," [Chemical abstract 120601u, 86(17):486 (1977)]: *Purasuchikkusu*, 27(7):48-49 (1976).
Fukui, S., Hirayama, T., Shindo, H., and M. Nohara. "Photochemical Reaction of Biphenyl (BP) and *o*-Phenylphenol (OPP) with Nitrogen Monoxide (1)," *Chemosphere*, 9(12):771-775 (1980).
Fuller, B.B. *Air Pollution Assessment of Tetrachloroethylene* (McLean, VA: The Mitre Corp., 1976), 87 p.
Funasaki, N., Hada, S., and S. Neya. "Partition Coefficients of Aliphatic Ethers - Molecular Surface Area Approach," *J. Phys. Chem.*, 89(14):3046-3049 (1985).
Gäb, S., Parlar, H., Nitz, S., Hustert, K., and F. Korte. "Beitrage zur Okologischen

Chemie. LXXXI. Photochemischer Abbau von Aldrin, Dieldrin and Photodieldrin als Festkorper im Sauerstoffstrom," *Chemosphere*, 3(5):183-186 (1974).

Gaffney, J.S., Streit, G.E., Spall, W.D., and J.H. Hall. "Beyond Acid Rain," *Environ. Sci. Technol.*, 21(6):519-524 (1987).

Galassi, S., Mingazzini, M., Viganò, L., Cesareo, D., and M.L. Tosato. "Approaches to Modeling Toxic Responses of Aquatic Organisms to Aromatic Hydrocarbons," *Ecotoxicol. Environ. Safety*, 16(2):158-169 (1988).

Gälli, R., and P.L. McCarty. "Biotransformation of 1,1,1- Trichloroethane, Trichloromethane, and Tetrachloromethane by a *Clostridium* sp.," *Appl. Environ. Microbiol.*, 55(4):837-844 (1989).

Garbarini, D.R., and L.W. Lion. "Influence of the Nature of Soil Organics on the Sorption of Toluene and Trichloroethylene," *Environ. Sci. Technol.*, 20(12):1263-1269 (1986).

Garten, C.T., and J.R. Trabalka. "Evaluation of Models for Predicting Terrestrial Food Chain Behavior of Xenobiotics," *Environ. Sci. Technol.*, 17(10):590-595 (1983).

Gauthier, T.D., Shane, E.C., Guerin, W.F., Seitz, W.R., and C.L. Grant. "Fluorescence Quenching Method for Determining Equilibrium Constants for Polycyclic Aromatic Hydrocarbons Binding to Humic Materials," *Environ. Sci. Technol.*, 20(11):1162-1166 (1986).

Gay, B.W., Jr., Hanst, P.L., Bufalini, J.J., and R.C. Noonan. "Atmospheric Oxidation of Chlorinated Ethylenes," *Environ. Sci. Technol.*, 10(1):58-67 (1976).

"General Industry Standards for Toxic and Hazardous Substances," U.S. Code of Federal Regulations 1910, Subpart Z Section 1910.1000 (July 1982).

Georgacakis, E., and M.A.Q. Khan. "Toxicity of the Photoisomers of Cyclodiene Insecticides to Freshwater Animals," *Nature*, 233(5315):120-121 (1971).

Geyer, H., Kraus, A.S., Klein, W., Richter, E., and F. Korte. "Relationship between Water Solubility and Bioaccumulation Potential of Organic Chemicals in Rats," *Chemosphere*, 9(5/6):277-294 (1980).

Geyer, H., Politzki, G., and D. Freitag. "Prediction of Ecotoxicological Behaviour of Chemicals: Relationship between *n*-Octanol/Water Partition Coefficient and Bioaccumulation of Organic Chemicals by Alga *Chlorella*," *Chemosphere*, 13(2):269-284 (1984).

Geyer, H., Sheehan, P., Kotzias, D., Freitag, D., and F. Korte. "Prediction of Ecotoxicological Behaviour of Chemicals: Relationship between Physico-Chemical Properties and Bioaccumulation of Organic Chemicals in the Mussell *Mytilus edulis*," *Chemosphere*, 11(11):1121-1134 (1982).

Geyer, H.J., Scheunert, I., and F. Korte. "Correlation between the Bioconcentration Potential of Organic Environmental Chemicals in Humans and Their *n*-Octanol/Water Partition Coefficients," *Chemosphere*, 16(1):239-252 (1987).

Gherini, S.A., Summers, K.V., Munson, R.K., and W.B. Mills. *Chemical Data for Predicting the Fate of Organic Compounds in Water, Volume 2: Database*, (Lafayette, CA: Tetra Tech, Inc., 1988), 433 p.

Giam, C.S., Atlas, E., Chan, H.S., and G.S. Neff. "Phthalate Esters, PCB and DDT Residues in the Gulf of Mexico Atmosphere," *Atmos. Environ.*, 14(1):65-69

(1980).

Gibbard, H.F., and J.L. Creek. "Vapor Pressure of Methanol from 288.15 to 337.65K," *J. Chem. Eng. Data*, 19(4):308-310 (1974).

Gibson, D.T. "Microbial Degradation of Aromatic Compounds," *Science*, 161(3846):1093-1097 (1968).

Gile, J.D., and J.W. Gillett. "Fate of Selected Fungicides in a Terrestrial Laboratory Ecosystem," *J. Agric. Food Chem.*, 27(6):1159-1164 (1979).

Ginnings, P.M., Plonk, D., and E. Carter. "Aqueous Solubilities of Some Aliphatic Ketones," *J. Am. Chem. Soc.*, 62(8):1923-1924 (1940).

Gledhill, W.E., Kaley, R.G., Adams, W.J., Hicks, O., Michael, P.R., and V.W. Saeger. "An Environmental Safety Assessment of Butyl Benzyl Phthalate," *Environ. Sci. Technol.*, 14(3):301-305 (1980).

Godsy, E.M., Goerlitz, D.F., and G.G. Ehrlich. "Methanogenesis of Phenolic Compounds by a Bacterial Consortium from a Contaminated Aquifer in St. Louis Park, Minnesota," *Bull. Environ. Contam. Toxicol.*, 30(3):261-268 (1983).

Gordon, A.J., and R.A. Ford. *The Chemist's Companion* (New York: John Wiley and Sons, Inc., 1972), 551 p.

Gordon, J.E., and R.L. Thorne. "Salt Effects on the Activity Coefficient of Naphthalene in Mixed Aqueous Electrolyte Solutions. I. Mixtures of Two Salts," *J. Phys. Chem.*, 71(13):4390-4399 (1967).

Gossett, J.M. "Measurement of Henry's Law Constants for C_1 and C_2 Chlorinated Hydrocarbons," *Environ. Sci. Technol.*, 21(2):202-208 (1987).

Graveel, J.G., Sommers, L.E., and D.W. Nelson. "Decomposition of Benzidine, α-Naphthylamine, and *p*-Toluidine in Soils," *J. Environ. Qual.*, 15(1):53-59 (1986).

Green, W.J., Lee, G.F., Jones, R.A., and T. Palit. "Interaction of Clay Soils with Water and Organic Solvents: Implications for the Disposal of Hazardous Wastes," *Environ. Sci. Technol.*, 17(5):278-282 (1983).

Greene, S., Alexander, M., and D. Leggett. "Formation of *N*-Nitrosodimethylamine During Treatment of Municipal Waste Water by Simulated Land Application," *J. Environ. Qual.*, 10(3):416-421 (1981).

Grosjean, D. "Atmospheric Reactions of Ortho Cresol: Gas Phase and Aerosol Products," *Atmos. Environ.*, 19(8):1641-1652 (1984).

Grosjean, D. "Atmospheric Reactions of Styrenes and Peroxybenzoyl nitrate," *Sci. Tot. Environ.*, 50:41-59 (1985).

Grosjean, D. "Photooxidation of Methyl Sulfide, Ethyl Sulfide, and Methanethiol," *Environ. Sci. Technol.*, 18(6):460-468 (1984).

Gross, F.C., and J.A. Colony. "The Ubiquitous Nature and Objectionable Characteristics of Phthalate Esters in Aerospace Industry," *Environ. Health Perspect.*, (January 1973), pp 37-48.

Gross, P. "The Determination of the Solubility of Slightly Soluble Liquids in Water and the Solubilities of the Dichloro- Ethanes and -Propanes," *J. Am. Chem. Soc.*, 51(8):2362-2366 (1929).

Gross, P.M., and J.H. Saylor. "The Solubilities of Certain Slightly Soluble Organic Compounds in Water," *J. Am. Chem. Soc.*, 53(5):1744-1751 (1931).

Gross, P.M., Saylor, J.H., and M.A. Gorman. "Solubility Studies. IV. The Solubilities of Certain Slightly Soluble Organic Compounds in Water," *J. Am.*

Chem. Soc., 55(2):650-652 (1933).
Groves, F., Jr. "Solubility of Cycloparaffins in Distilled Water and Salt Water," *J. Chem. Eng. Data*, 33(2):136-138 (1988).
"Guidelines Establishing Test Procedures for the Analysis of Pollutants," U.S. Code of Federal Regulations, 40 CFR 136, 44(233):69464-69575.
Gunther, F.A., and J.D. Gunther. "Residues of Pesticides and Other Foreign Chemicals in Foods and Feeds," *Res. Rev.*, 36:69-77 (1971).
Gunther, F.A., Westlake, W.E., and P.S. Jaglan. "Reported Solubilities of 738 Pesticide Chemicals in Water," *Res. Rev.*, 20:1-148 (1968).
Guseva, A.N. and E.I. Parnov. "Isothermal Cross-Sections of the Systems Cyclanes-Water," *Vestn. Mosk. Univ., Ser. II Khim.*, 19:77 (1964).
Guswa, J.H., Lyman, W.J., Donigan, A.S., Jr., Lo, T.Y.R., and E.W. Shanahan. *Groundwater Contamination and Emergency Response Guide* (Park Ridge, NJ: Noyes Publications, 1984), 490 p.
Hagin, R.D., Linscott, D.L., and J.E. Dawson. "2,4-D Metabolism in Resistant Grasses," *J. Agric. Food Chem.*, 18:848-850 (1970).
Haigler, B.E., Nishino, S.F., and J.C. Spain. "Degradation of 1,2-Dichlorobenzene by a *Pseudomonas* sp.," *Appl. Environ. Microbiol.*, 54(2):294-301 (1988).
Hakuta, T., Negishi, A., Goto, T., Kato, J., and S. Ishizaka. "Vapor-Liquid Equilibria of Some Pollutants in Aqueous and Saline Solutions," *Desalination*, 21(1):11-21 (1977).
Hallas, L.E., and M. Alexander. "Microbial Transformation of Nitroaromatic Compounds in Sewage Effluent," *Appl. Environ. Microbiol.*, 45(4):1234-1241 (1983).
Hammers, W.E., Meurs, G.J., and C.L. de Ligny. "Correlations between Liquid Chromatographic Capacity Ratio Data on Lichrosorb RP-18 and Partition Coefficients in the Octanol-Water System," *J. Chromatogr.*, 247:1-13 (1982).
Hansch, C., and S.M. Anderson. "The Effect of Intramolecular Hydrophobic Bonding on Partition Coefficients," *J. Org. Chem.*, 32(8):2583-2586 (1967).
Hansch, C., and T. Fujita. "ρ-σ-π Analysis. A Method for the Correlation of Biological Activity and Chemical Structure," *J. Am. Chem. Soc.*, 86(8):1616-1617 (1964).
Hansch, C., and A. Leo. *Substituent Constants for Correlation Analysis in Chemistry and Biology* (New York: John Wiley and Sons, Inc., 1979), 339 p.
Hansch, C., Leo, A., and D. Nikaitani. "On the Additive-Constitutive Character of Partition Coefficients," *J. Org. Chem.*, 37(20):3090-3092 (1972).
Hansch, C., Quinlan, J.E., and G.L. Lawrence. "The Linear Free-Energy Relationship between Partition Coefficients and Aqueous Solubility of Organic Liquids," *J. Org. Chem.*, 33(1):347-350 (1968).
Hansch, C., Vittoria, A., Silipo, C., and P.Y.C. Jow. "Partition Coefficients and the Structure-Activity Relationship of the Anesthetic Gases," *J. Med. Chem.*, 18(6):546-548 (1975).
Hanst, P.L. "Noxious Trace Gases in the Air, Part II: Halogenated Pollutants," *Chemistry*, 51(2):6-12.
Hanst, P.L., and B.W. Gay Jr. "Atmospheric Oxidation of Hydrocarbons: Formation of Hydroperoxides and Peroxyacids," *Atmos. Environ.*, 17(11):2259-

2265 (1983).

Hanst, P.L., and B.W. Gay, Jr. "Photochemical Reactions Among Formaldehyde, Chlorine, and Nitrogen Dioxide in Air," *Environ. Sci. Technol.*, 11(12):1105-1109 (1977).

Hanst, P.L., Spence, J.W., and M. Miller. "Atmospheric Chemistry of *N*-Nitrosodimethylamine," *Environ. Sci. Technol.*, 11(4):403-405 (1977).

Haque, R., Ed. *Dynamics, Exposure and Hazardous Assessment of Toxic Chemicals* (Ann Arbor, MI: Ann Arbor Science Publishers, Inc., 1980), 496 p.

Haque, R., Schmedding, D.W., and V.H. Freed. "Aqueous Solubility, Adsorption, and Vapor Behavior of Polychlorinated Biphenyl Aroclor 1254," *Environ. Sci. Technol.*, 8(2):139-142 (1974).

Harbison, K.G., and R.T. Belly. "The Biodegradation of Hydroquinone," *Environ. Toxicol. Chem.*, 1(1):9-15 (1982).

Harnisch, M., Mockel, H.J., and G. Schulze. "Relationship between Log P_{ow} Shake-Flask Values and Capacity Factors Derived from Reversed-Phase High Performance Liquid Chromatography for *n*-Alkylbenzene and some OECD Reference Substances," *J. Chromatogr.*, 282:315 (1983).

Harrison, R.M., Perry, R., and R.A. Wellings. "Polynuclear Aromatic Hydrocarbons in Raw, Potable and Waste Waters," *Water Res.*, 9:331-346 (1975).

Hassett, J.J., Means, J.C., Banwart, W.L., and S.G. Wood. "Sorption Properties of Sediments and Energy-Related Pollutants," Office of Research and Development, U.S. EPA Report-600/3-80-041 (1980), 150 p.

Hatakeyama, S., Ohno, M., Weng, J., Takagi, H., and H. Akimoto. "Mechanism for the Formation of Gaseous and Particulate Products from Ozone-Cycloalkene Reactions in Air," *Environ. Sci. Technol.*, 21(1):52-57 (1987).

Hatch, L.F., and L.E. Kidwell, Jr. "Preparation and Properties of 1-Bromo-1-propyne, 1,3-Dibromopropyne and 1-Bromo-3-chloro-1-propyne," *J. Am. Chem. Soc.*, 76(1):289-290 (1954).

Hawker, D.W., and D.W. Connell. "Influence of Partition Coefficient of Lipophilic Compounds on Bioconcentration Kinetics with Fish," *Water Res.*, 22(6):701-707 (1988).

Hawley, G.G. *The Condensed Chemical Dictionary* (New York: Van Nostrand Reinhold Co., 1981), 1135 p.

Hawthorne, S.B., Sievers, R.E., and R.M. Barkley. "Organic Emissions from Shale Oil Wastewaters and Their Implications for Air Quality," *Environ. Sci. Technol.*, 19(10):992-997 (1985).

Hayduk, W., and H. Laudie. "Vinyl Chloride Gas Compressibility and Solubility in Water and Aqueous Potassium Laurate Solutions," *J. Chem. Eng. Data*, 19(3):253-257 (1974).

Hayes, W.J. *Pesticides Studied in Man* (Baltimore, MD: The Williams and Wilkens Co., 1982), pp 234-247.

Healy, J.B., and L.Y. Young. "Anaerobic Biodegradation of Eleven Aromatic Compounds to Methane," *Appl. Environ. Microbiol.*, 38(1):85-89 (1979).

Heitkamp, M.A., Freeman, J.P., Miller, D.W., and C.C. Cerniglia. "Pyrene Degradation by a Mycobacterium sp.: Identification of Ring Oxidation and Ring Fission Products," *Appl. Environ. Microbiol.*, 54(10):2556-2565 (1988).

Henderson, G.L., and D.G. Crosby. "The Photodecomposition of Dieldrin Residues in Water," *Bull. Environ. Contam. Toxicol.*, 3(3):131-134 (1968).

Heritage, A.D., and I.C. MacRae. "Degradation of Lindane by Cell-free Preparations of *Clostridium sphenoides*," *Appl. Environ. Microbiol.*, 34(2):222-224 (1977).

Heritage, A.D., and I.C. MacRae. "Identification of Intermediates Formed during the Degradation of Hexachlorocyclohexanes by *Clostridium sphenoides*," *Appl. Environ. Microbiol.*, 33(6):1295-1297 (1977).

Hermann, R.B. "Theory of Hydrophobic Bonding. II. The Correlation of Hydrocarbon Solubility in Water with Solvent Cavity Surface Area," *J. Phys. Chem.*, 76(19):2754-2759 (1972).

Herzel, F., and A.S. Murty. "Do Carrier Solvents Enhance the Water Solubility of Hydrophobic Compounds?," *Bull. Environ. Contam. Toxicol.*, 32(1):53-58 (1984).

Hill, A.E., and R. Macy. "Ternary Systems. II. Silver Perchlorate, Aniline and Water," *J. Am. Chem. Soc.*, 46:1132-1143 (1924).

Hill, E.A. "On a System of Indexing Chemical Literature; Adopted by the Classification Division of the U.S. Patent Office," *J. Am. Chem. Soc.*, 22(8):478-494 (1900).

Hill, J., Kollig, H.P., Paris, D.F., Wolfe, N.L., and R.G. Zepp. "Dynamic Behavior of Vinyl Chloride in Aquatic Ecosystems," U.S. EPA Report-600/3-76-001 (1976), 59 p.

Hine, J., Haworth, H.W., and O.B. Ramsey. "Polar Effects on Rates and Equilibria. VI. The Effect of Solvent on the Transmission of Polar Effects," *J. Am. Chem. Soc.*, 85(10):1473-1475 (1963).

Hine, J., and P.K. Mookerjee. "The Intrinsic Hydrophilic Character of Organic Compounds. Correlations in Terms of Structural Contributions," *J. Org. Chem.*, 40(3):292-298 (1975).

Hinga, K.R., and M.E.Q. Pilson. "Persistence of Benz[*a*]anthracene Degradation Products in an Enclosed Marine Ecosystem," *Environ. Sci. Technol.*, 21(7):648-653 (1987).

Hinga, K.R., Pilson, M.E.Q., Lee, R.F., Farrington, J.W., Tjessem, K., and A.C. Davis. "Biogeochemistry of Benzanthracene in an Enclosed Marine Ecosystem," *Environ. Sci. Technol.*, 14(9):1136-1143 (1980).

Hirzy, J.W., Adams, W.J., Gledhill, W.E., and J.P. Mieure. "Phthalate Esters: The Environmental Issues," unpublished seminar document, Monsanto Industrial Chemicals Co. (1978), 54 p.

Hodgman, C.D., Weast, R.C., Shankland, R.S., and S.M. Selby. *Handbook of Chemistry and Physics* (Cleveland, OH: Chemical Rubber Publishing Co., 1961).

Hodson, J., and N.A. Williams. "The Estimation of the Adsorption Coefficient (K_{oc}) for Soils by High Performance Liquid Chromatography," *Chemosphere*, 19(1):67-77 (1988).

Hollifield, H.C. "Rapid Nephelometric Estimate of Water Solubility of Highly Insoluble Organic Chemicals of Environmental Interest," *Bull. Environ. Contam. Toxicol.*, 23(4/5):579-586 (1979).

Houser, J.J., and B.A. Sibbio. "Liquid-Phase Photolysis of Dioxane," *J. Org. Chem.*, 42(12):2145-2151 (1977).

Howard, P.H., and P.G. Deo. "Degradation of Aryl Phosphates in Aquatic Environments," *Bull. Environ. Contam. Toxicol.*, 22(3):337-344 (1979).
Howard, P.H., Banerjee, S., and K.H. Robillard. "Measurement of Water Solubilities, Octanol/Water Partition Coefficients and Vapor Pressures of Commercial Phthalate Esters," *Environ. Toxicol. Chem.*, 4:653-661 (1985).
Howard, P.H., and P.R. Durkin. "Sources of Contamination, Ambient Levels, and Fate of Benzene in the Environment," Office of Toxic Substances, U.S. EPA Report-560/5-75-005 (1974), 73 p.
Howard, P.H. *Handbook of Environmental Fate and Exposure Data for Organic Chemicals - Volume I. Large Production and Priority Pollutants* (Chelsea, MI: Lewis Publishers, Inc., 1989), 574 p.
Howard, P.H. *Handbook of Environmental Fate and Exposure Data for Organic Chemicals - Volume II. Solvents* (Chelsea, MI: Lewis Publishers, Inc., 1990), 546 p.
Hsu, C.C., and J.J. McKetta. "Pressure-Volume-Temperature Properties of Methyl Chloride," *J. Chem. Eng. Data*, 9(1):45-51 (1964).
Huntress, E.H., and S.P. Mulliken. *Identification of Pure Organic Compounds - Tables of Data on Selected Compounds of Order I* (New York: John Wiley and Sons, Inc., 1941), 691 p.
Hustert, K., Kotzias, D., and F. Korte. "Photokatalytischer Abbau von Chlornitrobenzolen an TiO_2 in Wäßriger Phase," *Chemosphere*, 16(4):809-812 (1987).
Hustert, K., and P.N. Moza. "Photokatalytischer Abbau von Phthalaten an Titandioxid in Wässriger Phase," *Chemosphere*, 17(9):1751-1754 (1988).
Hutzinger, O., Safe, S., and V. Zitko. *The Chemistry of PCB's* (Boca Raton, FL: CRC Press, Inc., 1974), 269 p.
Hwang, H.-M., Hodson, R.E., and R.F. Lee. "Degradation of Phenol and Chlorophenols by Sunlight and Microbes in Estuarine Water," *Environ. Sci. Technol.*, 20(10):1002-1007 (1986).
IARC Monographs on the Evaluation of Carcinogenic Risk of Chemicals to Man. Certain Polycyclic Aromatic Hydrocarbons and Heterocyclic Compounds, Volume 3 (Lyon, France: International Agency for Research on Cancer, 1973), 271 p.
IARC Monographs on the Evaluation of Carcinogenic Risk of Chemicals to Man. Some Halogenated Hydrocarbons, Volume 20 (Lyon, France: International Agency for Research on Cancer, 1979), 609 p.
IARC Monographs on the Evaluation of Carcinogenic Risk of Chemicals to Man. Some N-Nitroso Compounds Volume 17 (Lyon, France: International Agency for Research on Cancer, 1978), 365 p.
IARC Monographs on the Evaluation of Carcinogenic Risk of Chemicals to Man. Some Organochlorine Pesticides, Volume 4 (Lyon, France: International Agency for Research on Cancer, 1974), 241 p.
IARC Monographs on the Evaluation of the Carcinogenic Risk of Chemicals to Humans. Polynuclear Aromatic Compounds, Part 1, Chemical, Environmental and Experimental Data, Volume 32 (Lyon, France: International Agency for Research on Cancer, 1983), 477 p.
Ibusuki, T., and K. Takeuchi. "Toluene Oxidation on U.V.-Irradiated Titanium

Dioxide With and Without O_2, NO_2 or H_2O at Ambient Temperature," *Atmos. Environ.*, 29(9):1711-1715 (1986).

Instruction Manual - Model ISP1 101: Intrinsically Safe Portable Photoionization Analyzer (Newton, MA: HNU Systems, Inc., 1986), 86 p.

Isaac, R.A., and J.C. Morris. "Transfer of Active Chlorine from Chloramine to Nitrogenous Organic Compounds. 1. Kinetics," *Environ. Sci. Technol.*, 17(12):738-742 (1983).

Isaacson, P.J., and C.R. Frink. "Nonreversible Sorption of Phenolic Compounds by Sediment Fractions: The Role of Sediment Organic Matter," *Environ. Sci. Technol.*, 18(1):43-48 (1984).

Isensee, A.R., Holden, E.R., Woolson, E.A., and G.E. Jones. "Soil Persistence and Aquatic Bioaccumulation Potential of Hexachlorobenzene," *J. Agric. Food Chem.*, 24(6):1210-1214 (1976).

Ishikawa, S., and K. Baba. "Reaction of Organic Phosphate Esters with Chlorine in Aqueous Solution," *Bull. Environ. Contam. Toxicol.*, 41(1):143-150 (1988).

Isnard, S., and S. Lambert. "Estimating Bioconcentration Factors from Octanol-Water Partition Coefficient and Aqueous Solubility," *Chemosphere*, 17(1):21-34 (1988).

Ivie, G.W., and J.E. Casida. "Enhancement of Photoalteration of Cyclodiene Insecticide Chemical Residues by Rotenone," *Science*, 167:1620-1622 (1970).

Ivie, G.W., and J.E. Casida. "Photosensitizers for the Accelerated Degradation of Chlorinated Cyclodienes and Other Insecticide Chemicals Exposed to Sunlight on Bean Leaves," *J. Agric. Food Chem.*, 19(3):410-416 (1971).

Ivie, G.W., and J.E. Casida. "Sensitized Photodecomposition and Photosensitizer Activity of Pesticide Chemicals Exposed to Sunlight of Silica Gel Chromatoplates," *J. Agric. Food Chem.*, 19(3):405-409 (1971).

Ivie, G.W., Knox, J.R., Khalifa, S., Yamamoto, I., and J.E. Casida. "Novel Photoproducts of Heptachlor Epoxide, *Trans*-Chlordane, and *Trans*-Nonachlor," *Bull. Environ. Contam. Toxicol.*, 7(6):376-383 (1972).

Iwasaki, M., Ibuki, T., and Y. Takezaki. "Primary Processes of the Photolysis of Ethylenimine at Xe and Kr Resonance Lines," *J. Chem. Phys.*, 59(12):6321-6327 (1973).

Jacobson, S.N., and M. Alexander. "Enhancement of the Microbial Dehalogenation of a Model Chlorinated Compound," *Appl. Environ. Microbiol.*, 42(6):1062-1066 (1981).

Jafvert, C.T., Westall, J.C., Grieder, E., and R.P. Schwarzenbach. "Distribution of Hydrophobic Ionogenic Organic Compounds between Octanol and Water: Organic Acids," *Environ. Sci. Technol.*, 24(12):1795-1803 (1990).

Jafvert, C.T., and N.L. Wolfe. "Degradation of Selected Halogenated Ethanes in Anoxic Sediment-Water Systems," *Environ. Toxicol. Chem.*, 6(11):827-837 (1987).

James, T.L., and C.A. Wellington. "Thermal Decomposition of β-Propiolactone in the Gas Phase," *J. Am. Chem. Soc.*, 91(27):7743-7746 (1969).

Janssen, D.B., Jager, D., and B. Wilholt. "Degradation of *n*-Haloalkanes and α,ω-Dihaloalkanes by Wild Type and Mutants of *Acinetobacter* sp. Strain GJ70," *Appl. Environ. Microbiol.*, 53(3):561-566 (1987).

Jeffers, P.M., Ward, L.M., Woytowitch, L.M., and L.N. Wolfe. "Homogeneous Hydrolysis Rate Constants for Selected Chlorinated Methanes, Ethanes, Ethenes, and Propanes," *Environ. Sci. Technol.*, 23(8):965-969 (1989).

Jerina, D.M., Daly, J.W., Jeffrey, A.M., and D.T. Gibson. "*cis*-1,2-Dihydroxy-1,2-dihydronaphthalene: A Bacterial Metabolite from Naphthalene," *Arch. Biochem. Biophys.*, 142:394-396 (1971).

Jewett, D., and J.G. Lawless. "Formate Esters of 1,2-Ethanediol: Major Decomposition Products of *p*-Dioxane During Storage," *Bull. Environ. Contam. Toxicol.*, 25(1):118-121 (1980).

Jigami, Y., Omori, T., and Y. Minoda. "The Degradation of Isopropylbenzene and Isobutylbenzene by *Pseudomonas* sp." *Agric. Biol. Chem.*, 39(9):1781-1788 (1975).

Johnsen, S., Gribbestad, I.S., and S. Johansen. "Formation of Chlorinated PAH - A Possible Health Hazard from Water Chlorination," *Sci. Tot. Environ.*, 81/82:231-238 (1989).

Johnson, C.A., and J.C. Westall. "Effect of pH and KCl Concentration on the Octanol-Water Distribution of Methylanilines," *Environ. Sci. Technol.*, 24(12):1869-1875 (1990).

Jordan, T.E. *Vapor Pressures of Organic Compounds* (New York: Interscience Publishers, Inc., 1954), 266 p.

Joschek, H.I., and S.T. Miller. "Photooxidation of Phenol, Cresols, and Dihydroxybenzenes," *J. Am. Chem Soc.*, 88:3273-3281 (1966).

Jury, W.A., Focht, D.D., and W.J. Farmer. "Evaluation of Pesticide Pollution Potential from Standard Indices of Soil-Chemical Adsorption and Biodegradation," *J. Environ. Qual.*, 16(4):422-428 (1987).

Jury, W.A., Spencer, W.F., and W.J. Farmer. "Behavior Assessment Model for Trace Organics in Soil: III. Application of Screening Model," *J. Environ. Qual.*, 13(4):573-579 (1984).

Jury, W.A., Spencer, W.F., and W.J. Farmer. "Use of Models for Assessing Relative Volatility, Mobility, and Persistence of Pesticides and Other Trace Organics in Soil Systems," in *Hazard Assessment of Chemicals, Volume 2*, J. Saxena, Ed. (New York: Academic Press, Inc., 1983), pp 1-43.

Kaars Sijpesteijn, A., Dekhuijzen, H.M., and J.W. Vonk. "Biological Conversion of Fungicides in Plants and Microorganisms" in *Antifungal Compounds*, (New York: Marcel Dekker, 1977), pp 91-147.

Kaiser, K.L.E., and I. Valdmanis. "Apparent Octanol/Water Partition Coefficients of Pentachlorophenol as a Function of pH," *Can. J. Chem.*, 60(16):2104-2106 (1982).

Kaiser, K.L., and P.T.S. Wong. "Bacterial Degradation of Polychlorinated Biphenyls. I. Identification of Some Metabolic Products from Aroclor 1242," *Bull. Environ. Contam. Toxicol.*, 11(3):291-296 (1974).

Kallos, G.J., and J.C. Tou. "Study of Photolytic Oxidation and Chlorination Reactions of Dimethyl Ether and Chlorine in Ambient Air," *Environ. Sci. Technol.*, 11(12):1101-1105 (1977).

Kamlet, M.J., Abraham, M.H., Doherty, R.M., and R.W. Taft. "Solubility Properties in Polymers and Biological Media. 4. Correlation of Octanol/Water

Partition Coefficients with Solvatochromic Parameters," *J. Am. Chem. Soc.*, 106(2):464-466 (1984).

Kamlet, M.J., Doherty, R.M., Abraham, M.H., Carr, P.W., Doherty, R.F., and R.W. Taft. "Linear Solvation Energy Relationships. 41. Important Differences between Aqueous Solubility Relationships for Aliphatic and Aromatic Solutes," *J. Phys. Chem.*, 91(7):1996-2004 (1987).

Kanazawa, J. "Relationship between the Soil Sorption Constants for Pesticides and Their Physicochemical Properties," *Environ. Toxicol. Chem.*, 8(6):477-484 (1989).

Kanno, S., and K. Nojima. "Studies on Photochemistry of Aromatic Hydrocarbons. V. Photochemical Reaction of Chlorobenzene with Nitrogen Oxides in Air," *Chemosphere*, 8(4):225-232 (1979).

Kapoor, I.P., Metcalf, R.L., Hirwe, A.S., Coats, J.R., and M.S. Khalsa. "Structure Activity Correlations of Biodegradability of DDT Analogs," *J. Agric. Food Chem.*, 21(2):310-315 (1973).

Kapoor, I.P., Metcalf, R.L., Nystrom, R.F., and G.K. Sanghua. "Comparative Metabolism of Methoxychlor, Methiochlor, and DDT in Mouse, Insects, and in a Model Ecosystem," *J. Agric. Food Chem.*, 18(6):1145-1152 (1970).

Karickhoff, S.W. "Correspondence - On the Sorption of Neutral Organic Solutes in Soils," *J. Agric. Food Chem.*, 29(2):425-426 (1981).

Karickhoff, S.W., Brown, D.S., and T.A. Scott. "Sorption of Hydrophobic Pollutants on Natural Sediments," *Water Res.*, 13:241-248 (1979).

Kawamoto, K., and K. Urano. "Parameters for Predicting Fate of Organochlorine Pesticides in the Environment (I) Octanol-Water and Air-Water Partition Coefficients," *Chemosphere*, 18(9/10):1987-1996 (1989).

Kazano, H., Kearney, P.C., and D.D. Kaufman. "Metabolism of Methylcarbamate Insecticides in Soils," *J. Agric. Food Chem.*, 20(5):975-979 (1972).

Kearney, P.C., and D.D. Kaufman. *Herbicides: Chemistry, Degradation and Mode of Action* (New York: Marcel Dekker, Inc., 1976), 1036 p.

Keck, J., Sims, R.C., Coover, M., Park, K., and B. Symons. "Evidence for Cooxidation of Polynuclear Aromatic Hydrocarbons in Soil," *Wat. Res.*, 23(12):1467-1476 (1989).

Keely, D.F., Hoffpauir, M.A., and J.R. Meriwether. "Solubility of Aromatic Hydrocarbons in Water and Sodium Chloride Solutions of Different Ionic Strengths: Benzene and Toluene," *J. Chem. Eng. Data*, 33(2):87-89 (1988).

Keen, R., and C.R. Baillod. "Toxicity to *Daphnia* of the End Products of Wet Oxidation of Phenol and Substituted Phenols," *Water Res.*, 19(6):767-772 (1985).

Kenaga, E.E. "Correlation of Bioconcentration Factors of Chemicals in Aquatic and Terrestrial Organisms with Their Physical and Chemical Properties," *Environ. Sci. Technol.*, 14(5):553-556 (1980).

Kenaga, E.E. *Environmental Dynamics of Pesticides* (New York: Plenum Press, 1975), 243 p.

Kenaga, E.E. "Predicted Bioconcentration Factors and Soil Sorption Coefficients of Pesticides and Other Chemicals," *Ecotoxicol. Environ. Safety*, 4(1):26-38 (1980).

Kenaga, E.E., and C.A.I. Goring. "Relationship between Water Solubility, Soil Sorption, Octanol-Water Partitioning and Concentration of Chemicals in Biota," in *Aquatic Toxicology, ASTM STP 707*, Eaton, J.G., Parrish, P.R., and A.C.

Hendricks, Eds. (Philadelphia, PA: American Society for Testing and Materials 1980), pp 78-115.

Kenley, R.A., Davenport, J.E., and D.G. Hendry. "Hydroxyl Radical Reactions in the Gas Phase. Products and Pathways for the Reaction of OH with Toluene," *J. Phys. Chem.*, 82(9):1095-1096 (1978).

Khaledi, M.G., and E.D. Breyer. "Quantitation of Hydrophobicity with Micellar Liquid Chromatography," *Anal. Chem.*, 61(9):1040-1047 (1989).

Khalil, M.A.K., and R.A. Rasmussen. "Global Sources, Lifetimes and Mass Balances of Carbonyl Sulfide (OCS) and Carbon Disulfide (CS_2) in the Earth's Atmosphere," *Atmos. Environ.*, 18(9):1805-1813 (1984).

Kieatiwong, S., Nguyen, L.V., Hebert, V.R., Hackett, M., Miller, G.C., Miille, M.J., and R. Mitzel. "Photolysis of Chlorinated Dioxins in Organic Solvents and on Soils," *Environ. Sci. Technol.*, 24(10):1575-1580 (1990).

Kilzer, L., Scheunert, I., Geyer, H., Klein, W., and F. Korte. "Laboratory Screening of the Volatilization Rates of Organic Chemicals from Water and Soil," *Chemosphere*, 8(10):751-761 (1979).

Kim, I.-Y., and F.Y. Saleh. "Aqueous Solubilities and Transformations of Tetrahalogenated Benzenes and Effects of Aquatic Fulvic Acids," *Bull. Environ. Contam. Toxicol.*, 44(6):813-818 (1990).

Kim, Y.-H., Woodrow, J.E., and J.N. Seiber. "Evaluation of a Gas Chromatographic Method for Calculating Vapor Pressures with Organophosphorus Pesticides," *J. Chromatogr.*, 314:37-53 (1984).

Kirchnerová, J., and G.C.B. Cave. "The Solubility of Water in Low-Dielectric Solvents," *Can. J. Chem.*, 54(24):3909-3916 (1976).

Kishi, H., Kogure, N., and Y. Hashimoto. "Contribution of Soil Constituents in Adsorption Coefficient of Aromatic Compounds, Halogenated Alicyclic and Aromatic Compounds to Soil," *Environ. Sci. Technol.*, 21(7):867-876 (1990).

Kishi, H., and Y. Hashimoto. "Evaluation of the Procedures for the Measurement of Water Solubility and *n*-Octanol/Water Partition Coefficient of Chemicals Results of a Ring Test in Japan," *Chemosphere*, 18(9/10):1749-1749 (1989).

Klein, R.G. "Calculations and Measurements on the Volatility of *N*-Nitrosoamines and Their Aqueous Solutions," *Toxicology*, 23:135-147 (1982).

Klein, W., Kördel, W., Weiβ, M., and H.J. Poremski. "Updating of the OECD Test Guideline 107 'Partition Coefficient *n*-Octanol/Water': OECD Laboratory Intercomparison Test of the HPLC Method," *Chemosphere*, 17(2):361-386 (1988).

Kleopfer, R.D., Easley, D.M., Haas, B.B., Jr., Deihl, T.G., Jackson, D.E., and C.J. Wurrey. "Anaerobic Degradation of Trichloroethylene in Soil," *Environ. Sci. Technol.*, 19(3):277-280 (1985).

Klevens, H.B. "Solubilization of Polycyclic Hydrocarbons," *J. Phys. Colloid Chem.*, 54(2):283-298 (1950).

Klöpffer, W., Haag, F., Kohl, E-G., and R. Frank. "Testing of the Abiotic Degradation of Chemicals in the Atmosphere: The Smog Chamber Approach," *Ecotoxicol. Environ. Safety*, 15(3):298-319 (1988).

Klöpffer, W., Kaufman, G., Rippen, G., and H.-P. Poremski. "A Laboratory Method for Testing the Volatility from Aqueous Solution: First Results and

Comparison with Theory," *Ecotoxicol. Environ. Safety*, 6(6):545-559 (1982).

Knight, E.V., Novick, N.J., Kaplan, D.L., and J.R. Meeks. "Biodegradation of 2-Furaldehyde under Nitrate-Reducing and Methanogenic Conditions," *Environ. Toxicol. Chem.*, 9(6):725-730 (1990).

Knoevenagel, K., and R. Himmelreich. "Degradation of Compounds Containing Carbon Atoms by Photo-Oxidation in the Presence of Water," *Arch. Environ. Contam. Toxicol.*, 4:324-333 (1976).

Knuutinen, J., Palm, H., Hakala, H., Haimi, J., Huhta, V., and J. Salminen. "Polychlorinated Phenols and Their Metabolites in Soil and Earthworms of Sawmill Environment," *Chemosphere*, 20(6):609-623 (1990).

Kobayashi, H., and B.E. Rittman. "Microbial Removal of Hazardous Organic Compounds," *Environ. Sci. Technol.*, 16(3):170A-183A (1982).

Koester, C.J., and R.A. Hites. "Calculated Physical Properties of Polychlorinated Dibenzo-*p*-dioxins and Dibenzofurans," *Chemos-phere*, 17(12):2355-2362 (1988).

Kohring, G.-W., Rogers, J.E., and J. Wiegel. "Anaerobic Biodegradation of 2,4-Dichlorophenol in Freshwater Lake Sediments at Different Temperatures," *Appl. Environ. Microbiol.*, 55(2):348-353 (1989).

Könemann, H., Zelle, R., and F. Busser. "Determination of Log P_{oct} Values of Chloro-Substituted Benzenes, Toluenes, and Anilines by High-Performance Liquid Chromatography on ODS-Silica," *J. Chromatogr.*, 178:559-565 (1979).

Konietzko, H. "Chlorinated Ethanes: Sources, Distribution, Environmental Impact, and Health Effects," in *Hazard Assessment of Chemicals, Volume 3*, J. Saxena, Ed. (New York: Academic Press, Inc., 1984), pp 401-448.

Kopczynski, S.L., Altshuller, A.P., and F.D. Sutterfield. "Photochemical Reactivities of Aldehyde-Nitrogen Oxide Systems," *Environ. Sci. Technol.*, 8(10):909-918 (1974).

Korfmacher, W.A., Wehry, E.L., Mamantov, G., and D.F.S. Natusch. "Resistance to Photochemical Decomposition of Polycyclic Aromatic Hydrocarbons Vapor-Adsorbed on Coal Fly Ash," *Environ. Sci. Technol.*, 14(9):1094-1099 (1980).

Koskinen, W.C., Oliver, J.E., Kearney, P.C., and C.G. McWhorter. "Effect of Trifluralin Soil Metabolites on Cotton Growth and Yield," *J. Agric. Food Chem.*, 32(6):1246-1248 (1984).

Krasnoshchekova, R.Y., and M. Gubergrits. "Solubility of *n*-Alkylbenzene in Fresh and Salt Waters," [Chemical Abstracts 83(16):136583p]: *Vodn. Resur.*, 2:170-173 (1975).

Krijgsheld, K.R., and A. van der Gen. "Assessment of the Impact of the Emission of Certain Organochlorine Compounds on the Aquatic Environment - Part I: Monochlorophenols and 2,4-Dichlorophenol," *Chemosphere*, 15(7):825-860 (1986).

Krijgsheld, K.R., and A. van der Gen. "Assessment of the Impact of the Emission of Certain Organochlorine Compounds on the Aquatic Environment - Part II. Allyl chloride, 1,3- and 2,3-Dichloro-propene," *Chemosphere*, 15(7):861-880 (1986).

Kronberger, H., and J. Weiss. "Formation and Structure of Some Organic Molecular Compounds. III. The Dielectric Polarization of Some Solid Crystalline Molecular Compounds," *J. Chem. Soc. (London)*, (1944), pp 464-469.

Krzyzanowska, T., and J. Szeliga. "A Method for Determining the Solubility of Individual Hydrocarbons," *Nafta*, 28:414-417 (1978).

Kuchta, J.M., Furno, A.L., Bartkowiak, A., and G.H. Martindill. "Effect of Pressure and Temperature on Flammability Limits of Chlorinated Hydrocarbons in Oxygen-Nitrogen and Nitrogen Tetroxide-Nitrogen Atmospheres," *J. Chem. Eng. Data*, 13(3):421-428 (1968).

Kudchadker, A.P., Kudchadker, S.A., Shukla, R.P., and P.R. Patnaik. "Vapor Pressures and Boiling Points of Selected Halomethanes," *J. Phys. Chem. Ref. Data*, 8(2):499-517 (1979).

Kuo, P.P.K., Chian, E.S.K., and B.J. Chang. "Identification of End Products Resulting from Ozonation and Chlorination of Organic Compounds Commonly Found in Water," *Environ. Sci. Technol.*, 11(13):1177-1181 (1977).

Kurihara, N., Uchida, M., Fujita, T., and M. Nakajima. "Studies on BHC Isomers and Related Compounds. V. Some Physicochemical Properties of BHC Isomers," *Pestic. Biochem. Physiol.*, 2(4):383-390 (1973).

LaFleur, K.S. "Sorption of Pesticides by Model Soils and Agronomic Soils: Rates and Equilibria," *Soil Sci.*, 127(2):94-101 (1979).

Lagas, P. "Sorption of Chlorophenols in the Soil," *Chemosphere*, 17(2):205-216 (1988).

Lamparski, L.L., Stehl, R.H., and R.L. Johnson. "Photolysis of Pentachlorophenol-Treated Wood. Chlorinated Dibenzo-*p*-dioxin Formation," *Environ. Sci. Technol.*, 14(2):196-200 (1980).

Lande, S.S., and S. Banerjee. "Predicting Aqueous Solubility of Organic Nonelectrolytes from Molar Volumes," *Chemosphere*, 10:751:759 (1981).

Lande, S.S., Hagen, D.F., and A.E. Seaver. "Computation of Total Molecular Surface Area from Gas Phase Ion Mobility Data and its Correlation with Aqueous Solubilities of Hydrocarbons," *Environ. Toxicol. Chem.*, 4(3):325-334 (1985).

Landrum, P.F., Nihart, S.R., Eadie, B.J., and W.S. Gardner. "Reverse-Phase Separation Method for Determining Pollutant Binding to Aldrich Humic Acid and Dissolved Organic Carbon of Natural Waters," *Environ. Sci. Technol.*, 19(3):187-192 (1984).

Laplanche, Martin, G., and F. Tonnard. "Ozonation Schemes of Organophosphorous Pesticides. Application in Drinking Water Treatment," *Ozone: Sci. Engrg.*, 6:207-219 (1984).

Lau, Y.L., Oliver, B.G., and B.G. Krishnappan. "Transport of Some Chlorinated Contaminants by the Water, Suspended Sediments, and Bed Sediments in the St. Clair and Detroit Rivers," *Environ. Toxicol. Chem.*, 8(4):293-301 (1989).

Leadbetter, E.R., and J.W. Foster. "Oxidation Products formed from Gaseous Alkanes by the Bacterium *Pseudomonas methanica*," *Arch. Biochem. Biophys.*, 82:491-492 (1959).

Leavitt, D.D., and M.A. Abraham. "Acid-Catalyzed Oxidation of 2,4-Dichlorophenoxyacetic Acid by Ammonium Nitrate in Aqueous Solution," *Environ. Sci. Technol.*, 24(4):566-571 (1990).

Leenheer, J.A., Malcolm, R.L., and W.R. White. "Investigation of the Reactivity and Fate of Certain Organic Components of an Industrial Waste after Deep-

Well Injection," *Environ. Sci. Technol.*, 10(5):445-451 (1976).

Leffingwell, J.T. "The Photolysis of DDT in Water," PhD Thesis, University of California, Davis, CA (1975).

Leinster, P., Perry, R., and R.J. Young. "Ethylenebromide in Urban Air," *Atmos. Environ.*, 12:2382-2398 (1978).

Leland, H.V., Bruce, W.N., and N.F. Shimp. "Chlorinated Hydrocarbon Insecticides in Sediments in Southern Lake Michigan," *Environ. Sci. Technol.*, 7(9):833-838 (1973).

Lenchitz, C., and R.W. Velicky. "Vapor Pressure and Heat of Sublimation of Three Nitrotoluenes," *J. Chem. Eng. Data*, 15(3):401-403 (1970).

Leo, A., Hansch, C., and D. Elkins. "Partition Coefficients and Their Uses," *Chem. Rev.*, 71(6):525-616 (1971).

Lesage, S., Jackson, R.E., Priddle, M.W., and P.G. Riemann. "Occurrence and Fate of Organic Solvent Residues in Anoxic Groundwater at the Gloucester Landfill, Canada," *Environ. Sci. Technol.*, 24(4):559-566 (1990).

Leuenberger, C., Ligocki, M.P., and J.F. Pankow. "Trace Organic Compounds in Rain. 4. Identities, Concentrations, and Scavenging Mechanisms for Phenols in Urban Air and Rain," *Environ. Sci. Technol.*, 19(11):1053-1058 (1985).

Leyder, F., and P. Boulanger. "Ultraviolet Absorption, Aqueous Solubility and Octanol-Water Partition for Several Phthalates," *Bull. Environ. Contam. Toxicol.*, 30(2):152-157 (1983).

Lieberman, M.T., and M. Alexander. "Microbial and Nonenzymatic Steps in the Decomposition of Dichlorvos (2,2-Dichlorovinyl *O,O*-Dimethyl Phosphate)," *J. Agric. Food Chem.*, 31(2):265-267 (1983).

Liss, P.S., and P.G. Slater. "Flux of Gases across the Air-Sea Interface," *Nature*, 247(5438):181-184 (1974).

Lin, S., and R.M. Carlson. "Susceptibility of Environmentally Important Heterocycles to Chemical Disinfection: Reactions with Aqueous Chlorine, Chlorine Dioxide, and Chloramine," *Environ. Sci. Technol.*, 18(10):743-748 (1984).

Lin, S., Lukasewycz, M.T., Liukkonen, R.J., and R.M. Carlson. "Facile Incorporation of Bromine into Aromatic Systems under Conditions of Water Chlorination," *Environ. Sci. Technol.*, 18(12):985-986 (1984).

Liu, D. "Biodegradation of Aroclor 1221 Type PCBs in Sewage Wastewater," *Bull. Environ. Contam. Toxicol.*, 27(5):695-703 (1981).

Lloyd, A.C., Atkinson, R., Lurmann, F.W., and B. Nitta. "Modeling Potential Ozone Impacts from Natural Hydrocarbons - I. Development and Testing of a Chemical Mechanism for the NO_x-Air Photooxidations of Isoprene and α-Pinene under Ambient Conditions," *Atmos. Environ.*, 17(10):1931-1950 (1983).

Lodge, K.B. "Solubility Studies Using a Generator Column for 2,3,7,8-Tetrachlorodibenzo-*p*-dioxin," *Chemosphere*, 18(1-6):933-940 (1989).

Løkke, H. "Sorption of Selected Organic Pollutants in Danish Soils," *Ecotoxicol. Environ. Safety*, 8(5):395-409 (1984).

Low, G.K.-C., McEvoy, S.R., and R.W. Matthews. "Formation of Nitrate and Ammonium Ions in Titanium Dioxide Mediated Photocatalytic Degradation of Organic Compounds Containing Nitrogen Atoms," *Environ. Sci. Technol.*,

25(3):460-467 (1991).

Lu, P.-Y., and R.L. Metcalf. "Environmental Fate and Biodegradability of Benzene Derivatives as Studied in a Model Ecosystem," *Environ. Health Perspect.* 10:269-284 (1975).

Lu, P.-Y., Metcalf, R.L., Hirwe, A.S., and J.W. Williams. "Evaluation of Environmental Distribution and Fate of Hexachlorocyclopenta-diene, Chlordene, Heptachlor, and Heptachlor Epoxide in a Model Ecosystem," *J. Agric. Food Chem.*, 23(5):967-973 (1975).

Lyman, W.J., Reehl, W.F., and D.H. Rosenblatt. *Handbook of Chemical Property Estimation Methods: Environmental Behavior of Organic Compounds* (New York: McGraw-Hill, Inc., 1982).

Lyons, C.D., Katz, S.E., and R. Bartha. "Fate of Herbicide-derived Aniline Residues During Ensilage," *Bull. Environ. Contam. Toxicol.*, 35(5):704-710 (1985).

Lyons, C.D., Katz, S., and R. Bartha. "Mechanisms and Pathways of Aniline Elimination from Aquatic Environments," *Appl. Environ. Microbiol.*, 48(3):491-496 (1984).

Mabey, W., and T. Mill. "Critical Review of Hydrolysis of Organic Compounds in Water Under Environmental Conditions," *J. Phys. Chem. Ref. Data*, 7(2):383-415 (1978).

Mabey, W.R., Smith, J.H., Podoll, R.T., Johnson, H.L., Mill, T., Chou, T.-W., Gates, J., Partridge, I.W., Jaber, H., and D. Vandenberg. "Aquatic Fate Process Data for Organic Priority Pollutants - Final Report," Office of Regulations and Standards, U.S. EPA Report-440/4-81-014 (1982), 407 p.

MacIntyre, W.G., and P.O. deFur. "The Effect of Hydrocarbon Mixtures on Adsorption of Substituted Naphthalenes by Clay and Sediment from Water," *Chemosphere*, 14(1):103-111 (1985).

Mackay, D. "Correlation of Bioconcentration Factors," *Environ. Sci. Technol.*, 16(5):274-278 (1982).

Mackay, D., Bobra, A., Chan, D.W., and W.Y. Shiu. "Vapor Pressure Correlations for Low-Volatility Environmental Chemicals," *Environ. Sci. Technol.*, 16(10):645-649 (1982).

Mackay, D., Bobra, A., Shiu, W.-Y., and S.H. Yalkowsky. "Relationships between Aqueous Solubility and Octanol-Water Partition Coefficients," *Chemosphere*, 9:701-711 (1980).

Mackay, D., and P.J. Leinonen. "Notes - Rate of Evaporation of Low-Solubility Contaminants from Water Bodies to Atmosphere," *Environ. Sci. Technol.*, 9(13):1178-1180 (1975).

Mackay, D., and S. Paterson. "Calculating Fugacity," *Environ. Sci. Technol.*, 15(9):1006-1014 (1981).

Mackay, D., and W.-Y. Shiu. "A Critical Review of Henry's Law Constants for Chemicals of Environmental Interest," *J. Phys. Chem. Ref. Data*, 10(4):1175-1199 (1981).

Mackay, D., and W.-Y. Shiu. "Aqueous Solubility of Polynuclear Aromatic Hydrocarbons," *J. Chem. Eng. Data*, 22(4):399-402 (1977).

Mackay, D., and W.-Y. Shiu. "The Determination of the Solubility of

Hydrocarbons in Aqueous Sodium Chloride Solutions," *Can. J. Chem. Eng.*, 53:239-242 (1975).

Mackay, D., Shiu, W.-Y., and R.P. Sutherland. "Determination of AirWater Henry's Law Constants for Hydrophobic Pollutants," *Environ. Sci. Technol.*, 13(3):333-337 (1979).

Mackay, D., and A.W. Wolkoff. "Rate of Evaporation of Low-Solubility Contaminants from Water Bodies to Atmosphere," *Environ. Sci. Technol.*, 7(7):611-614 (1973).

Mackay, D., and A.T.K. Yeun. "Mass Transfer Coefficient Correlations for Volatilization of Organic Solutes from Water," *Environ. Sci. Technol.*, 17(4):211-217 (1983).

Macknick, A.B., and J.M. Prausnitz. "Vapor Pressures of High Molecular Weight Hydrocarbons," *J. Chem. Eng. Data*, 24(3):175-178 (1979).

MacLeod, H., Jourdain, L., Poulet, G., and G. LeBras. "Kinetic Study of Reactions of Some Organic Sulfur Compounds with OH Radicals," *Atmos. Environ.*, 18(12):2621-2626 (1984).

MacMichael, G.J., and L.R. Brown. "Role of Carbon Dioxide in Catabolism of Propane by "*Nocardia paraffinicum*" (*Rhodococcus rhodochrous*)," *Appl. Environ. Microbiol.*, 53(1):65-69 (1987).

MacRae, I.C. "Microbial Metabolism of Pesticides and Structurally Related Compounds," *Rev. Environ. Contam. Toxicol.*, 109:2-87 (1989).

MacRae, I.C., Raghu, K., and E.M. Bautista. "Anaerobic Degradation of the Insecticide Lindane by *Clostridium sp.*," *Nature*, 221:859-860 (1969).

MacRae, I.C., Raghu, K., and T.F. Castro. "Persistence and Biodegradation of Four Common Isomers of Benzene Hexachloride in Submerged Soils," *J. Agric. Food Chem.*, 15(5):911-914 (1967).

Madhun, Y.A., and V.H. Freed. "Degradation of the Herbicides Bromacil, Diuron and Chlortoluron in Soil," *Chemosphere*, 16(5):1003-1011 (1987).

Maeda, N., Ohya, T., Nojima, K., and S. Kanno. "Formation of Cyanide Ion or Cyanogen Chloride through the Cleavage of Aromatic Rings by Nitrous Acid or Chlorine. IX. On the Reactions of Chlorinated, Nitrated, Carboxylated or Methylated Benzene Derivatives with Hypochlorous Acid in the Presence of Ammonium Ion," *Chemosphere*, 16(10-12):2249-2258 (1987).

Mahaffey, W.R., Gibson, D.T., and C.E. Cerniglia. "Bacterial Oxidation of Chemical Carcinogens: Formation of Polycyclic Aromatic Acids from Benz[*a*]anthracene," *Appl. Environ. Microbiol.*, 54(10):2415-2423 (1988).

Maksimov, Y.Y. "Vapor Pressures of Aromatic Nitro Compounds at Various Temperatures," *Zh. Fiz. Khim.*, 42(11):2921-2925 (1968).

Mallik, M.A.B., and K. Tesfai. "Transformation of Nitrosamines in Soil and *in Vitro* By Soil Microorganisms," *Bull. Environ. Contam. Toxicol.*, 27(1):115-121 (1981).

Mallon, B.J., and F.L. Harrison. "Octanol-Water Partition Coefficient of Benzo(*a*)pyrene: Measurement, Calculation and Environmental Implications," *Bull. Environ. Contam. Toxicol.*, 32(3):316-323 (1984).

Mansour, M., Feicht, E., and P. Méallier. "Improvement of the Photostability of Selected Substances in Aqueous Medium," *Toxicol. Environ. Chem.*, 20-21:139-

147 (1989).
Mansour, M., Thaller, S., and F. Korte. "Action of Sunlight on Parathion," *Bull. Environ. Contam. Toxicol.*, 30(3):358-364 (1983).
Marple, L., Berridge, B., and L. Throop. "Measurement of the Water-Octanol Partition Coefficient of 2,3,7,8-Tetrachlorodibenzo-*p*-dioxin," *Environ. Sci. Technol.*, 20(4):397-399 (1986).
Marple, L., Brunck, R., and L. Throop. "Water Solubility of 2,3,7,8-Tetrachlorodibenzo-*p*-dioxin," *Environ. Sci. Technol.*, 20(2):180-182 (1986).
Martens, R. "Degradation of [8,9-^{14}C]Endosulfan by Soil Micro-organisms," *Appl. Environ. Microbiol.*, 31(6):853-858 (1975).
Martens, R. "Degradation of Endosulfan-8,9-^{14}C in Soil Under Different Conditions," *Bull. Environ. Contam. Toxicol.*, 17(4):438-446 (1977).
Martin, G., Laplanche, A., Morvan, J., Wei, Y., and C. LeCloirec. "Action of Ozone on Organo-Nitrogen Products," in *Proceedings Symposium on Ozonization: Environmental Impact and Benefit* (Paris, France: International Ozone Association, 1983), pp. 379-393.
Martin, H., Ed. *Pesticide Manual*, 3rd ed. (Worcester, England: British Crop Protection Council, 1972).
Masterton, W.L., and T.P. Lee. "Effect of Dissolved Salts on Water Solubility of Lindane," *Environ. Sci. Technol.*, 6(10):919-921 (1972).
Masunaga, S., Urushigawa, Y., and Y. Yonezawa. "Biodegradation Pathway of *o*-Cresol By Heterogeneous Culture," *Wat. Res.*, 20(4):477-484 (1986).
Mathur, S.P., and J.W. Rouatt. "Utilization of the Pollutant Di-2-ethylhexyl Phthalate by a Bacterium," *J. Environ. Qual.*, 4(2):273-275 (1975).
Mathur, S.P., and J.G. Saha. "Degradation of Lindane-^{14}C in a Mineral Soil and in an Organic Soil," *Bull. Environ. Contam. Toxicol.*, 17(4):424-430 (1977).
Mathur, S.P., and J.G. Saha. "Microbial Degradation of Lindane-C^{14} in a Flooded Sandy Loam Soil," *Soil Sci.*, 120(4):301-307 (1975).
Matthews, R.W. "Photo-oxidation of Organic Material in Aqueous Suspensions of Titanium Dioxide," *Wat. Res.*, 20(5):569-578 (1986).
May, W.E., Wasik, S.P., and D.H. Freeman. "Determination of the Aqueous Solubility of Polynuclear Aromatic Hydrocarbons by a Coupled Column Liquid Chromatographic Technique," *Anal. Chem.*, 50(1):175-179 (1978).
May, W.E., Wasik, S.P., and D.H. Freeman. "Determination of the Solubility Behavior of Some Polycyclic Aromatic Hydrocarbons in Water," *Anal. Chem.*, 50(7):997-1000 (1978).
Mazzocchi, P.H., and M.W. Bowen. "Photolysis of Dioxane," *J. Org. Chem.*, 40(18):2689-2690 (1975).
McAuliffe, C. "Solubility in Water of Paraffin, Cycloparaffin, Olefin, Acetylene, Cycloolefin, and Aromatic Compounds," *J. Phys. Chem.*, 70(4):1267-1275 (1966).
McCall, P.J., Swann, R.L., Laskowski, D.A., Vrona, S.A., Unger, S.M., and H.J. Dishburger. "Prediction of Chemical Mobility in Soil from Sorption Coefficients," in *Aquatic Toxicology and Hazard Assessment, Fourth Conference, ASTM STP 737*, Branson, D.R., and K.L. Dickson, Eds. (Philadelphia, PA: American Society for Testing and Materials, 1981), pp 49-58.
McCall, P.J., Vrona, S.A., and S.S. Kelley. "Fate of Uniformly Carbon-14 Ring

Labeled 2,4,5-Trichlorophenoxyacetic Acid and 2,4-Dichlorophenoxy Acid," *J. Agric. Food Chem.*, 29(1):100-107 (1981).

McConnell, G., Ferguson, D.M., and C.R. Pearson. "Chlorinated Hydrocarbons and the Environment," *Endeavour*, 34(121):13-18 (1975).

McConnell, G., and H.I. Schiff. "Methyl Chloroform: Impact on Stratospheric Ozone," *Science*, 199:174-177 (1978).

McCrady, J.K., Johnson, D.E., and L.W. Turner. "Volatility of Ten Priority Pollutants from Fortified Avian Toxicity Test Diets," *Bull. Environ. Contam. Toxicol.*, 34(5):634-644 (1985).

McDevit, W.F., and F.A. Long. "The Activity Coefficient of Benzene in Aqueous Salt Solutions," *J. Am. Chem. Soc.*, 74(7):1773-1777 (1952).

McMurry, P.H., and D. Grosjean. "Photochemical Formation of Organic Aerosols: Growth Laws and Mechanisms," *Atmos. Environ.*, 19(9):1445-1451 (1985).

McVeety, B.D., and R.A. Hites. "Atmospheric Deposition of Polycyclic Aromatic Hydrocarbons to Water Surfaces: A Mass Balance Approach," *Atmos. Environ.*, 22(3):511-536 (1988).

Means, J.C., Hassett, J.J., Wood, S.G., and W.L. Banwart. "Sorption Properties of Energy-Related Pollutants and Sediments," in *Polynuclear Aromatic Hydrocarbons, Third International Symposium on Chemistry, Biology, Carcinogenesis and Mutagenesis*, Jones, P.W., and P. Leber, Eds. (Ann Arbor, MI: Ann Arbor Science Publishers, Inc., 1979), pp 327-340.

Means, J.C., Wood, S.G., Hassett, J.J., and W.L. Banwart. "Sorption of Polynuclear Aromatic Hydrocarbons by Sediments and Soils," Environ. Sci. Technol., 14(2):1524-1528 (1980).

Medley, D.R., and E.L. Stover. "Effects of Ozone on the Biodegradability of Biorefractory Pollutants," *J. Water Poll. Control Fed.*, 55(5):489-494 (1983).

Meikle, R.W., and C.R. Youngson. "The Hydrolysis Rate of Chlorpyrifos, *O,O*-Diethyl-*O*-(3,5,6-trichloro-2-pyridyl)phosphorothioate and its Dimethyl Analog, Chloropyrifosmethyl in Dilute Aqueous Solution," *Arch. Environ. Contam. Toxicol.*, 7:349-357 (1978).

Meites, L., Ed. *Handbook of Analytical Chemistry*, 1st ed. (New York: McGraw-Hill, Inc., 1963), 1782 p.

Melnikov, N.N. *Chemistry of Pesticides* (New York: Springer-Verlag, Inc., 1971), 480 p.

Mercer, J.W., Skipp, D.C., and D. Giffin. "Basics of Pump-and-Treat Ground-Water Remediation Technology," U.S. EPA Report-600/8-90-003 (1990), 60 p.

Metcalf, R.L., Kapoor, I.P., Lu, P.-Y., Schuth, C.K., and P. Sherman. "Model Ecosystem Studies of the Environmental Fate of Six Organochlorine Pesticides," *Environ. Health Perspect.*, (June 1973), pp 35-44.

Mihelcic, J.R., and R.G. Luthy. "Microbial Degradation of Acenaphthene and Naphthalene under Denitrification Conditions in Soil-Water Systems," *Appl. Environ. Microbiol.*, 54(5):1188-1198 (1988).

Mikami, Y., Fukunaga, Y., Arita, M., Obi, Y., and T. Kisaki. "Preparation of Aroma Compounds by Microbial Transformation of Isophorone by *Aspergillus niger*," *Agric. Biol. Chem.*, 45(3):791-793 (1981).

Mikesell, M.D., and S.A. Boyd. "Reductive Dechlorination of the Pesticides 2,4-D,

2,4,5-T, and Pentachlorophenol in Anaerobic Sludges," *J. Environ. Qual.*, 14(3):337-3340 (1985).

Milano, J.C., Guibourg, A., and J.L. Vernet. "Non Biological Evolution, in Water, of Some Three- and Four-Carbon Atoms Organohalogenated Compounds: Hydrolysis and Photolysis," *Wat. Res.*, 22(12):1553-1562 (1988).

Miles, J.R.W., and P. Moy. "Degradation of Endosulfan and its Metabolites by a Mixed Culture of Soil Microorganisms," *Bull. Environ. Toxicol.*, 23(1/2):13-19 (1979).

Miller, G.C., and D.G. Crosby. "Photooxidation of 4-Chloroaniline and *N*-4-Chlorophenylbenzenesulfonamide to Nitroso Products and Nitro Compounds," *Chemosphere*, 12(9/10):1217-1228 (1983).

Miller, M.M., Ghodbane, S., Wasik, S.P., Tewari, Y.B., and D.E. Martire. "Aqueous Solubilities, Octanol/Water Partition Coefficients, and Entropies of Melting of Chlorinated Benzenes and Biphenyls," *J. Chem. Eng. Data*, 29(2):184-190 (1984).

Miller, M.M., Wasik, S.P., Huang, G.-L., Shiu, W.-Y., and D. Mackay. "Relationships between Octanol-Water Partition Coefficient and Aqueous Solubility," *Environ. Sci. Technol.*, 19(6):522-529 (1985).

Miller, G.C., and R.G. Zepp. "Photoreactivity of Aquatic Pollutants Sorbed on Suspended Sediments," *Environ. Sci. Technol.*, 13(7):860-863 (1979).

Mills, W.B., Porcella, D.B., Ungs, M.J., Gherini, S.A., Summers, K.V., Mok, L., Rupp, G.L., and G.L. Bowie. "Water Quality Assessment: A Screening Procedure for Toxic and Conventional Pollutants in Surface and Groundwater - Part I," Office of Research and Development, U.S. EPA Report-600/6-85-002a (1985), 638 p.

Minard, R.D., Russel, S., and J.-M. Bollag. "Chemical Transformation of 4-Chloroaniline to a Triazene in a Bacterial Culture Medium," *J. Agric. Food Chem.*, 25(4):841-844 (1977).

Mingelgrin, U., and Z. Gerstl. "Reevaluation of Partitioning as a Mechanism of Nonionic Chemicals Adsorption in Soils," *J. Environ. Qual.*, 12(1):1-11 (1983).

Mirvish, S.S., Issenberg, P., and H.C. Sornson. "Air-Water and Ether-Water Distribution of *N*-Nitroso Compounds: Implications for Laboratory Safety, Analytic Methodology, and Carcinogenicity for the Rat Esophagus, Nose, and Liver," *J. Nat. Cancer Instit.*, 56(6):1125-1129 (1976).

Mitra, A., Saksena, R.K., and C.R. Mitra. "A Prediction Plot for Unknown Water Solubilities of Some Hydrocarbons and Their Mixtures," *Chem. Petro-Chem. J.*, 8:16-17 (1977).

Monsanto Industrial Chemicals Co. "PCBs-Aroclors," Technical Bulletin O/PL 306A, (1974).

Montgomery, J.H. *Groundwater Chemicals Desk Reference - Volume 2* (Chelsea, MI: Lewis Publishers, Inc., 1990), 944 p.

Montgomery, J.H., and L.M. Welkom. *Groundwater Chemicals Desk Reference* (Chelsea, MI: Lewis Publishers, Inc., 1990), 640 p.

Moore, J.W., and S. Ramamoorthy. *Organic Chemicals in Natural Waters - Applied Monitoring and Impact Assessment* (New York: Springer-Verlag, Inc., 1984), 289 p.

Moreale, A., and R. Van Bladel. "Soil Interactions of Herbicide-Derived Aniline Residues: A Thermodynamic Approach," *Soil Sci.*, 127(1):1-9 (1979).

Morehead, N.R., Eadie, B.J., Lake, B., Landrum, P.F., and D. Berner. "The Sorption of PAH onto Dissolved Organic Matter in Lake Michigan Waters," *Chemosphere*, 15(4):403-412 (1986).

Moriguchi, I. "Quantitative Structure-Activity Studies on Parameters Related to Hydrophobicity," *Chem. Pharm. Bull.*, 23:247-257 (1975).

Morita, M., Nakagawa, J., and C. Rappe. "Polychlorinated Dibenzofuran (PCDF) Formation from PCB Mixture by Heat and Oxygen," *Bull. Environ. Contam. Toxicol.*, 19(6):665-670 (1978).

Morrison, R.T., and R.N. Boyd. *Organic Chemistry* (Boston, MA: Allyn and Bacon, Inc., 1971), 1258 p.

Moza, P.N., and E. Feicht. "Photooxidation of Aromatic Hydrocarbons as Liquid Film on Water," *Toxicol. Environ. Chem.*, 20-21:135-138 (1989).

Moza, P.N., Fytianos, K., Samanidou, V., and F. Korte. "Photo-decomposition of Chlorophenols in Aqueous Medium in Presence of Hydrogen Peroxide," *Bull. Environ. Contam. Toxicol.*, 41(5):678-682 (1988).

Mulla, M.S., Mian, L.S., and J.A. Kawecki. "Distribution, Transport and Fate of the Insecticides Malathion and Parathion in the Environment," *Res. Rev.*, 81:116-125 (1981).

Munshi, H.B., Rama Rao, K.V.S., and R. M. Iyer. "Characterization of Products of Ozonolysis of Acrylonitrile in Liquid Phase," *Atmos. Environ.*, 23(9):1945-1948 (1989).

Munshi, H.B., Rama Rao, K.V.S., and R.M. Iyer. "Rate Constants of the Reactions of Ozone with Nitriles, Acrylates and Terpenes in Gas Phase," *Atmos. Environ.*, 23(9):1971-1976 (1989).

Munz, C., and P.V. Roberts. "Air-Water Phase Equilibria of Volatile Organic Solutes," *J. Am. Water Works Assoc.*, 79(5):62-69 (1987).

Murphy, T.J., Mullin, M.D., and J.A. Meyer. "Equilibrium of Polychlorinated Biphenyls and Toxaphene with Air and Water," *Environ. Sci. Technol.*, 21(2):155-162 (1987).

Murray, J.M., Pottie, R.F., and C. Pupp. "The Vapor Pressures and Enthalpies of Sublimation of Five Polycyclic Aromatic Hydrocarbons," *Can. J. Chem.* 52(4):557-563 (1974).

Murthy, N.B.K., Kaufman, D.D., and G.F. Fries. "Degradation of Pentachlorophenol (PCP) in Aerobic and Anaerobic Soil," *J. Environ. Sci. Health*, B14(1):1-14 (1979).

Murty, A.S. *Toxicity of Pesticides to Fish. Volume 1* (Boca Raton, FL: CRC Press, Inc., 1986), 178 p.

Musoke, G.M.S., Roberts, D.J., and M. Cooke. "Heterogeneous Hydrodechlorination of Chlordan," *Bull. Environ. Contam. Toxicol.*, 28(4):467-472 (1982).

Nash, R.G. "Comparative Volatilization and Dissipation Rates of Several Pesticides from Soil," *J. Agric. Food Chem.*, 31(2):210-217 (1983).

Nathan, M.F. "Choosing a Process for Chloride Removal," *Chem. Eng.*, 85:93-100 (1978).

Nathwani, J.S., and C.R. Phillips. "Adsorption-Desorption of Selected Hydrocarbons in Crude Oil on Soils," *Chemosphere*, 6(4):157-162 (1977).

Neely, W.B., and G.E. Blau, Eds. *Environmental Exposure from Chemicals. Volume 1* (Boca Raton, FL: CRC Press, Inc. 1985), 245 p.

Neely, W.B., Branson, D.R., and G.E. Blau. "Partition Coefficient to Measure Bioconcentration Potential of Organic Chemicals in Fish," *Environ. Sci. Technol.*, 8(13):1113-1115 (1974).

Nelson, N.H., and S.D. Faust. "Acidic Dissociation Constants of Selected Aquatic Herbicides," *Environ. Sci. Technol.*, 3(11):1186-1188 (1969).

Nicholson, B.C., Maguire, B.P., and D.B. Bursill. "Henry's Law Constants for the Trihalomethanes: Effects of Water Composition and Temperature," *Environ. Sci. Technol.*, 18(7):518-521 (1984).

Nielsen, T., Ramdahl, T., and A. Bjørseth. "The Fate of Airborne Polycyclic Organic Matter," *Environ. Health Perspect.*, 47:103-114 (1983).

Niessen, R., Lenoir, D., and P. Boule. "Phototransformation of Phenol Induced by Excitation of Nitrate Ions," *Chemosphere*, 17(10):1977-1984 (1988).

Nikolaou, K., Masclet, P., and G. Mouvier. "Sources and Chemical Reactivity of Polynuclear Aromatic Hydrocarbons in the Atmosphere - A Critical Review," *Sci. Tot. Environ.*, 32(2):103-132 (1984).

"NIOSH Pocket Guide to Chemical Hazards," U.S. Department of Health and Human Services, U.S. Government Printing Office (1987), 241 p.

Nirmalakhandan, N.N., and R.E. Speece. "Prediction of Aqueous Solubility of Organic Compounds Based on Molecular Structure," *Environ. Sci. Technol.*, 22(3):328-338 (1988).

Nirmalakhandan, N.N., and R.E. Speece. "Prediction of Aqueous Solubility of Organic Compounds Based on Molecular Structure. 2. Application to PNAs, PCBs, PCDDs, etc.," *Environ. Sci. Technol.*, 23(6):708-713 (1989).

Nirmalakhandan, N.N., and R.E. Speece. "QSAR Model for Predicting Henry's Constant," *Environ. Sci. Technol.*, 22(11):1349-1357 (1988).

Nisbet, I.C., and A.F. Sarofim. "Rates and Routes of Transport of PCBs in the Environment," *Environ. Health Perspect.*, (April 1972), pp 21-38.

Nkedi-Kizza, P., Rao, P.S.C., and A.G. Hornsby. "Influence of Organic Cosolvents on Sorption of Hydrophobic Organic Chemicals by Soils," *Environ. Sci. Technol.*, 19(10):975-979.

Nkedi-Kizza, P., Rao, P.S.C., and J.W. Johnson. "Adsorption of Diuron and 2,4,5-T on Soil Particle-Size Separates," *J. Environ. Qual.*, 12(2):195-197 (1983).

Nojima, K., Ikarigawa, T., and S. Kanno. "Studies on Photo-chemistry of Aromatic Hydrocarbons. VI. Photochemical Reaction of Bromobenzene with Nitrogen Oxides in Air," *Chemosphere*, 9(7/8):421-436 (1980).

Nojima, K., and S. Kanno. "Studies on Photochemistry of Aromatic Hydrocarbons. VII. Photochemical Reaction of *p*-Dichlorobenzene with Nitrogen Oxides in Air," *Chemosphere*, 9(7/8):437-440 (1980).

Nyholm, N., Lindgaard-Jørgensen, P. and N. Hansen. "Biodegradation of 4-Nitrophenol in Standardized Aquatic Degradation Tests," *Ecotoxicol. Environ. Safety*, 8(6):451-470 (1984).

Nyssen, G.A., Miller, E.T., Glass, T.F., Quinn II, C.R., Underwood, J., and D.J.

Wilson. "Solubilities of Hydrophobic Compounds in Aqueous-Organic Solvent Mixtures," *Environ. Monit. Assess.*, 9(1):1-11 (1987).

Ogata, M., Fujisawa, K., Ogino, Y., and E. Mano. "Partition Coefficients as a Measure of Bioconcentration Potential of Crude Oil in Fish and Sunfish," *Bull. Environ. Contam. Toxicol.*, 33(5):561-567 (1984).

Ofstad, E.B., and T. Sletten. "Composition and Water Solubility Determination of a Commercial Tricresylphosphate," *Sci. Tot. Environ.*, 43(3):233-241 (1985).

Ohe, T. "Mutagenicity of Photochemical Reaction Products of Polycyclic Aromatic Hydrocarbons with Nitrite," *Sci. Tot. Environ.*, 39(1/2):161-175 (1984).

O'Reilly, K.T., and R.L. Crawford. "Kinetics of *p*-Cresol Degradation by an Immobilized *Pseudomonas* sp.," *Appl. Environ. Microbiol.*, 55(4):866-870 (1989).

Oliver, B.G. "Desorption of Chlorinated Hydrocarbons from Spiked and Anthropogenically Contaminated Sediments," *Chemosphere*, 14(8):1087-1106 (1985).

Oliver, B.G., and J.H. Carey. "Photochemical Production of Chlorinated Organics in Aqueous Solutions," *Environ. Sci. Technol.*, 11(9):893-895 (1977).

Oliver, B.G., and M.N. Charlton. "Chlorinated Organic Contaminants on Settling Particulates in the Niagara River Vicinity of Lake Ontario," *Environ. Sci. Technol.*, 18(12):903-908 (1984).

Oliver, B.G., and A.J. Niimi. "Bioconcentration Factors of Some Halogenated Organics for Rainbow Trout: Limitations in Their Use for Prediction of Environmental Residues," *Environ. Sci. Technol.*, 19(9):842-849 (1985).

Onitsuka, S., Kasai, Y., and K. Yoshimura. "Quantitative Structure-Toxicity Activity Relationship of Fatty Acids and the Sodium Salts to Aquatic Organisms," *Chemosphere*, 18(7/8):1621-1631 (1989).

Opperhuizen, A., Serné, P., and J.M.D. Van der Steen. "Thermodynamics of Fish/Water and Octan-1-ol/Water Partitioning of Some Chlorinated Benzenes," *Environ. Sci. Technol.*, 22(3):286-298 (1988).

Osborn, A.G., and D.R. Douslin. "Vapor Pressures and Derived Enthalpies of Vaporization of Some Condensed-Ring Hydrocarbons," *J. Chem. Eng. Data*, 20(3):229-231 (1975).

Osborn, A.G., and D.R. Douslin. "Vapor-Pressure Relationships for 15 Hydrocarbons," *J. Chem. Eng. Data*, 19(2):114-117 (1974).

Ou, L.-T., and J.J. Street. "Monomethylhydrazine Degradation and Its Effect on Carbon Dioxide Evolution and Microbial Populations in Soil," *Bull. Environ. Contam. Toxicol.*, 41(3):454-460 (1988).

Ou, L.-T., Gancarz, D.H., Wheeler, W.B., Rao, P.S.C., and J.M. Davidson. "Influence of Soil Temperature and Soil Moisture on Degradation and Metabolism of Carbofuran in Soils," *J. Environ. Qual.*, 11(2):293-298 (1982).

Owens, J.W., Wasik, S.P., and H. DeVoe. "Aqueous Solubilities and Enthalpies of Solution of *n*-Alkylbenzenes," *J. Chem. Eng. Data*, 31(1):47-51 (1986).

Oyler, A.R., Llukkonen, R.J., Lukasewycz, M.T., Heikkila, K.E., Cox, D.A., and R.M. Carlson. "Chlorine 'Disinfection' Chemistry of Aromatic Compounds. Polynuclear Aromatic Hydrocarbons: Rates, Products, and Mechanisms," *Environ. Sci. Technol.*, 17(6):334-342 (1983).

Palit, S.R. "Electronic Interpretations of Organic Chemistry. II. Interpretation of the

Solubility of Organic Compounds," *J. Phys. Chem.*, 51(3):837-857 (1947).

Pankow, J.F., and M.E. Rosen. "Determination of Volatile Compounds in Water by Purging Directly to a Capillary Column with Whole Column Cryotrapping," *Environ. Sci. Technol.*, 22(4):398-405 (1988).

Paris, D.F., Lewis, D.L., and J.T. Barnett. "Bioconcentration of Toxaphene by Microorganisms," *Bull. Environ. Contam. Toxicol.*, 17(5):564-572 (1977).

Paris, D.F., Steen, W.C., and G.L. Baughman. "Role of the Physico-Chemical Properties of Aroclors 1016 and 1242 in Determining Their Fate and Transport in Aquatic Environments," *Chemosphere*, 7(4):319-325 (1978).

Paris, D.F., and N.L. Wolfe. "Relationship between Properties of a Series of Anilines and Their Transformation by Bacteria," *Appl. Environ. Microbiol.*, 53(5):911-916 (1987).

Park, K.S., and W.N. Bruce. "The Determination of the Water Solubility of Aldrin, Dieldrin, Heptachlor and Heptachlor Epoxide," *J. Econ. Entomol.*, 61(3):770-774 (1968).

Parlar, H. "Photoinduced Reactions of Two Toxaphene Compounds in Aqueous Medium and Adsorbed on Silica Gel," *Chemosphere*, 17(11):2141-2150 (1988).

Parr, J.F., and S. Smith. "Degradation of Trifluralin under Laboratory Conditions and Soil Anaerobiosis," *Soil Sci.*, 115(1):55-63 (1973).

Parris, G.E. "Covalent Binding of Aromatic Amines to Humates. 1. Reactions with Carbonyls and Quinones," *Environ. Sci. Technol.*, 14(9):1099-1106 (1980).

Parsons, F., and G.B. Lage. "Chlorinated Organics in Simulated Groundwater Environments," *J. Am. Water Works Assoc.*, 75(5):52-59 (1985).

Pearson, C.R., and G. McConnell. "Chlorinated C_1 and C_2 Hydrocarbons in the Marine Environment," *Proc. R. Soc. London*, B189(1096):305-332 (1975).

Pereira, W.E., Short, D.L., Manigold, D.B., and P.K. Roscio. "Isolation and Characterization of TNT and Its Metabolites in Groundwater by Gas Chromatograph-Mass Spectrometer-Computer Techniques," *Bull Environ. Contam. Toxicol.*, 21(4/5):554-562 (1979).

Perry, R.H., and C.H. Chilton. *Chemical Engineers Handbook*, 5th ed. (New York: McGraw-Hill, Inc., 1973), p. 3-60.

Peterson, M.S., Lion, L.W., and C.A. Shoemaker. "Influence of Vapor-Phase Sorption and Dilution on the Fate of Trichloroethylene in an Unsaturated Aquifer System," *Environ. Sci. Technol.*, 22(5):571-578 (1988).

Petrasek, A.C., Kugelman, I.J., Austern, B.M., Pressley, T.A., Winslow, L.A., and R.H. Wise. "Fate of Toxic Organic Compounds in Wastewater Treatment Plants," *J. Water Pollut. Control Fed.*, 55(10):1286-1296 (1983).

Peyton, T.O., Steel, R.V., and W.R. Mabey. "Carbon Disulfide, Carbonyl Sulfide: Literature Review and Environmental Assessment," U.S. EPA Report PB-257-947/2 (1976), 64 p.

Piacente, V., Scardala, P., Ferro, D., and R. Gigli. "Vaporization of Study *o*-, *m*-, and *p*-Chloroaniline by Torsion-Weighing Effusion Vapor Pressure Measurements," *J. Chem. Eng. Data*, 30(4):372-376 (1985).

Pierce, R.H., Olney, C.E., and G.T. Felbeck. "*p,p'*-DDT Adsorption to Suspended Particulate Matter in Sea Water," *Geochim. Cosmochim. Acta*, 38(7):1061-1073 (1974).

Pignatello, J.J. "Microbial Degradation of 1,2-Dibromoethane in Shallow Aquifer Materials," *J. Environ. Qual.*, 16(4):307-312 (1987).

Pinal, R., Rao, S.C., Lee, L.S., Cline, P.V., and S.H. Yalkowsky. "Cosolvency of Partially Miscible Organic Solvents on the Solubility of Hydrophobic Organic Chemicals," *Environ. Sci. Technol.*, 24(5):639-647 (1990).

Patil, K.C., Matsumura, F., and G.M. Boush. "Metabolic Transformation of DDT, Dieldrin, Aldrin, and Endrin by Marine Microorganisms," *Environ. Sci. Technol.*, 6(7):629-632 (1972).

Pitts, J.N., Jr., Atkinson, R., Sweetman, J.A., and B. Zielinska. "The Gas-phase Reaction of Naphthalene with N_2O_5 to Form Nitronaphthalenes," *Atmos. Environ.*, 19(5):701-705 (1985).

Pitts, J.N., Jr., Grosjean, D., Cauwenberghe, K.V., Schmid, J.P., and D.R. Fitz. "Photooxidation of Aliphatic Amines Under Simulated Atmospheric Conditions: Formation of Nitrosamines, Nitramines, Amides, and Photochemical Oxidant," *Environ. Sci. Technol.*, 12(8):946-953 (1978).

Pitts, J.N., Jr., Lokensgard, D.M., Ripley, P.S., van Cauwenberghe, K.A., van Vaeck, L., Schaffer, S.D., Thill, A.J., and W.L. Belser, Jr. "'Atmospheric' Epoxidation of Benzo[*a*]pyrene by Ozone: Formation of the Metabolite Benzo[*a*]pyrene-4,5-oxide," *Science*, 210:1347-1349 (1980).

Pitts, J.N., Jr., Zielinska, B., Sweetman, J.A., Atkinson, R., and A.M. Winer. "Reactions of Adsorbed Pyrene and Perylene with Gaseous N_2O_5 Under Simulated Atmospheric Conditions," *Atmos. Environ.*, 19(6):911-915 (1985).

Platford, R.F., Carey, J.H., and E.J. Hale. "The Environmental Significance of Surface Films: Part 1. Octanol-Water Partition Coefficients for DDT and Hexachlorobenzene," *Environ. Pollut. (Series B)*, 3(2):125-128 (1982).

Plimmer, J.R. "Photolysis of TCDD and Trifluralin on Silica and Soil," *Bull Environ. Contam. Toxicol.*, 20(1):87-92 (1978).

Plimmer, J.R., and U.J. Klingebiel. "Photolysis of Hexachlorobenzene," *J. Agric. Food Chem.*, 24(4):721-723 (1976).

Plimmer, J.R. "Photolysis of TCDD and Trifluralin on Silica and Soil," *Bull. Environ. Contam. Toxicol.*, 29(1):87-92 (1978).

Plimmer, J.R., Klingbiel, U.I., and B.E. Hummer. "Photooxidation of DDT and DDE," *Science*, 167:67-69 (1970).

Podoll, R.T., Irwin, K.C., and H.J. Parish. "Dynamic Studies of Naphthalene Sorption on Soil from Aqueous Solution," *Chemosphere*, 18(11/12):2399-2412 (1989).

Podoll, R.T., Jaber, H.M., and T. Mill. "Tetrachlorodioxin: Rates of Volatilization and Photolysis in the Environment," *Environ. Sci. Technol.*, 20(5):490-492 (1986).

Pogány, E., Wallnöfer, P.R., Ziegler. W., and W. Mücke. "Metabolism of *o*-Nitroaniline and Di-*n*-butyl Phthalate in Cell Suspension Cultures of Tomatoes," *Chemosphere*, 21(4/5):557-562 (1990).

Polak, J., and B.C.-Y. Lu. "Mutual Solubilities of Hydrocarbons and Water at 0 and 25 °C," *Can. J. Chem.*, 51(24):4018-4023 (1973).

Pollero, R., and S.C. dePollero. "Degradation of DDT by a Soil Amoeba," *Bull. Environ. Contam. Toxicol.*, 19(3):345-350 (1978).

"Preliminary Study of Selected Potential Environmental Contaminants -Optical

Brighteners, Methyl Chloroform, Trichloroethylene, Tetrachloroethylene, Ion Exchange Resins," Office of Toxic Substances, U.S. EPA Report-560/2-75-002, 286 p.

Price, K.S., Waggy, G.T., and R.A. Conway. "Brine Shrimp Bioassay and Seawater BOD of Petrochemicals," *J. Water Poll. Control Fed.*, 46(1):63-77 (1974).

Price, L.C. "Aqueous Solubility of Petroleum as Applied to its Origin and Primary Migration," *Am. Assoc. Pet. Geol. Bull.*, 60(2):213-244 (1976).

Probst, G.W., Golab, T., Herberg, R.J., Holzer, F.J., Parka, S.J., van der Schans, C., and J.B. Tepe. "Fate of Trifluralin in Soils and Plants," *J. Agric. Food Chem.*, 15(4):592-599 (1967).

Pruden, A.L., and D.F. Ollis. "Degradation of Chloroform by Photoassisted Heterogeneous Catalysis in Dilute Aqueous Suspensions of Titanium Dioxide," *Environ. Sci. Technol.*, 17(10):628-631 (1983).

Pupp, C., Lao, R.C., Murray, J.J., and R.F. Pottie. "Equilibrium Vapor Concentrations of Some Polycyclic Hydrocarbons, Arsenic Trioxide (As_4O_6) and Selenium Dioxide and the Collection Efficiencies of these Air Pollutants," *Atmos. Environ.*, 8:915-925 (1974).

Putnam, T.B., Bills, D.D., and L.M. Libbey. "Identification of Endo-sulfan Based on the Products of Laboratory Photolysis," *Bull. Environ. Contam. Toxicol.*, 13(6):662-665 (1975).

Quirke, J.M.E., Marei, A.S.M., and G. Eglinton. "The Degradation of DDT and its Degradative Products by Reduced Iron (III) Porphyrins and Ammonia," *Chemosphere*, 8(3):151-155 (1979).

Radding, S.B., Mill, T., Gould, C.W., Lia, D.H., Johnson, H.L., Bomberger, D.S., and C.V. Fojo. "The Environmental Fate of Selected Polynuclear Aromatic Hydrocarbons," Office of Toxic Substances, U.S. EPA Report-560/5-75-009 (1976), 122 p.

Rajagopalan, R., Vohra, K.G., and A.M. Mohan Rao. "Studies on Oxidation of Benzo[*a*]pyrene by Sunlight and Ozone," *Sci. Tot. Environ.*, 27(1):33-42 (1983).

Ramamoorthy, S. "Competition of Fate Processes in the Bioconcentration of Lindane," *Bull. Environ. Contam. Toxicol.*, 34(3):349-358 (1985).

Randahl, T., Becher, G., and A. Bjørseth. "Nitrated Polycyclic Aromatic Hydrocarbons in Urban Air Particles," *Environ. Sci. Technol.*, 16(12):861-865 (1982).

Randall, T.L., and P.V. Knopp. "Detoxification of Specific Organic Substances by Wet Oxidation," *J. Water Poll. Control Fed.*, 52(8):2117-2130 (1980).

Rao, P.S.C., and J.M. Davidson. "Estimation of Pesticide Retention and Transformation Parameters Required in Nonpoint Source Pollution Models," in *Environmental Impact of Nonpoint Source Pollution*, Overcash, M.R., and J.M. Davidson, Eds. (Ann Arbor, MI: Ann Arbor Science Publishers, Inc., 1980), pp 23-67.

Rappe, C., Choudhary, G., and L.H. Keith. *Chlorinated Dioxins and Dibenzofurans in Perspective* (Chelsea, MI: Lewis Publishers, Inc., 1987), 570 p.

Raha, P., and A.K. Das. "Photodegradation of Carbofuran," *Chemosphere*, 21(1/2):99-106 (1990).

Rav-Acha, C., and E. Choshen. "Aqueous Reactions of Chlorine Dioxide with

Hydrocarbons," *Environ. Sci. Technol.*, 21(11):1069-1074 (1987).

Reed, C.D., and J.J. McKetta. "Solubility of 1,3-Butadiene in Water," *J. Chem. Eng. Data*, 4(4):294-295 (1959).

"Registry of Toxic Effects of Chemical Substances," U.S. Department of Health and Human Services, National Institute for Occupational Safety and Health (1985), 2050 p.

Reinbold, K.A., Hassett, J.J., Means, J.C., and W.L. Banwart. "Adsorption of Energy-Related Organic Pollutants: A Literature Review," Office of Research and Development, U.S. EPA Report-600/3-79-086 (1979), 180 p.

"Report on the Problem of Halogenated Air Pollutants and Stratigraphic Ozone," Office of Research and Development, U.S. EPA Report-600/9-75-008 (1975), 55 p.

Richard, Y., and L. Bréner. "Removal of Pesticides from Drinking Water by Ozone," in *Handbook of Ozone Technology and Applications, Volume II. Ozone for Drinking Water Treatment*, Rice, A.G., and A. Netzer, Eds. (Montvale, MA: Butterworth Publishers, 1984), pp. 77-97.

Riddick, J.A., Bunger, W.B., and T.K. Sakano. *Organic Solvents - Physical Properties and Methods of Purification. Volume II* (New York: John Wiley and Sons, Inc., 1986), 1325 p.

Riederer, M. "Estimating Partitioning and Transport of Organic Chemicals in the Foliage/Atmosphere System: Discussion of a Fugacity-Based Model," *Environ. Sci. Technol.*, 24(6):829-837 (1990).

Rippen, G., Ilgenstein, M., and W. Klöpffer. "Screening of the Adsorption Behavior of New Chemicals: Natural Soils and Model Adsorbents," *Ecotoxicol. Environ. Safety*, 6(3):236-245 (1982).

Roark, R.C. "A Digest of Information on Chlordane," Bureau of Entomology and Plant Quarantine, U.S. Dept. of Agric. Report E-817 (1951), 132 p.

Robb, I.D. "Determination of the Aqueous Solubility of Fatty Acids and Alcohols," *Aust. J. Chem.*, 18:2281-2285 (1966).

Robeck, G.G., Dostal, K.A., Cohen, J.M., and J.F. Kreissl. "Effectiveness of Water Treatment Processes in Pesticide Removal," *J. Am. Water Works Assoc.*, 57(2):181-200 (1965).

Roberts, A.L., and P.M. Gschwend. "Mechanism of Pentachloroethane Dehydrochlorination to Tetrachloroethylene," *Environ. ScJohnsoni. Technol.*, 25(1):76-86 (1991).

Roberts, P.V. "Nature of Organic Contaminants in Groundwater and Approaches to Treatment," in *Proceedings, AWWA Seminar on Organic Chemical Contaminants in Groundwater: Transport and Removal* (St. Louis, MO: American Water Works Association, 1981), pp 47-65.

Roberts, P.V., and P.G. Dändliker. "Mass Transfer of Volatile Organic Contaminants from Aqueous Solution to the Atmosphere During Surface Aeration," *Environ. Sci. Technol.*, 17(8):484-489 (1983).

Roberts, P.V., Munz, C., and P. Dändliker. "Modeling Volatile Organic Solute Removal by Surface and Bubble Aeration," *J. Water Pollut. Control Fed.*, 56(2):157-163 (1985).

Robertson, R.E., and S.E. Sugamori. "The Hydrolysis of Dimethyl Sulfate and

Diethyl Sulfate in Water," *Can. J. Chem.*, 44(14):1728-1730 (1966).
Robinson, J., Richardson, A., Bush, B., and K.E. Elgar. "A Photo-Isomerization Product of Dieldrin," *Bull. Environ. Contam. Toxicol.*, 1(4):127-132 (1966).
Rordorf, B.F. "Prediction of Vapor Pressures, Boiling Points and Enthalpies of Fusion for Twenty-Nine Halogenated Dibenzo-*p*-dioxins and Fifty-Five Dibenzofurans by a Vapor Pressure Correlation Method," *Chemosphere*, 18(1-6):783-788 (1989).
Rodorf, B.F. "Thermodynamic and Thermal Properties of Polychlorinated Compounds: The Vapor Pressures and Flow Tube Kinetics of Ten Dibenzo-*para*-dioxines," *Chemosphere*, 14(7):885-892 (1985).
Roeder, K.D., and E.A. Weiant. "The Site of Action of DDT in the Cockroach," *Science*, 103(2671):304-305 (1946).
Rogers, R.D., and J.C. McFarlane. "Sorption of Carbon Tetrachloride, Ethylene Dibromide, and Trichloroethylene on Soil and Clay," *Environ. Monitor. Assess.*, 1(2):155-162 (1981).
Rogers, R.D., McFarlane, J.C., and A.J. Cross. "Adsorption and Desorption of Benzene in Two Soils and Montmorillonite Clay," *Environ. Sci. Technol.*, 14(4):457-461 (1980).
Rondorf, B. "Thermal Properties of Dioxins, Furans and Related Compounds," *Chemosphere*, 15:1325-1332 (1986).
Rontani, J.F., Rambeloarisoa, E., Bertrand, J.C., and G. Giusti. "Favourable Interaction between Photooxidation and Bacterial Degradation of Anthracene in Sea Water," *Chemosphere*, 14(11/12):1909-1912 (1985).
Roy, W.R., Ainsworth, C.C., Chou, S.F.J., Griffin, R.A., and I.G. Krapac. "Development of Standardized Batch Adsorption Procedures: Experimental Considerations," in *Proceedings of the Eleventh Annual Research Symposium of the Solid and Hazardous Waste Research Division, April 29-May 1, 1985* (Cincinnati, OH: U.S. Environmental Protection Agency, 1985), U.S. EPA Report-600/9-85-013, pp 169-178.
Rosen, J.D., and W.F. Carey. "Preparation of the Photoisomers of Aldrin and Dieldrin," *J. Agric. Food Chem.*, 16(3):536-537 (1968).
Rosen, J.D., and D.J. Sutherland. "The Nature and Toxicity of the Photoconversion Products of Aldrin," *Bull. Environ. Contam. Toxicol.*, 2(1):1-9 (1967).
Rosen, J.D., Sutherland, J., and G.R. Lipton. "The Photochemical Isomerization of Dieldrin and Endrin and Effects on Toxicity," *Bull. Environ. Contam. Toxicol.*, 1(4):133-140 (1966).
Ross, R.D., and D.G. Crosby. "The Photooxidation of Aldrin in Water," *Chemosphere*, 4(5):277-282 (1975).
Rossi, S.S., and W.H. Thomas. "Solubility Behavior of Three Aromatic Hydrocarbons in Distilled Water and Natural Seawater," *Environ. Sci. Technol.*, 15(6):715-716 (1981).
Ruepert, C., Grinwis, A., and H. Govers. "Prediction of Partition Coefficients of Unsubstituted Polycyclic Aromatic Hydrocarbons from C_{18} Chromatographic and Structural Properties," *Chemosphere*, 14(3/4):279-291 (1985).
Russell, D.J., and B. McDuffie. "Chemodynamic Properties of Phthalate Esters: Partitioning and Soil Migration," *Chemosphere*, 15(8):1003-1021 (1986).

Sabljić, A. "On the Prediction of Soil Sorption Coefficients of Organic Pollutants from Molecular Structure: Application of Molecular Topology Model," *Environ. Sci. Technol.*, 21(4):358-366 (1987).

Sabljić, A. "Predictions of the Nature and Strength of Soil Sorption of Organic Pollutants by Molecular Topology," *J. Agric. Food Chem.*, 32(2):243-246 (1984).

Sabljić, A., and M. Protić. "Relationship between Molecular Connectivity Indices and Soil Sorption Coefficients of Polycyclic Aromatic Hydrocarbons," *Bull. Environ. Contam. Toxicol.*, 28(2):162-165 (1982).

Sadek, P.C., Carr, P.W., Doherty, R.M., Kamlet, M.J., Taft, R.W., and M.H. Abraham. "Study of Retention Processes in Reversed-Phase High-Performance Liquid Chromatography by the Use of the Solvatochromic Comparison Method," *Anal. Chem.*, 57(14):2971-2978 (1985).

Saeger, V.W., Hicks, O., Kaley, R.G., Michael, P.R., Mieure, J.P., and E.S. Tucker. "Environmental Fate of Selected Phosphate Esters," *Environ. Sci. Technol.*, 13(7):840-844 (1979).

Sahyun, M.R.V. "Binding of Aromatic Compounds to Bovine Serum Albumin," *Nature*, 209(5023):613-614 (1966).

Saleh, F.Y., Dickson, K.L., and J.H. Rodgers, Jr. "Fate of Lindane in the Aquatic Environment: Rate Constants of Physical and Chemical Processes," *Environ. Toxicol. Chem.*, 1(4):289-297 (1982).

Saltzman, S., Kliger, L., and B. Yaron. "Adsorption-Desorption of Parathion as Affected by Soil Organic Matter," *J. Agric. Food Chem.*, 20(6):1224-1226 (1972).

Sanborn, J.R., Metcalf, R.L., Bruce, W.N., and P.-Y. Lu. "The Fate of Chlordane and Toxaphene in a Terrestrial-Aquatic Model Ecosystem," *Environ. Entomol.*, 5(3):533-538 (1976).

Sanemasa, I., Araki, M., Deguchi, T., and H. Nagai. "Solubility Measurements of Benzene and the Alkylbenzenes in Water Making Use of Solute Vapor," *Bull. Chem. Soc. Jpn.*, 55(4):1054-1062 (1982).

Saravanja-Bozanic, V., Gäb, S., Hustert, K., and F. Korte. "Reacktionen von Aldrin, Chlorden, und 2,2-Dichlorobiphenyl mit $O(^3P)$," *Chemosphere*, 6(1):21-26 (1977).

Sarna, L.P., Hodge, P.E., and G.R.B. Webster. "Octanol-Water Partition Coefficients of Chlorinated Dioxins and Dibenzofurans by Reversed-Phase HPLC Using Several C_{18} Columns," *Chemosphere*, 13(9):975-983 (1984).

Sato, A., and T. Nakajima. "A Structure-Activity Relationship of Some Chlorinated Hydrocarbons," *Arch. Environ. Health*, 34(2):69-75 (1979).

Sax, N.I. *Dangerous Properties of Industrial Materials* (New York: Van Nostrand Reinhold Co., 1984), 3124 p.

Sax, N.I., Ed. *Dangerous Properties of Industrial Materials Report* (New York: Van Nostrand Reinhold Co., 1984), 4(1,3,6).

Sax, N.I., Ed. *Dangerous Properties of Industrial Materials Report* (New York: Van Nostrand Reinhold Co., 1985), 5(3,4).

Scala, A.A., and D. Salomon. "The Gas Phase Photolysis and γ Radiolysis of Ethylenimine," *J. Chem. Phys.*, 65(11):4455-4461 (1976).

Schellenberg, K., Leuenberger, C., and R.P. Schwarzenbach. "Sorption of Chlorinated Phenols by Natural Sediments and Aquifer Materials," *Environ. Sci.*

Technol., 18(9):652-657 (1984).
Scheunert, I., Vockel, D., Schmitzer, J., and F. Korte. "Biomineralization Rates of ^{14}C-Labelled Organic Chemicals in Aerobic and Anaerobic Suspended Soil," *Chemosphere*, 16(5):1031-1041 (1987).
Schroy, J.M., Hileman, F.D., and S.C. Cheng. "Physical/Chemical Properties of 2,3,7,8-TCDD," *Chemosphere*, 14(6/7):877-880 (1985).
Schultz, T.W., Wesley, S.K., and L.L. Baker. "Structure-Activity Relationships for Di and Tri Alkyl and/or Halogen Substituted Phenols," *Bull. Environ. Contam. Toxicol.*, 43(2):192-198 (1989).
Schumacher, H.G., Parlar, H., Klein, W., and F. Korte. "Beitrage zur Okologischen Chemie. LV. Photochemische Reaktion von Endosulfan," *Chemosphere*, 3(2):65-70 (1974).
Schwarz, F.P. "Determination of Temperature Dependence of Solubilities of Polycyclic Aromatic Hydrocarbons in Aqueous Solutions by a Fluorescence Method," *J. Chem. Eng. Data*, 22(3):273-277 (1977).
Schwarz, F.P. "Measurement of the Solubilities of Slightly Soluble Organic Liquids in Water by Elution Chromatography," *Anal. Chem.*, 52(1):10-15 (1980).
Schwarz, F.P., and S.P. Wasik. "Fluorescence Measurements of Benzene, Naphthalene, Anthracene, Pyrene, Fluoranthene, and Benzo[*e*]pyrene in Water," *Anal. Chem.*, 48(3):524-528 (1976).
Schwarzenbach, R.P., Giger, W., Hoehn, E., and J.K. Schneider. "Behavior of Organic Compounds during Infiltration of River Water to Groundwater. Field Studies," *Environ. Sci. Technol.*, 17(8):472-479 (1983).
Schwarzenbach, R.P., Giger, W., Schaffner, C., and O. Wanner. "Groundwater Contamination by Volatile Halogenated Alkanes: Abiotic Formation of Volatile Sulfur Compounds under Anaerobic Conditions," *Environ. Sci. Technol.*, 19(4):322-327 (1985).
Schwarzenbach, R., Stierli, R., Folsom, B.R., and J. Zeyer. "Compound Properties Relevant for Assessing the Environmental Partitioning of Nitrophenols," *Environ. Sci. Technol.*, 22(1):83-92 (1988).
Schwarzenbach, R.P., and J. Westall. "Transport of Nonpolar Organic Compounds from Surface Water to Groundwater. Laboratory Sorption Studies," *Environ. Sci. Technol.*, 15(11):1360-1367 (1981).
Schwille, F. *Dense Chlorinated Solvents* (Chelsea, MI: Lewis Publishers, Inc., 1988), 146 p.
Seip, H.M., Alstad, J., Carlberg, G.E., Matinsen, K., and R. Skaane. "Measurement of Mobility of Organic Compounds in Soils," *Sci. Tot. Environ.*, 50(1):87-101 (1986).
Shafer, D. *Hazardous Materials Training Handbook* (Madison, CT: Bureau of Law and Business, Inc., 1987), 195 p.
Sharma, R.K., and H.B. Palmer. "Vapor Pressure of Biphenyl Near Fusion Temperature," *J. Chem. Eng. Data*, 19(1):6-8 (1974).
Sharom, M.S., Miles, J.R.W., Harris C.R., and F.L. McEwen. "Behaviour of 12 Pesticides in Soil and Aqueous Suspensions of Soil and Sediment," *Water Res.*, 14:1095-1100 (1980).
Shelton, D.R., Boyd, S.A., and J.M. Tiedje. "Anaerobic Biodegradation of Phthalic

Acid Esters in Sludge," *Environ. Sci. Technol.*, 18(2):93-97 (1984).

Shevchenko, M.A., Taran, P.N., and P.V. Marchenko. "Technology of Water Treatment and Demineralization," *Soviet J. Water Chem. Technol.*, 4(4):53-71 (1982).

Shiaris, M.P., and G.S. Sayler. "Biotransformation of PCB by Natural Assemblages of Freshwater Microorganisms," *Environ. Sci. Technol.*, 16(6):367-369 (1982).

Shinozuka, N., Lee, C., and S. Hayano. "Solubilizing Action of Humic Acid from Marine Sediment," *Sci. Tot. Environ.*, 62:311-314 (1987).

Shiraishi, H., Pilkington, N.H., Otsuki, A., and K. Fuwa. "Occurrence of Chlorinated Polynuclear Aromatic Hydrocarbons in Tap Water," *Environ. Sci. Technol.*, 19(7):585-590 (1985).

Shriner, C.R., Drury, J.S., Hammons, A.S., Towill, L.E., Lewis, E.B., and D.M. Opresko. "Reviews of the Environmental Effects of Pollutants: II. Benzidine," Office of Research and Development, U.S. EPA Report-600/1-78-024 (1978), 157 p.

Sims, R.C., Doucette, W.C., McLean, J.E., Grenney, W.J., and R.R. Dupont. "Treatment Potential for 56 EPA Listed Hazardous Chemicals in Soil," U.S. EPA Report-600/6-88-001 (1988), 105 p.

Singh, J. "Conversion of Heptachloride to its Epoxide," *Bull. Environ. Contam. Toxicol.*, 4(2):77-79 (1969).

Singh, A.K., and P.K. Seth. "Degradation of Malathion by Microorganisms Isolated from Industrial Effluents," *Bull. Environ. Contam. Toxicol.*, 43(1):28-35 (1989).

Singmaster, J.A., III. "Environmental Behavior of Hydrophobic Pollutants in Aqueous Solutions," PhD Thesis, University of California, Davis CA (1975).

Siragusa, G.R., and D.D. DeLaune. "Mineralization and Sorption of *p*-Nitrophenol in Estuarine Sediment," *Environ. Toxicol. Chem.*, 5(1):175-178 (1986).

Sisler, H.D., and C.E. Cox. "Effects of Tetramethyl Thiuram Disulfide on Metabolism of *Fusarium roseum*," *Am. J. Bot.*, 41:338 (1954).

Sittig, M. *Handbook of Toxic and Hazardous Chemicals and Carcinogens* (Park Ridge, NJ: Noyes Publications, 1985), 950 p.

Smith, A.E. "Identification of 2,4-Dichloroanisole and 2,4-Dichlorophenol as Soil Degradation Products of Ring-Labelled [^{14}C]2,4-D," *Bull. Environ. Contam. Toxicol.*, 34(2):150-157 (1985).

Smith, L.R., and J. Dragun. "Degradation of Volatile Chlorinated Aliphatic Priority Pollutants in Groundwater," *Environ. Int.*, 19(4):291-298 (1984).

Smith, R.V., and J.P. Rosazza. "Microbial Models of Mammalian Metabolism. Aromatic Hydroxylation," *Arch. Biochem. Biophys.*, 161(2):551-558 (1974).

Smith, J.H., Mabey, W.R., Bohonos, N., Holt, B.R., Lee, S.S., Chou, T.-W., Bomberger, D.C., and T. Mill. "Environmental Pathways of Selected Chemicals in Freshwater Systems. Part II: Laboratory Studies," U.S. EPA Report-600/7-78-074 (1978), 406 p.

Snider, E.H., and F.C. Alley. "Kinetics of the Chlorination of Biphenyl under Conditions of Waste Treatment Processes," *Environ. Sci. Technol.*, 13(10):1244-1248 (1979).

Snider, J.R., and G.A. Dawson. "Tropospheric Light Alcohols, Carbonyls, and

Acetonitrile: Concentrations in the Southwestern United States and Henry's Law Data," *J. Geophys. Res., D: Atmos.*, 90(D2):3797-3805 (1985).

Socha, S.B., and R. Carpenter. "Factors Affecting the Pore Water Hydrocarbon Concentrations in Puget Sound Sediments," *Geochim. Cosmochim. Acta*, 51(5):1273-1284 (1987).

Sonnefeld, W.J., Zoller, W.H., and W.E. May. "Dynamic Coupled-Column Liquid Chromatographic Determination of Ambient Temperature Vapor Pressures of Polynuclear Aromatic Hydrocarbons," *Anal. Chem.*, 55(2):275-280 (1983).

Southworth, G.R. "The Role of Volatilization in Removing Polycyclic Aromatic Hydrocarbons from Aquatic Environments," *Bull. Environ. Contam. Toxicol.*, 21(4/5):507-514 (1979).

Southworth, G.R., and J.L. Keller. "Hydrophobic Sorption of Polar Organics by Low Organic Carbon Soils," *Water, Air, Soil Poll.*, 28:239-248 (1986).

Special Occupational Hazard Review for DDT (Rockville, MD: National Institute for Occupational Safety and Health, 1978), 205 p.

Spence, J.W., Hanst, P.L., and B.W. Gay, Jr. "Atmospheric Oxidation of Methyl Chloride, Methylene Chloride, and Chloroform," *J. Air Pollut. Control Assoc.*, 26(10):994-996 (1976).

Spencer, W.F., Adams, J.D., Shoup, T.D., and R.C. Spear. "Conversion of Parathion to Paraoxon on Soil Dusts and Clay Minerals as Affected by Ozone and UV Light," *J. Agric. Food Chem.*, 28(2):366-371 (1980).

Spencer, W.F., and M.M. Cliath. "Vapor Density and Apparent Vapor Pressure of Lindane (γ-BHC)," *J. Agric. Food Chem.*, 18(3):529-530 (1970).

Spencer, W.F., and M.M. Cliath. "Vapor Density of Dieldrin," *Environ. Sci. Technol.*, 3(7):670-674 (1969).

Spencer, W.F., and M.M. Cliath. "Volatility of DDT and Related Compounds," *J. Agric. Food Chem.*, 20(3):645-649 (1972).

Spencer, W.F., Cliath, M.M., Jury, W.A., and L.-Z. Zhang. "Volatilization of Organic Chemicals from Soil as Related to Their Henry's Law Constants," *J. Environ. Qual.*, 17(3):504-509 (1988).

Standen, A., Ed. *Kirk-Othmer Encyclopedia of Chemical Technology, Volumes 1,3,4,5,12,13,15,19,21,22*, 2nd ed. (New York: John Wiley and Sons, Inc., 1963).

Stauffer, T.B., and W.G. MacIntyre. "Sorption of Low-Polarity Organic Compounds on Oxide Minerals and Aquifer Material," *Environ. Toxicol. Chem.*, 5(11):949-955 (1986).

Stearns, R.S., Oppenheimer, H., Simon, E., and W.D. Harkins. "Solubilization of Long-Chain Colloidal Electrolytes," *J. Chem. Phys.*, 15(7):496-507 (1947).

Steinwandter, H. "Experiments on Lindane Metabolism in Plants III. Formation of β-HCH," *Bull. Environ. Contam. Toxicol.*, 20(4):535-536 (1978).

Stephen, H., and T. Stephen. *Solubilities of Inorganic and Organic Compounds - Part 1, Volume 1* (London: Pergamon Printing and Art Services, Ltd., 1963), 960 p.

Stevens, A.A., Slocum, C.J., Seeger, D.R., and G.G. Robeck. "Chlorination of Organics in Drinking Water," *J. Am. Water Works Assoc.*, 68(11):615-620 (1976).

Streitweiser, A. Jr., and L.L. Nebenzahl. "Carbon Acidity. LII. Equilibrium Acidity of Cyclopentadiene in Water and Cyclohexylamine," *J. Am. Chem. Soc.*,

98(8):2188-2190 (1976).
Struif, B., Weil, L., and K.-E. Quentin. "Verhalten Herbizider Phenoxyalkan-carbonäuren bei der Wasseraufbereitung mit Ozon," *Zeit. fur Wasser und Abwasser-Forschung*, 3/4:118-127 (1978).
Stucki, G., and M. Alexander. "Role of Dissolution Rate and Solubility in Biodegradation of Aromatic Compounds," *Appl. Environ. Microbiol.*, 53(2):292-297 (1987).
Su, F., Calvert, J.G., and J.H. Shaw. "Mechanism of the Photooxidation of Gaseous Formaldehyde," *J. Phys. Chem.*, 83(25):3185-3191 (1979).
Sudhakar, B., and N. Sethunathan. "Biological Hydrolysis of Parathion in Natural Ecosystems," *J. Environ. Qual.*, 7(3):346-348 (1978).
Suffet, I.H., Faust, S.D., and W.F. Carey. "Gas-Liquid Chromatographic Separation of Some Organophosphate Pesticides, Their Hydrolysis Products, and Oxons," *Environ. Sci. Technol.*, 1(8):639-643 (1967).
Suflita, J.M., Stout, J., and J.M. Tiedje. "Dechlorination of (2,4,5-Trichlorophenoxy)acetic acid by Anaerobic Microorganisms," *J. Agric. Food Chem.*, 32(2):218-221 (1984).
Sumer, K.M., and A.R. Thompson. "Refraction, Dispersion, and Densities of Benzene, Toluene, and Xylene Mixtures," *J. Chem. Eng. Data*, 13(1):30-34 (1968).
Summers, L. "The *alpha*-Haloalkyl Ethers," *Chem. Rev.*, 55:301-353 (1955).
Sunshine, I., Ed. *Handbook of Analytical Toxicology* (Cleveland, OH: The Chemical Rubber Co., 1969), 1081 p.
Sutton, C., and J.A. Calder. "Solubility of Alkylbenzenes in Distilled Water and Seawater at 25 °C," *J. Chem. Eng. Data*, 20(3):320-322 (1975).
Sutton, C., and J.A. Calder. "Solubility of Hiher-Molecular-Weight *n*-Paraffins in Distilled Water and Seawater," *Environ. Sci. Technol.*, 8(7):654-657 (1974).
Suzuki, J., Hagino, T., and S. Suzuki. "Formation of 1-Nitropyrene by Photolysis of Pyrene in Water Containing Nitrite Ion," *Chemosphere*, 16(4):859-867 (1987).
Swain, C.G., and E.R. Thornton. "Initial-state and Transition-state Isotope Effects of Methyl Halides in Light and Heavy Water," *J. Am. Chem. Soc.*, 84:822-826 (1962).
Swann, R.L., Laskowski, D.A., McCall, P.J., Vander Kuy, K., and H.J. Dishburger. "A Rapid Method for the Estimation of the Environmental Parameters Octanol/Water Partition Coefficient, Soil Sorption Constant, Water to Air Ratio, and Water Solubility," *Res. Rev.*, 85:17-28 (1983).
Szabo, G., Prosser, S.L., and R.A. Bulman. "Adsorption Coefficient (K_{oc}) and HPLC Retention Factors of Aromatic Hydrocarbons," *Chemosphere*, 21(4/5):495-505 (1990).
Szabo, G., Prosser, S.L., and R.A. Bulman. "Determination of the Adsorption Coefficient (K_{oc}) of Some Aromatics for Soil by RP-HPLC on Two Immobilized Humic Acid Phases," *Chemosphere*, 21(6):777-788 (1990).
Szabo, G., Prosser, S.L., and R.A. Bulman. "Prediction of the Adsorption Coefficient (K_{oc}) for Soil by a Chemically Immobilized Humic Acid Column using RP-HPLC," *Chemosphere*, 21(6):729-739 (1990).
Tables of Useful Information (Houston, TX: Exxon Corp., 1974), 70 p.

Takagi, H., Hatakeyama, S., Akimoto, H., and S. Koda. "Formation of Methyl Nitrite in the Surface Reaction of Nitrogen Dioxide and Methanol. 1. Dark Reaction," *Environ. Sci. Technol.*, 20(4):387-393 (1986).

Takahashi, N. "Ozonation of Several Organic Compounds having Low Molecular Weight under Ultraviolet Irradiation," *Ozone Sci. Eng.*, 12(1):1-18 (1990).

Takeuchi, K., Yazawa, T., and T. Ibusuki. "Heterogeneous Photocatalytic Effect of Zinc Oxide on Photochemical Smog Formation Reaction of C_4H_8-NO_2-Air," *Atmos. Environ.*, 17(11):2253-2258 (1983).

Tan, C.K., and T.C. Wang. "Reduction of Trihalomethanes in a Water-Photolysis System," *Environ. Sci. Technol.*, 21(5):508-511 (1987).

Taymaz, K., Williams, D.T., and F.M. Benoit. "Chlorine Dioxide Oxidation of Aromatic Hydrocarbons Commonly Found in Water," *Bull. Environ. Contam. Toxicol.*, 23(3):398-404 (1979).

Tewari, Y.B., Miller, M.M., Wasik, S.P., and D.E. Martire. "Aqueous Solubility and Octanol/Water Partition Coefficient of Organic Compounds at 25.0 °C," *J. Chem. Eng. Data*, 27(4):451-454 (1982).

Threshold Limit Values and Biological Exposure Indices for 1987-1988 (Cincinnati, OH: American Conference of Governmental Hygienists, 1987), 114 p.

Tinsley, L.J. *Chemical Concepts in Pollutant Behavior* (New York: John Wiley and Sons, Inc., 1979), 265 p.

Tou, J.C., and G.J. Kallos. "Study of Aqueous HCl and Formaldehyde Mixtures for Formation of Bis(chloromethyl)ether," *J. Am. Ind. Hyg. Assoc.*, 35(7):419-422 (1974).

Toxic and Hazardous Industrial Chemicals Safety Manual for Handling and Disposal with Toxicity and Hazard Data (Tokyo, Japan: International Technical Information Institute, 1986), 700 p.

Travis, C.C., and A.D. Arms. "Bioconcentration of Organics in Beef, Milk and Vegetation," *Environ. Sci. Technol.*, 22(3):271-274 (1988).

"Treatability Manual - Volume 1: Treatability Data," Office of Research and Development, U.S. EPA Report-600/8-80-042a (1980), 1035 p.

Tu, C.M. "Utilization and Degradation of Lindane by Soil Micro-organisms," *Arch. Microbiol.*, 108(3):259-263 (1976).

Tuazon, E.C., Atkinson, R., Winer, A.M., and J.N. Pitts, Jr. "A Study of the Atmospheric Reactions of 1,3-Dichloropropene and Other Selected Organochlorine Compounds," *Arch. Environ. Contam. Toxicol.*, 13(6):691-700 (1984).

Tuazon, E.C., Carter, W.P.L., Atkinson, R., Winer, A.M., and J.N. Pitts, Jr. "Atmospheric Reactions of *N*-Nitrosodimethylamine and Dimethylnitramine," *Environ. Sci. Technol.*, 18(1):49-54 (1984).

Tuazon, E.C., Leod, H.M., Atkinson, R., and W.P.L. Carter. "α-Dicarbonyl Yields from the NO_x-Air Photooxidations of a Series of Aromatic Hydrocarbons in Air," *Environ. Sci. Technol.*, 20(4):383-387 (1986).

Tuazon, E.C., Winer, A.M., Graham, R.A., Schmid, J.P., and J.N. Pitts, Jr. "Fourier Transform Infrared Detection of Nitramines in Irradiated Amine-NO_x Systems," *Environ. Sci. Technol.*, 12(8):954-958 (1978).

Tundo, P., Facchetti, S., Tumiatti, W., and U.G. Fortunati. "Chemical Degradation of 2,3,7,8-TCDD by Means of Polyethylene-glycols in the Presence of Weak Bases and an Oxidant," *Chemosphere*, 14(5):403-410 (1985).

Uchrin, C.G., and G. Michaels. "Chloroform Sorption to New Jersey Coastal Plain Ground Water Aquifer Solids," *Environ. Toxicol. Chem.*, 5(4):339-343 (1986).

Ungnade, H.E., and E.T. McBee. "The Chemistry of Perchlorocyclo-pentadienes and Cyclopentadienes," *Chem. Rev.*, 58:249-320 (1957).

Ursin, C. "Degradation of Organic Chemicals at Trace Levels in Seawater and Marine Sediment. The Effect of Concentration on the Initial Turnover Rate," *Chemosphere*, 14(10):1539-1550 (1985).

U.S. Department of Health and Human Services. *Hazardous Substances Data Bank*. National Library of Medicine, Toxnet File (Bethedsa, MD: National Institute of Health, 1989).

Uyeta, M., Taue, S., Chikasawa, K., and M. Mazaki. "Photoformation of Polychlorinated Biphenyls from Chlorinated Benzenes," *Nature*, 264:583-584 (1976).

Valsaraj, K.T. "On the Physio-Chemical Aspects of Partitioning of Non-Polar Hydrophobic Organics at the Air-Water Interface," *Chemosphere*, 17(5):875-887 (1988).

Valvani, S.C., Yalkowsky, S.H., and T.J. Roseman. "Solubility and Partitioning IV: Aqueous Solubility and Octanol-Water Partition Coefficients of Liquid Nonelectrolytes," *J. Pharm. Sci.*, 70(5):502-506 (1981).

van Gestel, C.A.M., and W.-C. Ma. "Toxicity and Bioaccumulation of Chlorophenols in Earthworms, in Relation to Bioavailability in Soil," *Ecotoxicol. Environ. Safety*, 15(3):289-297 (1988).

van Ginkel, C.G., Welten, H.G.J., and J.A.M. de Bont. "Oxidation of Gaseous and Volatile Hydrocarbons by Selected Alkene-Utilizing Bacteria," *Appl. Environ. Microbiol.*, 53(12):2903-2907 (1987).

Veith, G.D., Call, D.J., and L.T. Brooke. "Structure-Toxicity Relationships for the Fathead Minnow *Pimephales promelas*: Narcotic Industrial Chemicals," *Can. J. Fish. Aquat. Sci.*, 40(6):743-748 (1983).

Veith, G.D., Macek, K.J., Petrocelli, S.R., and J. Carroll. "An Evaluation of Using Partition Coefficients and Water Solubility to Estimate Bioconcentration Factors for Organic Chemicals in Fish," in *Aquatic Toxicology, ASTM STP 707*, Eaton, J.G., Parrish, P.R., and A.C. Hendricks, Eds. (Philadelphia, PA: American Society for Testing and Materials 1980), pp 116-129.

Venkateswarlu, K., and N. Sethunnathan. "Degradation of Carbofuran by *Azospirillium lipoferum* and *Streptomyces* spp. Isolated from Flooded Alluvial Soil," *Bull. Environ. Contam. Toxicol.*, 33(5):556-560 (1984).

Verschueren, K. *Handbook of Environmental Data on Organic Chemicals* (New York: Van Nostrand Reinhold Co., 1983), 1310 p.

Vogel, T.M., Criddle, C.S., and P.L. McCarty. "Transformations of Halogenated Aliphatic Compounds," *Environ. Sci. Technol.*, 21(8):722-736 (1987).

Vogel, T.M., and P.L. McCarty. "Biotransformation of Tetrachloro-ethylene to Trichloroethylene, Dichloroethylene, Vinyl Chloride, and Carbon Dioxide Under Methanogenic Conditions," *Appl. Environ. Microbiol.*, 49(5):1080-1083

(1985).

Vogel, T.M., and M. Reinhard. "Reaction Products and Rates of Disappearance of Simple Bromoalkanes, 1,2-Dibromopropane, and 1,2-Dibromoethane in Water," *Environ. Sci. Technol.*, 20(10):992-997 (1986).

Voice, T.C., and W.J. Weber, Jr. "Sorbent Concentration Effects in Liquid/Solid Partitioning," *Environ. Sci. Technol.*, 19(9):789-796 (1985).

Vowles, P.D., and R.F.C. Mantoura. "Sediment-Water Partition Coefficients and HPLC Retention Factors of Aromatic Hydrocarbons," *Chemosphere*, 16(1):109-116 (1987).

Wahid, P.A., Ramakrishna, C., and N. Sethunathan. "Instantaneous Degradation of Parathion in Anaerobic Soils," *J. Environ. Qual.*, 9(1):127-130 (1980).

Wahid, P.A., and N. Sethunathan. "Sorption-Desorption of *Alpha*, *Beta* and *Gamma* Isomers of BHC in Soils," *J. Agric. Food Chem.*, 27(5):1050-1053 (1979).

Wajon, J.E., Rosenblatt, D.H., and E.P. Burrows. "Oxidation of Phenol and Hydroquinone by Chlorine Dioxide," *Environ. Sci. Technol.*, 16(7):396-402 (1982).

Walters, R.W., and R.G. Luthy. "Equilibrium Adsorption of Polycyclic Aromatic Hydrocarbons from Water onto Activated Carbon," *Environ. Sci. Technol.*, 18(6):395-403 (1984).

Walters, R.W., Ostazeski, S.A., and A. Guiseppi-Elie. "Sorption of 2,3,7,8-Tetrachlorodibenzo-*p*-dioxin from Water by Surface Soils," *Environ. Sci. Technol.*, 23(4):480-484 (1989).

Walton, B.T., Anderson, T.A., Hendricks, M.S., and S.S. Talmage. "Physicochemical Properties as Predictors of Organic Chemical Effects on Soil Microbial Respiration," *Environ. Toxicol. Chem.*, 8(1):53-63 (1989).

Walton, W.C. *Practical Aspects of Ground Water Modeling* (Worthington, OH: National Water Well Association, 1985), 587 p.

Wams, T.J. "Diethylhexylphthalate as an Environmental Contaminant - A Review," *Sci. Tot. Environ.*, 66:1-16 (1987).

Wang, Y.-T. "Methanogenic Degradation of Ozonation Products of Bio-refractory or Toxic Aromatic Compounds," *Wat. Res.*, 24(2):185-190 (1990).

Wang, T.C., and C.K. Tan. "Enhanced Degradation of Halogenated Hydrocarbons in a Water-Photolysis System," *Bull. Environ. Contam. Toxicol.*, 40(1):60-65 (1988).

Wang, T.C., Tan, C.K., and M.C. Liou. "Degradation of Bromoform and Chlorodibromomethane In a Catalyzed H_2-Water System," *Bull. Environ. Contam. Toxicol.*, 41(4):563-568 (1988).

Wannstedt, C., Rotella, D., and J.F. Siuda. "Chloroperoxidase Mediated Halogenation of Phenols," *Bull. Environ. Contam. Toxicol.*, 44(2):282-287 (1990).

Warner, H.P., Cohen, J.M., and J.C. Ireland. "Determination of Henry's Law Constants of Selected Priority Pollutants," Office of Science and Development, U.S. EPA Report-600/D-87/229 (1987), 14 p.

Warne, M.St.J., Connell, D.W., Hawker, D.W., and G. Schüürmann. "Prediction of Aqueous Solubility and the Octanol-Water Partition Coefficient for Lipophilic Organic Compounds Using Molecular Descriptors and Physicochemical Properties," *Chemosphere*, 21(7):877-888 (1990).

Wasik, S.P., Miller, M.M., Tewari, Y.B., May, W.E., Sonnefeld, W.J., DeVoe, H., and W.H. Zoller. "Determination of the Vapor Pressure, Aqueous Solubility, and Octanol/Water Partition Coefficient of Hydrophobic Substances by Coupled Generator Column/Liquid Chromatographic Methods," *Res. Rev.*, 85:29-42 (1983).

Wasik, S.P., Schwarz, F.P., Tewari, Y.B., and M.M. Miller. "A Head-Space Method for Measuring Activity Coefficients, Partition Coefficients, and Solubilities of Hydrocarbons in Saline Solutions," *NBS J. Res.*, 89(3):273-277 (1984).

Watanabe, I., Kashimoto, T., and R. Tatsukawa. "Brominated Phenol Production from the Chlorination of Wastewater Containing Bromide Ions," *Bull. Environ. Contam. Toxicol.*, 33(4):395-399 (1984).

Watanabe, N., Sato, E., and Y. Ose. "Adsorption and Desorption of Polydimethylsiloxane, Cadmium Nitrate, Copper Sulfate, Nickel Sulfate and Zinc Nitrate by River Surface Sediments," *Sci. Tot. Environ.*, 41(2):163-161 (1985).

Watarai, H., Tanaka, M., and N. Suzuki. "Determination of Partition Coefficients of Halobenzenes in Heptane/Water and 1-Octanol/Water Systems and Comparison with the Scaled Particle Calculation," *Anal. Chem.*, 54(4):702-705 (1982).

Watson, E., and C.F. Parrish. "Laser Induced Decomposition of 1,4-Dioxane," *J. Chem. Phys.*, 54(3):1427-1428 (1971).

Wauchope, R.D., and F.W. Getzen. "Temperature Dependence of Solubilities in Water and Heats of Fusion of Solid Aromatic Hydrocarbons," *J. Chem. Eng. Data*, 17(1):38-41 (1972).

Wauchope, R.D., and R. Haque. "Effects of pH, Light and Temperature on Carbaryl in Aqueous Media," *Bull. Environ. Contam. Toxicol.*, 9(5):257-261 (1973).

Weast, R.C., Ed. *CRC Handbook of Chemistry and Physics*, 67th ed. (Boca Raton, FL: CRC Press, Inc., 1986), 2406 p.

Weast, R.C., and M.J. Astle., Eds. *CRC Handbook of Data on Organic Compounds - 2 Volumes* (Boca Raton, FL: CRC Press, Inc. 1986).

Webb, R.F., Duke, A.J., and L.S.A. Smith. "Acetals and Oligoacetals. Part I. Preparation and Properties of Reactive Oligoformals," *J. Chem. Soc. (London)*, (1962), pp 4307-4319.

Weber, J.B. "Interaction of Organic Pesticides with Particulate Matter in Aquatic and Soil Systems," in *Fate of Organic Pesticides in the Aquatic Environment, Advances in Chemistry Series*, R.F. Gould, Ed. (Washington, DC: American Chemical Society, 1972), pp 55-120.

Webster, G.R.B., Friesen, K.J., Sarna, L.P., and D.C.G. Muir. "Environmental Fate Modelling of Chlorodioxins: Determination of Physical Constants," *Chemosphere*, 14(6/7):609-622 (1985).

Weil, L., Dure, G., and K.E. Quentin. "Solubility in Water of Insecticide Chlorinated Hydrocarbons and Polychlorinated Biphenyls in View of Water Pollution," *Z. Wasser Forsch.*, 7(6):169-175 (1974).

Weiss, G. *Hazardous Chemicals Data Book* (Park Ridge, NJ: Noyes Data Corp., 1986), 1069 p.

Westcott, J.W., Simon, C.G., and T.F. Bidleman. "Determination of Polychlorinated Biphenyl Vapor Pressures by a Semimicro Gas Saturation Method," *Environ. Sci. Technol.*, 15(11):1375-1378 (1981).

Whitbeck, M. "Photo-oxidation of Methanol," *Atmos. Environ.*, 17(1):121-126 (1983).

Wiese, C.S., and D.A. Griffin. "The Solubility of Aroclor 1254 in Seawater," *Bull. Environ. Contam. Toxicol.*, 19(4):403-411 (1978).

Wilhoit, R.C., and B.J. Zwolinski. *Handbook of Vapor Pressure and Heats of Vaporization of Hydrocarbons and Related Compounds*, Publication 101 (College Station, TX: Thermodynamics Research Station, 1971), 329 p.

Wilson, B.H., Smith, G.B., and J.F. Rees. "Biotransformations of Selected Alkylbenzenes and Halogenated Aliphatic Hydrocarbons in Methanogenic Aquifer Material: A Microcosm Study," *Environ. Sci. Technol.*, 20(10):997-1002 (1986).

Wilson, J.T., Enfield, C.G., Dunlap, W.J., Cosby, R.L., Foster, D.A., and L.B. Baskin. "Transport and Fate of Selected Organic Pollutants in a Sandy Soil," *J. Environ. Qual.*, 10(4):501-506 (1981).

Windholz, M., Budavari, S., Blumetti, R.F., and E.S. Otterbein, Eds. *The Merck Index*, 10th ed. (Rahway, NJ: Merck and Co., 1983), 1463 p.

Wolfe, N.L., Steen, W.C., and L.A. Burns. "Phthalate Ester Hydrolysis: Linear Free Energy Relationships," *Chemosphere*, 9(7/8):403-408 (1980).

Wolfe, L.N., Zepp, R.G., Gordon, J.A., Baughman, G.L., and D.M. Cline. "Kinetics of Chemical Degradation of Malathion in Water," *Environ. Sci. Technol.*, 11(1):88-93 (1977).

Wolfe, N.L., Zepp, R.G., Paris, D.F., Baughman, G.L., and R.C. Hollis. "Methoxychlor and DDT Degradation in Water: Rates and Products," *Environ. Sci. Technol.*, 11(12):1077-1081 (1977).

Wong, A.S., and D.G. Crosby. "Photodecomposition of Pentachlorophenol in Water," *J. Agric. Food Chem.*, 29(1):125-130 (1981).

Woodrow, J.E., Crosby, D.G., Mast, T., Moilanen, K.W., and J.N. Seiber. "Rates of Transformation of Trifluralin and Parathion Vapors in Air," *J. Agric. Food Chem.*, 26(6):1312-1316 (1978).

Worthing, C.R., and R.J. Hance, Eds. *The Pesticide Manual - A World Compendium*, 9th ed. (Great Britain: British Crop Protection Council, 1991), 1141 p.

Worthing, C.R., and S.B. Walker, Eds. *The Pesticide Manual - A World Compendium*, 7th ed. (Lavenham, Suffolk, Great Britain: The Lavenham Press Ltd., 1983), 695 p.

Yagi, O., and R. Sudo. "Degradation of Polychlorinated Biphenyls by Microorganisms," *J. Water Poll. Control Fed.*, 52(5)1035-1043 (1980).

Yalkowsky, S.H., Orr, R.J., and S.C. Valvani. "Solubility and Partitioning. 3. The Solubility of Halobenzenes in Water," *Indust. Eng. Chem. Fund.*, 18(4):351-353 (1979).

Yalkowsky, S.H., and S.C. Valvani. "Solubilities and Partitioning 2. Relationships between Aqueous Solubilities, Partition Coefficients, and Molecular Surface Areas of Rigid Aromatic Hydrocarbons," *J. Chem. Eng. Data*, 24(2):127-129

(1979).
Yalkowsky, S.H., Valvani, S.C., and D. Mackay. "Estimation of the Aqueous Solubility of Some Aromatic Compounds," *Res. Rev.*, 85:43-55 (1983).
Yasuhara, A., and M. Morita. "Formation of Chlorinated Compounds in Pyrolysis of Trichloroethylene," *Chemosphere*, 21(4/5):479-486 (1990).
Yokley, R.A., Garrison, A.A., Mamantov, G., and E.L. Wehry. "The Effect of Nitrogen Dioxide on the Photochemical and Nonphoto-chemical Degradation of Pyrene and Benzo[*a*]pyrene Adsorbed on Coal Fly Ash," *Chemosphere*, 14(11/12):1771-1778 (1985).
Yoshida, K., Shigeoka, T., and F. Yamauchi. "Non-Steady State Equilibrium Model for the Preliminary Prediction of the Fate of Chemicals in the Environment," *Ecotoxicol. Environ. Safety*, 7(2):179-190 (1983).
Yoshida, K., Tadayoshi, S., and F. Yamauchi. "Relationship between Molar Refraction and *n*-Octanol/Water Partition Coefficient," *Ecotoxicol. Environ. Safety*, 7(6):558-565 (1983).
Young, L.Y., and M.D. Rivera. "Methanogenic Degradation of Four Phenolic Compounds," *Water Res.*, 19(10):1325-1332 (1985).
Zabik, M.J., Schuetz, R.D., Burton, W.L., and B.E. Pape. "Photochemistry of Bioreactive Compounds. Studies of a Major Product of Endrin," *J. Agric. Food Chem.*, 19(2):308-313 (1971).
Zafiriou, O.C. "Reaction of Methyl Halides with Seawater and Marine Aerosols," *J. Mar. Res.*, 33(1):75-81 (1975).
Zepp, R.G., and P.F. Scholtzhauer. "Photoreactivity of Selected Aromatic Hydrocarbons in Water," in *Polynuclear Aromatic Hydrocarbons, 3rd International Symposium on Chemistry, Biology, Carcinogenesis and Mutagenesis*, Jones, P.W., and P. Leber, Eds. (Ann Arbor, MI: Ann Arbor Science Publishers, Inc., 1979), pp 141-158.
Zepp, R.G., Wolfe, N.L., Baughman, G.L., Schlotzhauer, P.F., and J.N. MacAllister. "Dynamics of Processes Influencing the Behavior of Hexachlorocyclopentadiene in the Aquatic Environment," 178th Meeting of the American Chemical Society, Washington, DC (September 1979).
Zepp, R.G., Wolfe, N.L., Gordon, J.A., and R.C. Fincher. "Light-Induced Transformations of Methoxychlor in Aquatic Systems," *J. Agric. Food Chem.*, 24(4):727-733 (1976).
Zeyer, J., and P.C. Kearney. "Degradation of *o*-Nitrophenol and *m*-Nitrophenol by a *Pseudomonas putida*," *J. Agric. Food Chem.*, 32(2):238-242 (1984).
Zeyer, J., and P.C. Kearney. "Microbial Metabolism of [^{14}C]Nitroanilines to [^{14}C]Carbon Dioxide," *J. Agric. Food Chem.*, 31(2):304-308 (1983).
Zhou, X., and K. Mopper. "Apparent Partition Coefficients of 15 Carbonyl Compounds between Air and Seawater and between Air and Freshwater; Implications for Air-Sea Exchange," *Environ. Sci. Technol.*, 24(12):1864-1869 (1990).